2661-C

OO

NG.

MICROELECTRONIC DEVICES

McGraw-Hill Series in Electrical Engineering

Consulting Editor

Stephen W. Director, *Carnegie-Mellon University*

CIRCUITS AND SYSTEMS
COMMUNICATIONS AND SIGNAL PROCESSING
CONTROL THEORY
ELECTRONICS AND ELECTRONIC CIRCUITS
POWER AND ENERGY
COMPUTER ENGINEERING
INTRODUCTORY
RADAR AND ANTENNAS
VLSI

Previous Consulting Editors

**Ronald N. Bracewell, Colin Cherry, James F. Gibbons, Willis W. Harman,
Hubert Heffner, Edward W. Herold, John G. Linvill, Simon Ramo, Ronald A. Rohrer,
Anthony E. Siegman, Charles Susskind, Frederick E. Terman, John G. Truxal,
Ernst Weber, and John R. Whinnery**

ELECTRONICS AND ELECTRONIC CIRCUITS

Consulting Editor

Stephen W. Director, *Carnegie-Mellon University*

Colclaser and Diehl-Nagle: *Materials and Devices for Electrical Engineers and
   Physicians*
Franco: *Design with Operational Amplifiers and Analog Integrated Circuits*
Grinich and Jackson: *Introduction to Integrated Circuits*
Hamilton and Howard: *Basic Integrated Circuits Engineering*
Hodges and Jackson: *Analysis and Design of Digital Integrated Circuits*
Millman and Grabel: *Microelectronics*
Millman and Halkias: *Integrated Electronics: Analog, Digital Circuits, and Systems*
Millman and Taub: *Pulse, Digital, and Switching Waveforms*
Schilling and Belove: *Electronic Circuits: Discrete and Integrated*
Smith: *Modern Communication Circuits*
Sze: *VSLI Technology*
Taub: *Digital Circuits and Microprocessors*
Taub and Schilling: *Digital Integrated Electronics*
Wait, Huelsman, and Korn: *Introduction to Operational and Amplifier Theory Applications*
Yang: *Microelectronic Devices*

MICROELECTRONIC DEVICES

Edward S. Yang

Department of Electrical Engineering
Columbia University

McGraw-Hill Book Company

New York St. Louis San Francisco Auckland Bogotá Caracas
Colorado Springs Hamburg Lisbon London Madrid Mexico
Milan Montreal New Delhi Oklahoma City Panama Paris
San Juan São Paulo Singapore Sydney Tokyo Toronto

This book was set in Times Roman by Eta Services Ltd.
The editors were Alar E. Elken and David Damstra;
the designer was Albert M. Cetta;
the production supervisor was Salvador Gonzales.
R. R. Donnelley & Sons Company was printer and binder.

MICROELECTRONIC DEVICES

234567890 DOCDOC 89321098

ISBN 0-07-072238-2

Library of Congress Cataloging-in-Publication Data
Yang, Edward S.
    Microelectronic devices/Edward S. Yang.
       p. cm.—(McGraw-Hill series in electrical engineering.
    Electronics and electronic circuits.)
    ISBN 0-07-072238-2
    1. Microelectronics. 2. Semiconductors. I. Title. II. Series.
TK7874.Y36 1988   621.381′73—dc19   87-19414 CIP

To
Stephen Kaung and Hokmo Chan

for their
profound influence and inspiration

CONTENTS

PREFACE

This book presents the principles of operation of microelectronic devices and their practical implementation. It is intended as an introductory text for students in electrical engineering, physics, or material science. The topical coverage is broad enough to be adapted for juniors or seniors depending on the selection of material. Some sections deal with new devices and technology, which may be useful for first-year graduate students. Most of the subjects have been simplified such that they can be understood by beginning students who have a background in differential equations and in the basic concepts of modern physics.

This book was originally planned as a second edition to *Fundamentals of Semiconductor Devices* (McGraw-Hill 1978). However, so much new material has been added and the revisions have been so extensive that a new title seems appropriate. The revisions include lowering the level by eliminating the more difficult subjects, expanding explanation sections, and adding illustrative examples. The organization has also been changed. The bipolar transistor is introduced before the MOS devices, and five chapters are devoted to MOS structures. Many new subjects have been added, for example, bipolar scaling, polyemitter, silicides, short-channel MOS transistors, LDD structure, CMOS technology, amorphous solar cells, photodetectors and laser diodes, to name a few.

The material has been used in the classroom for two different groups of students at Columbia. It has been tested in an introductory solid-state devices course for juniors covering chapters 1 to 5, 9, and 14. For a mixed class including seniors and graduate students, Chapters 1 to 9 and selected topics in later chapters have been used. In a different setting, Chapters 3, 7 to 9, 11, and 12 were covered in a field-effect transistor course for seniors at the University of California at Irvine.

Another possible one-year sequence is to have the bipolar transistor and opto-electronics in the first semester and field-effect devices in the second semester.

Many individuals have contributed in providing suggestions and criticisms. In particular, I am indebted to Professors Yannis Tsividis and Eric Fossum of Columbia, Farid Shoucair of Brown, and Chin Lee and James Mulligan of University of California at Irvine; Larry Burton, Virginia Polytechnic Institute; David Dumin, Clemson University; O. Eknoyan, Texas A&M University; Ronald Gutman, Rensselaer Polytechnic Institute; Shmuel Mardix, University of Rhode Island; Jeffrey Smith, Mississippi State University; John Uyemura, Georgia Institute of Technology; and Mustafa Yousef, University of Illinois for their critical review of portions of the manuscript. Special thanks are due to Drs. Stefan Lai of Intel, Morris Wu of Sharp, and R. Zuleeg of McDonnell-Douglas for preprints and valuable discussion. My appreciation is extended to my former students at Columbia, Ajay Nahata, Chung Ng, and Xu Wu for their contribution in adding problems as well as in improving the accuracy and content of the text. Dr. Wu did all the homework, prepared the solutions manual, and made many important suggestions. Thanks are also due to Betty Lim for assistance in typing. Editorial burdens were borne by David and Grace Yang who translated the unintelligible into the native tongue. Various versions of the manuscript were typed and prepared by my wife, Ruth, whose constant support in keeping track of the mass confusion created by the writing project made possible the completion of this work. My deep appreciation to my family for their cheerfulness in putting up with it all.

Edward S. Yang

MICROELECTRONIC DEVICES

1

SEMICONDUCTOR FUNDAMENTALS

In this chapter, we present the elementary physical theory and material properties of semiconductors that are crucial for microelectronic device development. Today, silicon is the material used in essentially all commercial integrated circuits, including memory and logic circuits, operational amplifiers, and power devices. However, gallium arsenide, a compound semiconductor, has recently emerged as an alternative to silicon in some special systems in microwave, optoelectronics, and high-speed circuits. These two materials will be emphasized to illustrate the basic semiconductor parameters.

1-1   CRYSTAL STRUCTURE

Solids may be classified by structural organization into *crystalline, polycrystalline*, and *amorphous* types. An amorphous solid does not have a well-defined structure; in fact, it is distinguished by its formlessness, as shown in Fig. 1-1*a*. In the past decade, amorphous silicon has received a great deal of attention, primarily due to its application in low-cost solar cells. Recently, amorphous silicon solar cells with an efficiency of greater than 10 percent have been realized, and lower-efficiency devices are being used in consumer electronics, e.g., hand-held calculators and cameras. Exploratory work is being performed to build amorphous field-effect transistors (FETs) for large-area displays and image sensors. Nevertheless, amorphous semiconductors are not expected to play an important role in microelectronics in the foreseeable future. Those who are interested in this new field may consult Ref. 1. It should be pointed out that silicon dioxide, an extremely

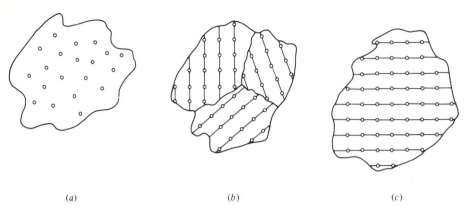

FIGURE 1-1
Three classes of solids: (*a*) amorphous, (*b*) polycrystalline, and (*c*) crystalline.

important material in semiconductor technology, is also an amorphous solid. However, it is used as an insulator, so its electrical conduction property is not of primary importance to us.

In a polycrystalline solid (Fig. 1-1*b*), there are many small regions, each having a well-organized structure but differing from its neighboring regions. Such a material can be produced inexpensively and is used extensively in microelectronics, e.g., polycrystalline silicon (poly-Si), which is used as a conductor, contact, or gate in transistors. Both amorphous and polycrystalline materials are structurally more complex, resulting in less well defined device physics that are not suitable for beginning undergraduates. In this text, we will pass over amorphous solids but will treat poly-Si as a poor cousin of its crystalline counterpart.

In a crystalline solid, atoms are arranged in an orderly array (Fig. 1-1*c*) that defines a periodic structure called the *lattice*. It is possible to specify a *unit cell* which, when repeating itself, produces the crystalline solid. Let us consider the two-dimensional (2-D) array of atoms shown in Fig. 1-2*a*. A unit cell may be an equal-sided square with an atom at the center (Fig. 1-2*b*) or with four atoms at the corners (Fig. 1-2*c*). In the latter case, only a quarter of each corner atom belongs to this unit cell. It is obvious that the 2-D array can be formed by arranging the same kind of unit cells together. A three-dimensional lattice can then be produced by stacking up the 2-D planes. A unit cell contains complete information regarding the arrangement of atoms, and hence the unit cell can be used to describe the crystal structure. Among the many structures of crystals, we shall limit ourselves to the five cubic structures relevant to our later discussion.

1. *Simple cubic (sc) crystal.* An sc crystal is shown in Fig. 1-3*a*. Each corner of the cubic lattice is occupied by an atom which is shared by eight neighboring unit cells. The dimension *a* in the cubic unit cell is called the *lattice constant*. Few crystals exhibit this structure, and polonium is the only element crystallized in this form.

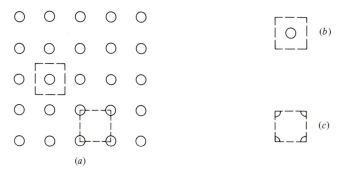

FIGURE 1-2
(a) A crystal lattice with two different unit cells (b) and (c).

2. *Body-centered cubic (bcc) crystal.* A bcc crystal is illustrated in Fig. 1-3b, where, in addition to the corner atoms, an atom is located at the center of the cube. Crystals exhibiting this structure include sodium, molybdenum, and tungsten.

3. *Face-centered cubic (fcc) crystal.* Figure 1-3c shows an fcc crystal, which contains one atom at each of the six cubic faces in addition to the eight corner atoms. A large number of elements exhibit this crystal form, including aluminum, copper, gold, silver, nickel, and platinum.

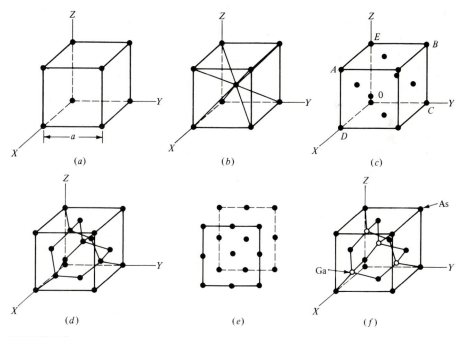

FIGURE 1-3
Unit cells of cubic crystals: (a) simple cubic, (b) body-centered cubic, (c) face-centered cubic, (d) diamond, (e) two penetrating fcc lattices in two dimensions, (f) zinc blende.

4. *Diamond structure.* A diamond unit cell is depicted in Fig. 1-3*d*. It may be seen as two interpenetrating fcc sublattices with one displaced from the other by one-quarter of the distance along a diagonal of the cube. The top view of such two penetrating fcc lattices is shown in Fig. 1-3*e* in two-dimensional form. Note that four atoms of the second fcc lattice are located within the first fcc lattice. These four atoms belong to the same unit cell of the first fcc lattice in the diamond structure. The atoms outside the original unit cell belong to the adjacent unit cells. Both silicon and germanium exhibit this structure. We shall come back to this lattice later.

5. *Zinc blende structure.* Gallium arsenide (GaAs) exhibits the zinc blende structure, as shown in Fig. 1-3*f*. It results from the diamond structure when Ga atoms are placed on one fcc lattice and As atoms on the other fcc lattice. Other materials of the zinc blende family include gallium phosphide, zinc sulfide, and cadmium sulfide.

Let us take a further look at the fcc crystal shown in Fig. 1-3*c*. We notice that there are six atoms in the *ABCD* plane and five atoms in the *AEOD* plane and that the atomic spacings are different for the two planes. For these reasons, the crystal properties along different directions are different; i.e., the crystal is *anisotropic*. Therefore it is necessary to find a convenient way to specify the orientation of crystal planes and directions. In practice, a crystal plane is specified by its *Miller indices*, which are obtained by using the following procedure:

1. Determine the intercepts of the plane on the three cartesian coordinates.
2. Measure the distances of the intercepts from the origin in multiples of the lattice constant.
3. Take the reciprocals of the intercepts and then multiply by the least common multiple of the intercepts.

> **Example.** Find the Miller indices for the crystal plane shown in Fig. 1-4. Assume the lattice constant is unity.

> **Solution** In Fig. 1-4, the plane intercepts the three crystal axes at (4, 0, 0), (0, 2, 0), and (0, 0, 2), or simply 4, 2, 2. The reciprocals of these intercepts are $\frac{1}{4}, \frac{1}{2}, \frac{1}{2}$. Multiplying by 4, these fractions are reduced to 1, 2, 2, which are the Miller indices. The plane is known as the (122) plane.

The direction of a crystal plane is normal to the plane itself and is designated by brackets of the same Miller indices. For example, the direction of the (111) plane is written as [111]. Some directions in a crystal lattice are equivalent because they depend only on the orientation of the axes. As an example, the directions of the crystal axes in a cubic lattice are designated [001], [010], and [100]. These directions are equivalent because the crystal planes they represent are the same and the only difference is the arbitrary naming of the axes. These equivalent direction indices are written as ⟨100⟩. Miller indices for three important lattice planes are shown in Fig. 1-5.

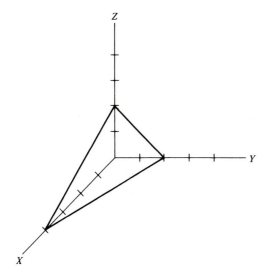

FIGURE 1-4
A crystal plane.

Let us return to examine the diamond structure a second time. The bonds between atoms are emphasized in Fig. 1-6. Notice that each atom bonds with its four nearest neighbors to form a tetrahedron. We have darkened the top corner section to provide a clearer picture. The tetrahedral bonds are very strong and directional, giving rise to specific mechanical, metallurgical, and electrical properties of the solid as in the case of diamond, silicon, germanium, and gallium arsenide. In GaAs, the tetrahedral is formed such that all the bonds are between Ga and As, as shown in the lower left corner of Fig. 1-3f. The tetrahedral configuration specifies the bond angles and lengths with a perfect symmetry. If, however, the bond angles are allowed to vary by less than 10° *randomly*, the silicon will become amorphous.

Example. The lattice constant a of silicon is 0.543 nm.
(a) Find the *bond length*, which is defined as the minimum distance between the adjacent neighbor atoms.
(b) Calculate the atomic radius if the atoms are packed together as rigid spheres.

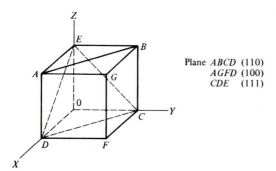

Plane *ABCD* (110)
AGFD (100)
CDE (111)

FIGURE 1-5
Important crystal planes represented by Miller indices.

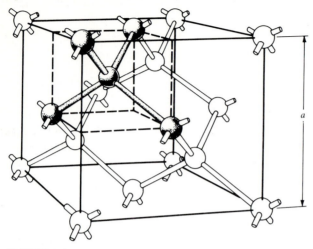

FIGURE 1-6
The diamond structure, showing how each atom forms four bonds with its nearest neighbors. (*After Shockley [2].*)

Solution

(*a*) Since the diagonal of a face is $a\sqrt{2}$, the diagonal of the cube is

$$D = \sqrt{a^2 + 2a^2} = \sqrt{3}a$$

The minimum distance between the adjacent neighbor is

$$\frac{D}{4} = \frac{\sqrt{3}a}{4}$$

(*b*) Assuming the same radius for all atoms, then

$$r = \frac{1}{2} \times \frac{D}{4} = \frac{\sqrt{3}a}{8}$$

One may now ask, "What is the bonding force between the two neighboring atoms?" Before we can answer this question, we have to step back to consider the physical model of an atom.

1-2 BOHR'S ATOM

The simplest model of an atom consists of a positively charged nucleus with negatively charged electrons orbiting around it, analogous to the relation between the sun and the planets. Bohr postulates that the angular momentum of an electron may assume only certain discrete values which in turn give rise to a set of discrete allowable energy levels. As shown in Appendix A [Eq. (A-8)], these energy levels are derived for the hydrogen atom:

$$E_n = -\frac{13.6}{n^2} \quad \text{eV} \tag{1-1}$$

where n is the principal quantum number and E_n is the electron binding energy in electronvolts.

For $n = 1$, 2, and 3, for example, the corresponding energies are

$$E_1 = -13.6 \text{ eV} \qquad E_2 = -3.41 \text{ eV} \qquad E_3 = -1.53 \text{ eV}$$

E_1, being the lowest energy level, is known as the *ground* state. Other higher-energy levels are known as *excited* states. Thus, the orbits of the electrons are specified by their energy levels. If an electron at E_2 desires to move to a higher energy E_3, it must gain an amount of energy exactly equal to $E_3 - E_2 = 1.88$ eV. On the other hand, the reverse transition from E_3 to E_2 is accompanied by the emission of a "quantum" of energy equal to 1.88 eV. The quantum is emitted in the form of light and is known as a *photon*. Its energy is given by

$$E_2 - E_1 = h\nu \tag{1-2}$$

for the transition between E_2 and E_1, where h is Planck's constant and ν is the frequency of the absorbed or emitted energy. The corresponding wavelength is given by

$$\lambda = \frac{c}{\nu} \tag{1-3}$$

where c is the velocity of light, 3×10^8 m/s.

Let us digress for a moment to say a few words about the electronvolt. In the mks system, the unit of energy is the joule (J). Unfortunately, a joule is too large to be used in semiconductor devices because each electron possesses only a tiny amount of energy. For this reason, we use the unit of energy called the electronvolt (eV), defined as

$$1 \text{ eV} = 1.6 \times 10^{-19} \text{ J}$$

A more familiar unit is the watt, which is defined as joules per second. The name "electronvolt" arises from the fact that if an electron falls through a potential of 1 V, its kinetic energy will increase by 1 eV. One may ask how an electronic device can handle a sizable power. The answer is that we have trillions and quadrillions of electrons in a real semiconductor device. A typical high-power rectifier can easily control kilowatts of energy.

Example. Calculate the wavelength of the photon that provides the necessary energy to excite an electron to the excited state E_4 of a hydrogen atom from any lower level.

Solution The energy required is given by

$$\Delta E = h\nu = \frac{hc}{\lambda}$$

so

$$\lambda = \frac{1.24}{\Delta E} \quad \mu m$$

Since

$$E_4 - E_1 = -0.87 + 13.6 = 12.7 \text{ eV}$$

$$E_4 - E_2 = 2.54 \text{ eV}$$

$$E_4 - E_3 = 0.66 \text{ eV}$$

the corresponding wavelengths are

$$0.098 \ \mu\text{m} \qquad 0.488 \ \mu\text{m} \qquad \text{and} \qquad 1.88 \ \mu\text{m}$$

These three wavelengths fall in the *ultraviolet*, *visible*, and *infrared* regions of light, respectively. Roughly speaking, the visible spectrum covers 0.4 to 0.72 μm, with the infrared in the long-wavelength side and the ultraviolet in the short-wavelength side.

When an atom has more than one electron, the lowest energy level is the first to be filled, followed by the filling up of each successively higher energy level. As shown in Appendix A, the number of energy states associated with each quantum number n is different, and they bunch together to form a shell. The electron shell associated with E_1 has two energy states, so it can accept two electrons. The shells associated with E_2, E_3, and E_4 can accommodate 8, 18, and 32 electrons, respectively. The unique distribution of electrons in any particular element gives rise to its specific chemical properties.

Let us consider the case of an isolated silicon atom. Its atomic number is 14. Since the first two shells can accommodate 2 and 8 electrons, we have 4 electrons occupying the E_3 shell. The completely filled shells, that is, E_1 and E_2 in silicon, represent tightly bound electrons that would not be disturbed by interatomic forces or chemical reaction. The four outermost electrons, known as *valence electrons*, are relatively free to participate in chemical reactions or interatomic bonding. One may therefore visualize that the nucleus and the two inner shells form the core with an effective charge of $+4q^\dagger$ and the valence electrons have a charge of $-4q$. This simplification is represented schematically in Fig. 1-7 and is the basis of the description of Si, Ge, and GaAs in the next section.

---

† The charge associated with the nucleus is $+14q$, and the 10 electrons occupying the inner two shells have $-10q$; thus, there is a net charge of $+4q$.

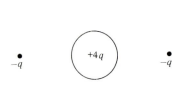

FIGURE 1-7
ıplified representation of a silicon atom with its four valence electrons.

1-3 VALENCE-BOND MODEL OF SOLID

In a crystal lattice, a positively charged nucleus is surrounded by negatively charged orbiting electrons in each constituent atom. If the atoms are closely packed, the orbits of the outer-shell electrons will overlap to produce strong interatomic forces. The outer electrons, i.e., valence electrons, are of primary importance in determining the electrical properties of the solid. In a metallic conductor such as aluminum or gold, the valence electrons are shared by all the atoms in the solid. These electrons are not bound to individual atoms and are free to contribute to the conduction of current upon the application of an electric field. The free-electron density of a metallic conductor is on the order of 10^{23} cm^{-3}, and the resulting resistivity is smaller than 10^{-5} $\Omega\cdot$cm. In an insulator such as quartz (SiO$_2$), almost all the valence electrons remain tightly bound to the constituent atoms and are not available for current conduction. As a result, the resistivity of silicon dioxide is greater than 10^{16} $\Omega\cdot$cm. In silicon, a crystalline semiconductor, each atom has four valence electrons to share with its four nearest-neighboring atoms, as shown schematically in two-dimensional form in Fig. 1-8a. The valence electrons are shared in a paired configuration called a *covalent bond*. At low temperatures, these electrons are bound, and they are not available for conduction. At high temperatures, the thermal energy enables some electrons to break the bond, and the liberated electrons are then free to contribute to current conduction. Thus, a semiconductor behaves like an insulator at low temperatures and a conductor at high temperatures. At room temperature, the resistivity of pure silicon is 2×10^5 $\Omega\cdot$cm, which is considerably higher than that of a good conductor.

Whenever a valence electron is liberated in a semiconductor, a vacancy is left behind in the covalent bond, as shown in Fig. 1-8b. This vacancy may be filled by one of its neighboring valence electrons, which would result in a shift of the vacancy location. One may then see the vacancy as moving inside the crystalline structure. As a result, this vacancy may be considered as a particle analogous to an electron. This fictitious particle is called a *hole*. It carries a positive charge and

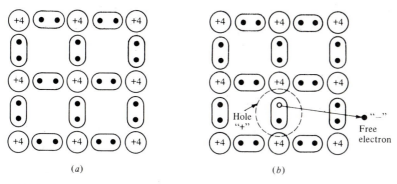

FIGURE 1-8
Two-dimensional schematic representations of crystal structure in silicon: (a) complete covalent bond and (b) broken covalent bond.

moves in the direction opposite that of an electron under an externally applied electric field.

1-4 ENERGY-BAND MODEL OF SOLIDS

In Sec. 1-2, we have considered the isolated atom, and the theory is applicable to elements in the gaseous state in which individual atoms are widely separated. In a solid, the atoms are closely packed, so they tend to interact. Figure 1-9 illustrates what would happen if two hydrogen atoms were brought together from isolation. The individual atoms have identical energy states, as shown in part (a). As we bring the two hydrogen atoms together to form molecular hydrogen, the electron orbits start to overlap, and each energy level splits into two discrete levels, as shown in Fig. 1-9b. Think about this as if you were observing the Indy 500. When the racing cars are widely separated, they all travel the same route—the shortest path. But when one car is attempting to pass another car, i.e., the two cars are close enough to interact, the two cars must travel two different paths. The two electrons react in a similar way. Notice that, in part (b), as the energy E' moves below E and E'' moves above E, the energy splitting is wider for E_2 because of more interaction. In the case of a solid, the atomic interaction involves more than two atoms. In fact, electrons in the outermost shell are shared by many atoms, so each energy level will split into a band of energy. The nature of the separation of bands and their properties are related to the atomic spacing of the atoms. Suppose we can adjust the atomic spacing of carbon at will; its energy bands will vary as shown in Fig. A-7, where the lattice constant a is an adjustable parameter.

Electrons are allowed to reside within the energy bands but not between the bands of allowed energy states. This is no different from the case of the isolated hydrogen atom, where the electron can only occupy the allowed discrete energy levels. Thus, certain bands of allowed energy levels are separated by gaps of forbidden energies in which an electron cannot exist. Each allowed energy band contains a limited number of states that can accommodate a definite number of

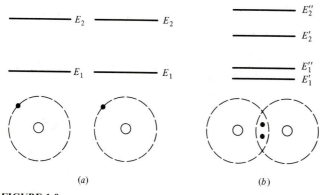

(a)   (b)

FIGURE 1-9
Two hydrogen atoms: (a) noninteracting and (b) interacting. Splitting of energy levels is illustrated for (b).

electrons. In a semiconductor the valence electrons group together to occupy a band of energy levels, called the *valence band*. The next higher band of allowed energy levels, called the *conduction band*, is separated from the former by a *forbidden gap of energy* E_g. This physical picture, called an *energy-band diagram*, is shown in Fig. 1-10 for three classes of solids. In metals, the conduction band is only partially filled by electrons, as shown in Fig. 1-10a; alternatively, the energy-band diagram for a metal may be seen as two bands which overlap so that there is no forbidden gap (Fig. 1-10b). In insulators (Fig. 1-10c) the lower band has just enough states in it to accommodate the number of valence electrons supplied by the atoms. The valence band is completely filled, and the conduction band is empty. From quantum-mechanical considerations, it is not possible to impart any net momentum to the electrons in a completely filled band. As an analogy, no one would be able to swim in a swimming pool filled with people. Thus, a completely filled band does not contribute to the current flow. In terms of the band diagram, energy added thermally or by an externally applied field must be equal to E_g for electrons in the filled valence band to jump to the conduction band in order to produce current flow. The probability of such transition is very low when E_g is large, as is true in an insulator in which $E_g > 5$ eV.

In a semiconductor (Fig. 1-10d) the forbidden-gap energy E_g is not too large, for example, $E_g = 1.12$ eV in silicon and 1.43 eV in GaAs. This corresponds to the not-so-tight bonding of the valence electrons. At room temperature, some electrons in the valence band may receive enough thermal energy to overcome the forbidden gap and reach the conduction band. For each successful transition of a valence electron to the conduction band, a hole is left behind in the valence band. In other words, in the energy-band picture, a hole is an unoccupied or empty energy level in the valence band. When an external electric field is applied, the electrons in the conduction band gain kinetic energy, which results in the flow of electrons, or current conduction. At the same time the holes in the valence band also gain kinetic energy from the applied field. Thus, electrical conduction in a semiconductor may take place by two distinct and independent mechanisms: transport of

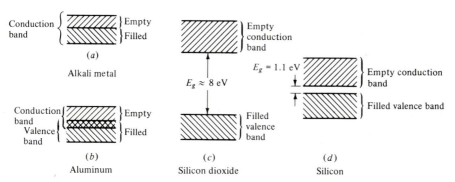

FIGURE 1-10
Energy-band diagrams for (*a* and *b*) metals, (*c*) insulator, and (*d*) semiconductor at 0 K. The vertical axis represents the energy, and the horizontal axis represents spatial position.

negatively charged electrons in the conduction band and transport of positively charged holes in the valence band.

1-5 THE CONCEPT OF EFFECTIVE MASS

Electrons and holes in a semiconductor are carriers with a charge $-q$ associated with an electron and a charge $+q$ associated with a hole, q being the electronic charge of 1.6×10^{-19} C. Each particle also carries a mass which is different from the electron mass m_o in free space. The mass of a carrier depends on the energy band it occupies, the momentum of the carrier, and the direction of the applied field. In most cases, the masses of the carriers are lighter than m_o, indicating that they behave differently from the classical particles. An important point is that the nonclassical (quantum-mechanical) model describes an electron by a traveling wave with a *wavelength* λ and a *crystal momentum p*. These two are related by the de Broglie relation

$$p = \frac{h}{\lambda} = \frac{hk}{2\pi} \tag{1-4}$$

where $k = 2\pi/\lambda$ and is known as the *wave number*.

To understand the mass of a carrier in a solid, let us consider the ideal case of an electron in a one-dimensional lattice at absolute zero temperature, as depicted in Fig. 1-11. The potential wells represent the decrease in energy as the electron passes by the positively charged atomic core. The electron wave in free space is a pure sinusoid, as illustrated by the dashed curve. However, it is disturbed in the solid by the crystal's periodic potential. The amplitude of the electron wave increases slightly in the neighborhood of the atomic cores because of coulombic attraction, but the electron propagates freely in spite of the interference. The difference between the solid and dashed curves in Fig. 1-11b illustrates an important property of the electron in a solid; i.e., its mass is different from the electron mass in free space. This is known as the *effective mass*. Physically, it means that we can

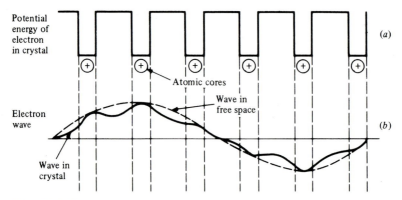

FIGURE 1-11
Representation of motion of electron wave in crystal potential. (*After Wolfendale [3]*.)

represent the perturbed sinusoid by an effectively pure sinusoid which differs from the free-space electron wave. By taking the averaged effect of the crystal field, we achieve an extremely important simplification; namely, the quantum-mechanical nature of electrons is summed up in the effective mass. We may now consider the electron in a solid as a classical particle, having the energy-momentum relationship

$$E = \frac{p^2}{2m_e} = \frac{h^2 k^2}{8\pi^2 m_e} \qquad (1\text{-}5)$$

where Eq. (1-4) has been used and E represents the kinetic energy. Because actually the energy and momentum in the solid are related in a complicated manner, we cannot use Eq. (1-5) directly to calculate the effective mass. From quantum-mechanical consideration, m_e is obtained from the following expression:

$$m_e = \frac{h^2/4\pi^2}{\partial^2 E/\partial k^2} \qquad (1\text{-}6)$$

Thus, the effective mass can be calculated if the E–k relationship is known. In practice, the effective mass is measured in a cyclotron experiment, as described in Prob. 1-11.

The energy-band diagram in Fig. 1-10d is a simplified representation of a very complex three-dimensional picture. Slightly more complicated energy-band diagrams are shown in Fig. 1-12, where Eq. (1-5) is used to plot energy vs. crystal momentum. Normally, the top of the valence band is taken as the reference level. Note that the lowest conduction-band minimum in GaAs is located at zero momentum, directly above the valence-band maximum. For this reason, GaAs is called a *direct* band-gap semiconductor. The lowest conduction-band minimum in Si is not at zero momentum, and silicon is called an *indirect* band-gap material. In either crystal, however, the band-gap energy E_g is the difference between the lowest conduction-band minimum and the valence-band maximum. These features

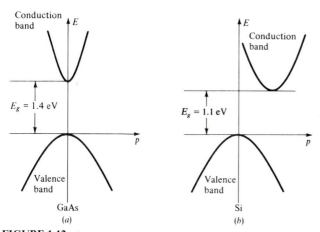

FIGURE 1-12
Energy-band diagram with energy vs. momentum for (*a*) GaAs (direct) and (*b*) Si (indirect).

are important in optoelectronic devices. With the $E–k$ curves shown in Fig. 1-12, we may make use of Eq. (1-6) to derive the effective masses for electrons in the conduction band and holes in the valence band, respectively. Since the second derivative $\partial^2 E / \partial k^2$ describes the curvature, the sharper conduction band of GaAs indicates a smaller effective mass for its electrons.

1-6 INTRINSIC AND EXTRINSIC SEMICONDUCTORS

In an absolutely perfect semiconductor, concentrations of electrons and holes are identical since the process of producing an electron in the conduction band creates a hole in the valence band. This is known as an *intrinsic* semiconductor where electrons and holes are equal. Let us designate the electron and hole concentrations per unit volume as n and p, respectively. Then, we have

$$n = p = n_i \tag{1-7}$$

where n_i is the *intrinsic* carrier concentration. The intrinsic carrier density of most semiconductors is very low and is not readily usable for device purposes. In practical devices, the carrier concentration is controlled by adding impurities. The impurity introduced is called a *dopant*, and the doped semiconductor is said to be *extrinsic*. Let us now introduce a pentavalent element from group V in the periodic table such as phosphorus, arsenic, or antimony to substitute for some silicon atoms in an otherwise perfect silicon crystal. Each new substitutional atom has one electron more than is necessary for making the covalent bonds in the lattice, as shown in Fig. 1-13a. Because they are not in a valence bond, the excess electrons are rather loosely bound to their constituent atoms and can easily be set free. Since each pentavalent atom *donates* an excess electron in silicon, the pentavalent element is called a *donor*. The free electrons carry a negative charge and leave behind a fixed positive charge with the substitutional impurity atoms. Thus, the charge-neutrality condition is preserved, and there is no vacancy in the lattice site to produce a hole. When the excess electrons are set free, the atoms are said to be *ionized*. At room temperature, the thermal energy is sufficient to ionize all impurity atoms, so the number of excess electrons equals the number of impurity atoms. In this type of extrinsic semiconductor, called *n type*, the concentration of electrons is much larger than that of holes. The carrier density of the crystal is controlled by the group V impurity concentration.

If a trivalent element, boron, aluminum, or gallium, is used as the substitutional impurity, one electron is missing from the bonding configuration, as illustrated in Fig. 1-13b. A hole is thus produced with a fixed negative charge remaining with the impurity atom. The number of holes is equal to the number of impurity atoms if ionization is complete. The trivalent impurity in silicon is known as an *acceptor* since it accepts an electron to produce a hole. This type of semiconductor is called *p type*.

Energy-band diagrams for both *n*- and *p*-type semiconductors are illustrated in Fig. 1-14, where E_c and E_v are the conduction and valence band-edge energies,

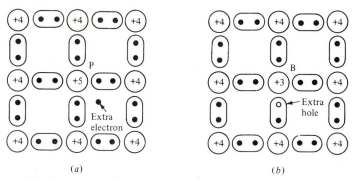

FIGURE 1-13
Crystal structure of silicon with a silicon atom displaced by (a) a pentavalent (donor) impurity atom and (b) a trivalent (acceptor) impurity atom.

respectively. The energy level E_a is the acceptor energy level, and $E_a - E_v$ is called the *ionization energy* of the acceptor impurity. The ionization energy is small because an acceptor impurity can readily accept an electron. The small ionization energy puts the impurity energy level near the valence-band edge and inside the forbidden gap. Similarly, E_d is the donor energy level, and $E_c - E_d$, the donor ionization energy, represents the small energy required to set the excess electron in a donor atom free. The ionization energies of some important impurities for Si, Ge, and GaAs are given in Fig. 1-15, where the levels in the lower half of the forbidden gap are measured from E_v and are acceptors unless indicated by D for a donor level. The levels in the upper half of the forbidden gap are measured from E_c and are donors unless indicated by A for an acceptor level. Although most impurity atoms introduce a single energy level, interaction between host and impurity atoms may give rise to multiple impurity levels in the case of gold or copper in silicon.

   Let us consider a simple way of estimating the ionization energy in an *n*-type silicon. It may be recognized that the ionization process is related to the strength of the bonding between the extra electron and the donor atom. This situation is, roughly speaking, the same as that of a hydrogen atom expressed in Eq. (1-1) except that the impurity atom is embedded in solid silicon instead of in a vacuum.

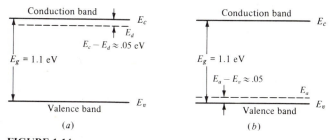

FIGURE 1-14
Energy-band diagram of (a) an *n*-type semiconductor and (b) a *p*-type semiconductor.

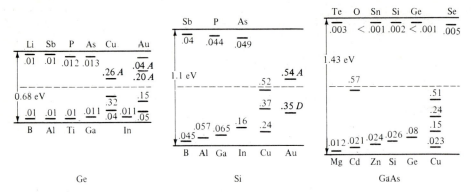

FIGURE 1-15
Ionization energies for various impurities in Ge, Si, and GaAs at 300 K. (*After Sze [4].*)

If we use the original form of Eq. (1-1), that is, Eq. (A-8), and put in the proper dielectric constant and effective mass, we have, for $n = 1$,

$$E = \frac{13.6(0.33)}{(11.8)^2} = 0.03 \text{ eV}$$

This is the required energy to free the extra electron. Compared to a hydrogen atom, it represents a much larger orbital with a smaller binding energy. The experimental values given in Fig. 1-15 range between 0.04 and 0.054 eV, reasonably close to the estimated value and much less than the band-gap energy of 1.12 eV.

1-7 FREE-CARRIER DENSITY IN SEMICONDUCTORS

To determine the electrical behavior of a semiconductor, we need to know the number of electrons and holes available for current conduction. We shall consider the electron density in the conduction band and shall apply the result to holes in the valence band by analogy. The electron density in the conduction band can be obtained if the *density-of-states function $N(E)$* and the *distribution function $f(E)$* are given. The first function describes the density of available energy states that may be occupied by an electron. The second function tells us the probability that a particular state is occupied. Thus, the density of electrons is the density of occupied states in the conduction band.

Energy and Density of States

In the energy-band diagram shown in Fig. 1-16, the lowest energy level in the conduction band, i.e., the band-edge energy E_c, is the potential energy of an electron at rest. When an electron gains energy, it moves up from the band edge to energy level E. In this picture, $E - E_c$ is the kinetic energy of the electron. Similarly, E_v

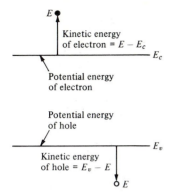

FIGURE 1-16
Schematic representation of kinetic and potential energy in the energy-band diagram.

is the potential energy of a hole at rest. As a hole's energy is increased, it moves down from the band edge to a level below E_v, $E_v - E$ being the kinetic energy of the hole.

The density $N(E)$ of available states as a function of the energy $E - E_c$ in the conduction band is derived in Appendix A:

$$N(E) = \frac{4\pi}{h^3} (2m_e)^{3/2}(E - E_c)^{1/2} \qquad (1\text{-}8)$$

where h is Planck's constant, the numerical value of which is given on the inside of the front cover, along with other important physical constants. The density of allowed energy states in the valence band is given by[1]

$$N(E) = \frac{4\pi}{h^3} (2m_h)^{3/2}(E_v - E)^{1/2} \qquad (1\text{-}9)$$

The density of states represents the number of states that could be occupied by electrons. However, not every state is occupied. To find the density of electrons, we must find out the probability of occupancy of these states, which is called the *distribution function*. We shall consider the evaluation of the electron density in the conduction band and then apply the result to holes in the valence band. This problem is analogous to finding the number of cars parked inside a municipal garage. The number of parking spots per floor is the density of states. The probability of a space being occupied depends on various factors, e.g., whether the spaces in the lower floors are filled. This probability is the distribution function. The number of cars per floor is the product of the density of states and the distribution function. The total number of cars is obtained simply by summing the cars on each floor.

---

[1] m_e and m_h in Eqs. (1-8) and (1-9) are known as the density-of-state effective masses, not to be confused with the conductivity effective masses to be described in Chap. 2.

The Distribution Function

The probability that an energy level E is occupied by an electron is given by the Fermi-Dirac distribution function

$$f(E) = \frac{1}{e^{(E-E_f)/kT} + 1} \tag{1-10}$$

where E_f is an important parameter called the *Fermi level*, k is Boltzmann's constant, and T is the temperature in kelvins. Figure 1-17 shows the probability function at 0, 100, 300, and 400 K. There are several interesting observations we can make regarding this figure. First, at 0 K, $f(E)$ is unity for $E < E_f$. This indicates that all energy levels below E_f are occupied and all energy levels greater than E_f are empty. Second, the probability of occupancy for $T > 0$ K is always $\frac{1}{2}$ at $E = E_f$, independent of temperature. Third, the function $f(E)$ is symmetrical with respect to E_f. Thus, the probability that the energy level $E_f + \Delta E$ is occupied equals the probability that the energy level $E_f - \Delta E$ is unoccupied. Since $f(E)$ gives the probability that a level is occupied, the probability that a level is not occupied by an electron is

$$1 - f(E) = \frac{1}{1 + e^{(E_f - E)/kT}} \tag{1-11}$$

Since a level not occupied by an electron in the valence band means that the level is occupied by a hole, Eq. (1-11) is useful for finding the density of holes in the valence band.

For all energy levels higher than approximately $3kT$ above E_f, the function $f(E)$ can be approximated by

$$f(E) = e^{-(E-E_f)/kT} \tag{1-12}$$

which is identical to the Maxwell-Boltzmann distribution function for classical gas particles. The electrons in the conduction band obey Eq. (1-12) if $E_c - E_f > 3kT$. For most device applications, the function in Eq. (1-12) is a good approximation

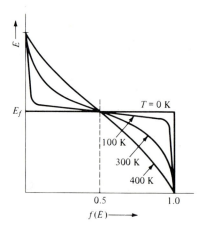

FIGURE 1-17
The Fermi-Dirac distribution function at different temperatures.

for $f(E)$. In this text, the Boltzmann function will be used throughout, and the Fermi-Dirac function will be referred to only in unusual situations.

Equilibrium Carrier Density

The electron density in the conduction band represents the number of free carriers available to participate in current conduction. It is, therefore, an important device parameter. It can be calculated by integrating the product of the density of states and the occupancy probability:

$$n = \int_{E_c}^{\infty} f(E)N(E)\, dE \tag{1-13}$$

The integration limits are from the lower conduction band edge E_c to the upper band edge. Since the upper band edge is usually very large, it has been replaced by infinity to simplify the definite integral. Performing the integration, we have[1]

$$n = N_c e^{-(E_c - E_f)/kT} \tag{1-14}$$

where

$$N_c = 2\left(\frac{2\pi m_e kT}{h^2}\right)^{3/2} \tag{1-15}$$

The quantity N_c is called the *effective density of states in the conduction band*. In silicon, $N_c = 2.8 \times 10^{19}\ \text{cm}^{-3}$ at 300 K (room temperature). Since N_c and E_c are fixed, the density of electrons is specified if the Fermi level is known. In other words, specifying E_f is the same as saying how many electrons are in the conduction band, and knowing one is equivalent to knowing the other. The use of the Fermi level frequently produces a more concise mathematical expression which is used in some advanced texts. However, using the density of electrons makes more sense to most beginning students, and we shall take this approach in this text.

The density of holes in the valence band is

$$p = \int_{-\infty}^{E_v} [1 - f(E)]N(E)\, dE \tag{1-16}$$

Substituting Eqs. (1-9) and (1-12) in Eq. (1-16) and integrating yields

$$p = N_v e^{-(E_f - E_v)/kT} \tag{1-17}$$

where

$$N_v = 2\left(\frac{2\pi m_h kT}{h^2}\right)^{3/2} \tag{1-18}$$

---

[1] Using the definite integral

$$\int_0^{\infty} x^{1/2} e^{-ax}\, dx = \frac{1}{2a}\sqrt{\frac{\pi}{a}}$$

The quantity N_v is called the *effective density of states in the valence band,* and $N_v = 10^{19}$ cm^{-3} for silicon at room temperature. Notice that the hole density like the electron density is related to the Fermi level exponentially.

What we have gone through mathematically is illustrated graphically in Fig. 1-18. Parts (a) and (b) depict the functions $N(E)$ and $f(E)$, respectively. The product $f(E)N(E)$ is shown in the upper half of part (c), and the function $[1 - f(E)]N(E)$ is sketched there in the lower half. The areas of these two halves are the electron and hole densities. In the present case, the semiconductor is intrinsic, so the electron density is equal to the hole density. Therefore, the two areas are the same as shown.

Using Eqs. (1-14) and (1-17), we obtain the product of electron and hole densities as follows:

$$np = N_c N_v e^{-E_g/kT} \tag{1-19}$$

where $E_g = E_c - E_v$, the energy needed to bridge the forbidden gap.

Equation (1-19) states that the np product is a constant in a semiconductor for a given temperature under *thermal equilibrium.* Thermal equilibrium is defined as the steady-state condition at a given temperature without external forces or excitation. This pn product depends only on the effective density of allowed energy states and the forbidden-gap energy, but it is independent of the impurity density or the position of the Fermi level.

Experimental data of silicon's band gap as a function of temperature is shown in Fig. 1-19. The intrinsic-carrier density, effective density of states, and band-gap energy, along with other important properties, are given for Ge, Si, and GaAs on the inside of the back cover. The band-gap energy is slightly temperature-dependent and can be described by the empirical relation

$$E_g = E_g(0) - \frac{\alpha T^2}{T + \beta} \tag{1-20}$$

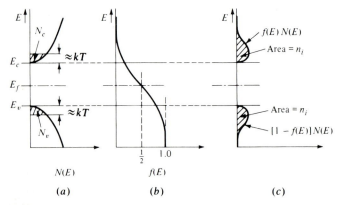

FIGURE 1-18
Graphic procedures for obtaining the intrinsic-carrier concentration: (a) the density-of-states function $N(E)$, (b) the Fermi-Dirac function $f(E)$, and (c) the density of carriers $N(E)f(E)$ at 300 K. The shaded areas in (a) correspond to the effective density of states; see Prob. 1-15 for the meaning of N_c and N_v in (a).

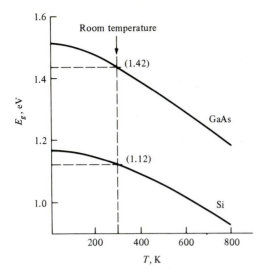

FIGURE 1-19
Energy band gaps of Si and GaAs.

where α and β are material constants, and $E_g(0)$ is the value of E_g at 0 K. For silicon we have $E_g(0) = 1.17$ eV, $\alpha = 4.73 \times 10^{-4}/\text{K}^2$, and $\beta = 636$ K.

1-8 CARRIER CONCENTRATION AND FERMI LEVEL

In an intrinsic semiconductor, the electron density is exactly equal to the hole density, as expressed by Eq. (1-7). But in an extrinsic semiconductor, the increase of one type of carriers tends to reduce the number of the other type through *recombination* (described in Chap. 2) such that the product of the two remains constant at a given temperature. This is known as the *mass-action law*:

$$np = n_i^2 \tag{1-21}$$

which is valid for both intrinsic and extrinsic materials under thermal equilibrium.

By using Eqs. (1-19) and (1-21) we can write the intrinsic-carrier density as

$$n_i = \sqrt{N_c N_v}\, e^{-E_g/2kT} \tag{1-22}$$

Near-exponential temperature dependence of n_i is obtained by using Eq. (1-22). Experimental data of the intrinsic-carrier density for Ge, Si, and GaAs as a function of temperature are shown in Fig. 1-20. From the slope of these plots, the band-gap energy can be calculated. Note that the intrinsic-carrier density is smaller for a semiconductor having a higher band-gap energy. For the same semiconductor, a higher temperature produces a higher carrier density. We can use Fig. 1-19 to obtain $E_g = 1.12$ eV and then find $n_i = 1.5 \times 10^{10}$ cm^{-3} in Fig. 1-20 for silicon at room temperature.

The Fermi level in an intrinsic semiconductor, denoted E_i, can be obtained

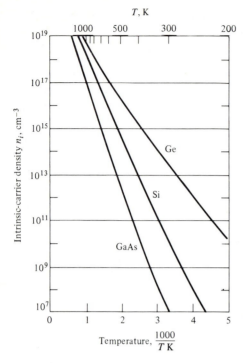

FIGURE 1-20
Experimental data of n_i for Ge, Si, and GaAs as a function of temperature. (*After Sze [4].*)

by equating Eqs. (1-14) and (1-17) and by setting $E_f = E_i$:

$$E_i = \tfrac{1}{2}(E_c + E_v) + \tfrac{3}{4}kT \ln \frac{m_h}{m_e} \qquad (1\text{-}23)$$

If the effective masses of electrons and holes are equal, the intrinsic Fermi level would be located at the middle of the forbidden gap. The small deviation of E_i from midgap due to the difference of effective masses is usually negligible. From here on, we shall use E_i to specify the midgap energy of semiconductor.

For an intrinsic semiconductor, we may set $E_f = E_i$ in Eqs. (1-14) and (1-17) and substitute them into Eq. (1-7). The result is

$$n_i = N_c e^{-(E_c - E_i)/kT} = N_v e^{-(E_i - E_v)/kT} \qquad (1\text{-}24)$$

Using the foregoing relationship, we can rewrite Eqs. (1-14) and (1-17) as

$$n = n_i e^{(E_f - E_i)/kT} \qquad (1\text{-}25)$$

$$p = n_i e^{(E_i - E_f)/kT} \qquad (1\text{-}26)$$

In Eqs. (1-25) and (1-26), we express the electron and hole concentrations in terms of the intrinsic concentration n_i and midgap energy E_i. These equations are valid for both intrinsic and extrinsic semiconductors and are sometimes more convenient to use than Eqs. (1-14) and (1-17).

With the doping of donor density N_d, the concentration of electrons in the conduction band increases from its intrinsic concentration. The result corresponds to an increase of the probability of occupancy in the conduction band. Under this condition, the Fermi level shifts upward from the midgap position. Figure 1-21 shows the graphical procedures in obtaining the carrier concentration in an n-type semiconductor. As the Fermi level moves up near the conduction band in (a), the whole curve of the distribution function $f(E)$ moves up, as shown in (b). Now, when we multiply, the $N(E)f(E)$ product increases dramatically, as is seen in the upper part of (c). At the same time, the wider space between E_f and E_v means higher occupancy of electrons or lower density of holes in the valence band, as shown in the lower part of (c). The product of np follows the mass-action law given by Eq. (1-21).

For numerical calculation, the new electron density and new position of the Fermi level can be obtained by using the following arguments. Let us assume that all donor atoms are ionized. The total number of electrons in the conduction band must equal the sum of electrons originated from the donor level and the valence band. Since each electron originated from the valence band leaves a hole behind, the electron density can be written as

$$n = p + N_d \qquad (1\text{-}27)$$

Equation (1-27) actually describes the condition of space-charge neutrality in an n-type semiconductor. When both donors N_d and acceptors N_a are present, the space-charge neutrality condition is generalized to

$$n + N_a = p + N_d \qquad (1\text{-}28)$$

By solving Eqs. (1-21) and (1-27) and by using the quadratic formula, we get the

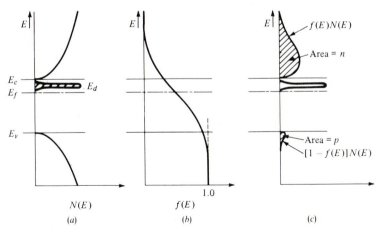

FIGURE 1-21
Graphical procedures for obtaining carrier concentrations in an n-type semiconductor.

equilibrium electron and hole concentrations in an n-type semiconductor:

$$n_n = \frac{\sqrt{N_d^2 + 4n_i^2} + N_d}{2} \tag{1-29}$$

$$p_n = \frac{\sqrt{N_d^2 + 4n_i^2} - N_d}{2} \tag{1-30}$$

The subscript n refers to the n-type semiconductor. Since the electron is the dominant carrier, it is called the *majority carrier*. The hole in the n-type semiconductor is called the *minority carrier*.

In most practical situations, the donor concentration is much higher than the intrinsic concentration; that is, $n_i/N_d \ll 1$. Therefore, we can use the binomial expansion[1] in Eqs. (1-29) and (1-30) to obtain

$$n_n = N_d + \frac{n_i^2}{N_d} \approx N_d \tag{1-31}$$

$$p_n = \frac{n_i^2}{N_d} \tag{1-32}$$

The location of the Fermi level is obtained by substituting Eq. (1-31) into Eq. (1-14) and solving for E_f. Thus,

$$E_f = E_c - kT \ln \frac{N_c}{N_d} \tag{1-33}$$

Example.   A silicon wafer is doped with 10^{15} phosphorus atoms/cm^{-3}. Find the carrier concentrations and Fermi level at room temperature (300 K).

Solution.   At room temperature, we can assume complete ionization of impurity atoms. Since $N_d = 10^{15}$ cm$^{-3} \gg n_i$, we have

$$n_n = N_d = 10^{15} \text{ cm}^{-3}$$

$$p_n = \frac{n_i^2}{N_d} = 2.25 \times 10^5 \text{ cm}^{-3}$$

The Fermi level is calculated by using Eq. (1-33):

$$E_c - E_f = 0.0258 \ln \frac{2.8 \times 10^{19}}{10^{15}} = 0.265 \text{ eV}$$

As explained in Sec. 1-7, the Fermi level represents the position in energy with an occupancy of one-half. Equation (1-33) shows that E_f is below E_c for $N_d < N_c$ in an n-type semiconductor. For a lighter doping, the Fermi level is farther away from the conduction band edge. In the above example, $E_c - E_f$ would become

---

[1] $(1 + x)^m = 1 + mx + \cdots \approx 1 + mx$ for $x \ll 1$.

quite small if N_d were greater than $10^{17}\,\text{cm}^{-3}$. In any case, the Fermi level is uniquely related to the doping concentration.

When the impurity concentration is very high and approaches the effective density of states, the Fermi level will coincide with E_c. The semiconductor is now *degenerate*. A more rigorous definition of a degenerate semiconductor is

$$E_c - E_f < 3kT \tag{1-34}$$

Under this condition, Eq. (1-33) is no longer valid since the assumption leading to Eq. (1-12) is no longer applicable. Instead of using the Boltzmann function, we must use the Fermi-Dirac distribution function for a degenerate semiconductor.

By analogy, the equilibrium carrier concentrations in a p-type semiconductor are given by

$$n_p = \frac{n_i^2}{N_a} \tag{1-35}$$

$$p_p = N_a \tag{1-36}$$

The Fermi level in a p-type semiconductor is, using Eqs. (1-17) and (1-36),

$$E_f = E_v + kT \ln \frac{N_v}{N_a} \tag{1-37}$$

Making use of Eqs. (1-25), (1-26), (1-29), and (1-30), we can obtain the position of the Fermi level of the semiconductor as a function of temperature for a given donor concentration. Calculated results are plotted in the upper half of the forbidden gap in Fig. 1-22. Similarly, we calculate the Fermi-level position in the p-type semiconductor and plot the results in the lower half of the forbidden gap

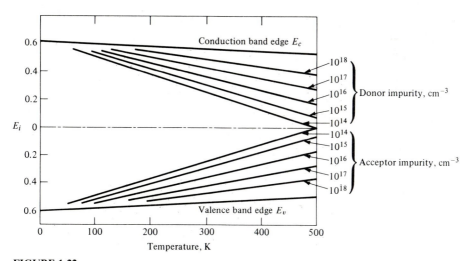

FIGURE 1-22
The Fermi level in silicon as a function of temperature for various impurity concentrations. (*After Grove [5].*)

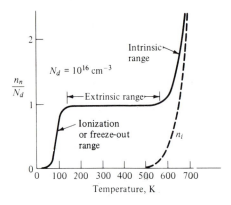

FIGURE 1-23
Electron density as a function of temperature in n-type silicon.

in the same figure. It is noted that the temperature dependence of the band-gap energy E_g is also shown. As the temperature increases, the Fermi level moves toward the midgap position; thus, the semiconductor becomes intrinsic at high temperatures.

For most of us, the electron concentration instead of the Fermi level is more informative. In Fig. 1-23, experimental data of the electron density as a function of temperature are shown for a donor density of 10^{16} cm^{-3}. At low temperatures, the thermal energy is not sufficient to ionize all the donor impurities. Therefore, the free-electron concentration is lower than the donor concentration, and the regime is known as the *freeze-out* regime, which is approximately below 100 K in this n-type silicon. As the temperature is increased, the electron density reaches the level of the donor density when all impurities are ionized. This is the *extrinsic regime* where the electron density is controlled by the doping concentration. In Fig. 1-23, this is shown between 150 and 550 K. Beyond this temperature range, the intrinsic-carrier density, being exponentially related to temperature, becomes comparable to or surpasses the impurity density. The semiconductor now behaves like an intrinsic crystal. Of course, the boundaries between different regions depend on the specific donor density. This experiment demonstrates that the carrier concentration may be very low at low temperature because of carrier freeze-out, an important consideration for low-temperature operation of devices. On the other hand, the same semiconductor may become intrinsic at high temperature even with a relatively high doping concentration. As will be explained in later chapters, a device's characteristics depend on our ability to selectively control the carrier density. When the material becomes intrinsic, the device can no longer function as a switch or an amplifier. For these reasons, both the freeze-out and the intrinsic regimes should be avoided in device operation.

REFERENCES

1. Pankove, J. I. (ed): "Amorphous Silicon," parts A–D, Academic, New York, 1984.
2. Shockley, W.: "Electrons and Holes in Semiconductors," Van Nostrand Reinhold, New York, 1950.
3. Wolfendale, E. (ed): "The Junction Transistor and Its Applications," MacMillan, New York, 1958.
4. Sze, S. M.: "Physics of Semiconductor Devices," 2d ed., Wiley, New York, 1981.
5. Grove, A. S.: "Physics and Technology of Semiconductor Devices," Wiley, New York, 1967.

ADDITIONAL READINGS

Addler, R. B., A. C. Smith, and R. L. Longini: "Introduction to Semiconductor Physics," SEEC Series, vol. 1, Wiley, New York, 1964.

Pierret, R. F.: "Semiconductor Fundamentals," Addison-Wesley, Reading, Mass., 1983.

Streetman, B. G.: "Solid-State Electronic Devices," 2d ed., Prentice-Hall, Englewood Cliffs, N.J., 1981.

PROBLEMS

1-1. Show that the bcc lattice can be formed by two interpenetrating sc lattices.

1-2. Sketch the crystal planes (010), (011), and (001) in a cubic lattice.

1-3. Determine the Miller indices for the crystal plane shown in Fig. P1-3.

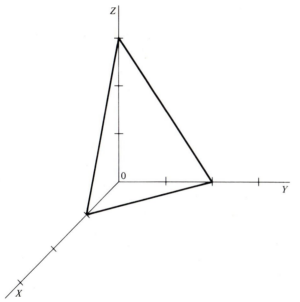

FIGURE P1-3

1-4. Sketch the two-dimensional diagram of a diamond lattice if you project all the atoms onto (*a*) the (111) plane and (*b*) the (110) plane.

1-5. Sketch a unit cube to represent a cubic lattice and draw the equivalent [111] directions in your cube.

1-6. (*a*) Determine the number of atoms in a unit cell of an fcc crystal.

  (*b*) What is the distance in units of the lattice constant *a* between two nearest-neighboring atoms?

  (*c*) If each atom is assumed to be a sphere and the spherical surface of each atom comes into contact with its nearest neighbors, what percentage of the total volume of the unit cell is being occupied?

1-7. The lattice constant of silicon is 0.543 nm. Calculate the density of valence electrons.

1-8. A light source contains a 10-W mercury-vapor lamp. Assuming that 0.1 percent of the electric energy is converted to the ultraviolet emission line of 253.7 nm, calculate the number of emitted photons per second at this wavelength.

1-9. The seven lowest energy levels of sodium vapor are 0, 2.10, 3.19, 3.60, 3.75, 4.10 and 4.26 eV. Upon absorption of a photon of wavelength 330 nm by an atom of the vapor, three photons are emitted. If one of these lines is 1138 nm, what are the other two emission lines? What energy states are involved to produce these lines?

1-10. A *mole* is defined as the quantity of a substance equal to its molecular weight in grams. In any substance, the number of molecules contained in a mole is called Avogadro's number ($N_A = 6.02 \times 10^{23}$ molecules/mole).

(a) Show that the number of atoms N per cubic centimeter is given by

$$N = \frac{N_A \rho}{\text{atomic weight} \times (1 \text{ g/mole})}$$

where ρ is the density in grams per cubic centimeter.

(b) Calculate N for silicon and germanium.

1-11. The electron effective mass is measured in the cyclotron resonance experiment by placing a crystal sample in a dc magnetic field B. An electron moves in a circular path which is defined by equating the centrifugal force to the magnetic force qvB, where v is the electron velocity.

(a) Derive the expression for the angular frequency of rotation.

(b) The frequency is measured by applying an ac electric field (normal to the magnetic field) whose energy is absorbed if the two frequencies are in resonance. If the resonant frequency is 5 GHz and $B = 2000$ G, find m_e in terms of the free-electron mass.

1-12. (a) Draw the energy-band diagram for a silicon wafer doped with 10^{15} boron atoms at 77 and 600 K and room temperature. Specify the band-gap energy. Assume constant ionization energy.

(b) At each temperature, calculate the electron and hole concentrations and the Fermi level.

1-13. In a semiconductor, the Fermi level is 250 meV below the conduction band. What is the probability of finding an electron in a state kT above the conduction-band edge at room temperature?

1-14. (a) In an n-type silicon, the donor concentration corresponds to 1 atom per 10^7 silicon atoms and the electron effective mass is $0.33m_o$. Calculate the Fermi level with respect to the conduction-band edge at room temperature.

(b) Repeat (a) if donor impurities are added to the ratio of 1 donor per 5×10^3 silicon atoms.

(c) Under what value of donor concentration will E_f coincide with E_c? Assume Boltzmann statistics is valid.

1-15. Show that the effective density of states N_c represents a total number of states in the conduction band $1.2\,kT$ wide near the edge of the conduction-band edge. Explain the physical meaning of your result.

1-16. Determine the donor or acceptor concentration that produces degenerate Si and GaAs.

1-17. Carry out the integration of Eq. (1-16) to derive Eq. (1-17).

1-18. Obtain the band-gap energy of Ge, Si, and GaAs at 300 K from Fig. 1-20. Compare your results with the data in the table on the inside back cover.

1-19. Using the data in the table on the inside back cover, calculate the deviation of E_i from the midgap energy for Ge and Si. What conclusion do you arrive at after these calculations?

1-20. (a) Calculate the Fermi level of silicon wafers doped with 10^{15}, 10^{18}, and 10^{20} arsenic atoms/cm^3 at room temperature by assuming complete ionization.

(b) Using the Fermi level obtained in (a), show whether the assumption of complete ionization is justified in each case.

CHAPTER
2

CARRIER
TRANSPORT
AND
RECOMBINATION

The movements of electrons and holes are fundamental to the operation of all semiconductor devices. In this chapter, we introduce the concepts of carrier scattering and mobility, followed by a discussion of drift and diffusion currents. The Hall effect is presented to demonstrate the magnetic effect and the method of determining the carrier concentration and mobility. The topic of adding excess carriers by means of generation and injection is discussed. It will be shown that, on average, the excess carriers will survive for a time before they are annihilated by recombination processes. During their lifetime, excess carriers can diffuse or drift for a useful distance to produce a current. The main object of this chapter is to establish the basic laws governing the recombination and transport of non-equilibrium electrons and holes. These basic laws will be used in later chapters to derive the current-voltage characteristics of diodes and transistors.

2-1 SCATTERING AND DRIFT OF ELECTRONS AND HOLES

Under thermal equilibrium , mobile electrons and atoms of the lattice are always in random thermal motion. From the theory of statistical mechanics, a carrier at a temperature T is estimated to have an average thermal energy of $3kT/2$. This thermal energy can be translated into an average velocity v_{th} by using the relationship

$$\tfrac{1}{2}m_e^* v_{th}^2 = \tfrac{3}{2}kT \tag{2-1}$$

The average velocity of thermal motion for electrons in silicon at room temperature is calculated to be approximately 10^7 cm/s. We may now visualize the electrons as moving rapidly in all directions. At the same time, the lattice atoms experience thermal vibrations that cause small periodic deviations from their mean positions. As a result, the electrons make frequent collisions with the vibrating atoms. Such a collision may be perceived as the interference of the electron wave (Fig. 1-11) by the lattice-vibration wave so that electrons may no longer propagate freely. Under thermal-equilibrium condition, the random motions of these electrons cancel out, and the average current in any direction is zero. It should be pointed out that the effect of the periodic field established by the lattice atoms has been taken into account by the effective mass (Sec. 1-5). Therefore, the lattice atoms do not deflect the conduction electrons. However, the effective mass of electrons does not account for the deviation of the atoms from their periodic positions caused by the thermal energy of the system. It is these deviations from a perfect periodic lattice that deflect or scatter the free electrons. The thermal vibrations may be treated quantum-mechanically as discrete particles called *phonons*,[1] and the collision of phonons with electrons and holes is called *lattice* or *phonon scattering*. Lattice scattering increases with increasing temperature because of the increased lattice vibration. It dominates other scattering processes at and above room temperature in lightly doped silicon. Most semiconductor devices are operated in this temperature range.

Besides lattice scattering, there are three other scattering mechanisms: (1) The ionized impurity atoms are charged centers that may deflect the free carriers. The effect of this mechanism depends on the temperature and the impurity concentration. (2) The neutral impurity atoms may introduce scattering if the concentration of these atoms is high. This scattering takes place when the free carriers are influenced by the atomic core and orbiting electrons separately. Usually, the effect of this scattering is negligible. (3) The Coulomb force between carriers (electron-electron, electron-hole) can cause scattering at high concentration of carrier densities.

Under thermal equilibrum, the random motion of electrons leads to zero current in any direction. This physical picture is shown in Fig. 2-1a. The average distance between collisions is called the *mean free path* l_m. Regardless of which scattering mechanism is dominating, the typical value of the mean free path is between 10^{-6} and 10^{-4} cm. With a velocity of 10^7 cm/s, the *mean free time* τ_m between collisions is about 1 ps (1 picosecond $= 10^{-12}$ s). Let us apply an external electric field across the crystal. The electrons are acted upon to move in the direction

---

[1] Like an electron, a phonon has the dual nature of wave and particle. Its frequency or wavelength of vibration is quantized corresponding to a discrete energy. In a solid, the discrete energies become phonon bands. The vibration or oscillation frequencies are in the acoustic and optical ranges so that they are known as *acoustic* and *optical phonons*.

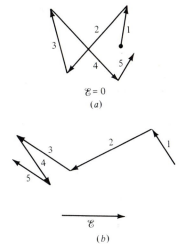

$\mathcal{E} = 0$
(a)

$\mathcal{E}$
(b)

FIGURE 2-1
Typical path of electron in crystal (a) without and (b) with electric field.

opposite that of the electric field. This external force superimposes on the random motion of electrons to produce activity such as that shown in Fig. 2-1b. We notice that the electrons now *drift* in the direction opposite the field. Thus, current flow is induced.

Unlike electrons moving in a vacuum, electrons in a semiconductor under an external field do not achieve constant acceleration. Instead, due to energy-losing collisions, they move with an average velocity called the *drift velocity*. The carriers are accelerated by the field and then decelerated by a collision. The drift velocity is a function of the applied field, and at low electric fields the magnitude of the velocity can be derived by using the following simple model.

Since the acceleration of an electron is given by $\bar{a} = -q\mathcal{E}/m_e^*$ (Newton's second law) and the increase of velocity between two consecutive collisions is $\Delta v = \bar{a}\tau_m$, the average drift velocity (if zero initial velocity is assumed) is

$$v_d = \frac{\Delta v}{2} = -\frac{q\tau_m}{2m_e^*}\mathcal{E} \tag{2-2}$$

In a more accurate model including the effect of statistical distribution, the factor 2 does not appear in the denominator of Eq. (2-2). The correct expression is[1]

$$v_d = -\frac{q\tau_m}{m_e^*}\mathcal{E} = -\mu_n\mathcal{E} \tag{2-3}$$

where $\mu_n \equiv q\tau_m/m_e^*$ is called the *electron mobility*. A similar equation can be written for the hole:

$$v_d = \mu_p\mathcal{E} \tag{2-4}$$

where μ_p is the *hole mobility*. The negative sign is missing in Eq. (2-4) because holes drift in the same direction as the electric field.

---

[1] m_e^* in Eq. (2-3) is known as the *conductivity effective mass* [1].

According to Eq. (2-3), the drift velocity is linearly related to the applied electric field. This linear relationship is valid when the electric field is low. By increasing the applied field, the drift velocity eventually approaches a limit very close to the thermal velocity at room temperature. The velocity does not increase with the electric field because at high electric field, the energy gained by the electron will be lost through the emission of optical phonons, i.e., generating a lattice vibration of optical mode as a result of the collision. Thus, the mobility decreases as the electric field is increased, giving rise to a nonlinear mobility. Experimental data showing the field-dependent drift velocity in silicon are plotted in Fig. 2-2, which can be approximated by the empirical expression

$$v_d = v_{th}(1 - e^{-\mathscr{E}/\mathscr{E}_c})\qquad(2\text{-}5)$$

where $\mathscr{E}_c$ is called the *critical field*. For *n*-type silicon the critical field is approximately 1.5×10^4 V/cm.

The drift velocity for electrons in GaAs is also shown in Fig. 2-2. The most striking feature of this curve is that the velocity reaches a peak and then decreases with further increase of the electric field. This unusual behavior arises from electron transport in a semiconductor with two separate valleys in the conduction band, as shown in Fig. 2-3. At a lower electric field, electrons reside in the lower valley where electrons travel faster because of the lower effective mass. As the electric field is increased, the resulting higher kinetic energy transfers some of these electrons to the upper valley where electron motion is slower. The decrease in velocity with increasing electric field gives rise to a differential negative mobility. Since a positive resistance consumes power and a negative resistance supplies power, a negative differential mobility may generate ac power, i.e., oscillation in the microwave regime. This is known as the *Gunn effect* and is the basis of operation of the Gunn or transfer-electron oscillator.

Since the mobility takes into account the effects of scattering, the contribution of different scattering mechanisms to the mobility variation should be readily

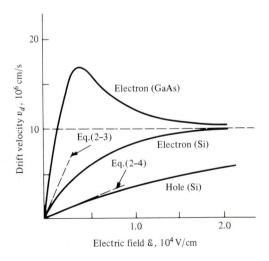

FIGURE 2-2
Drift velocity as a function of the electric field for electrons and holes.

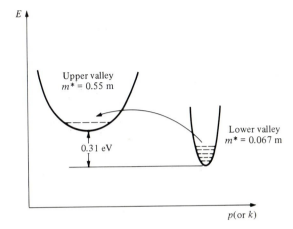

FIGURE 2-3
Schematic diagram of the two-valley conduction band in GaAs.

observable. It has already been mentioned that the lattice scattering increases with increasing temperature. Increased scattering results in a shorter mean free time, yielding a lower mobility. For a very pure sample, the impurity and carrier scattering mechanisms are negligible, and the temperature effect of the mobility is dominated by the lattice scattering. Experimental results show a temperature dependence ranging between $T^{-3/2}$ and $T^{-5/2}$ for the electron and hole mobilities in silicon, as demonstrated by the sample doped with 10^{13} cm^{-3} in Fig. 2-4. On the other hand, the effect of impurity scattering on the mobility is most pronounced for

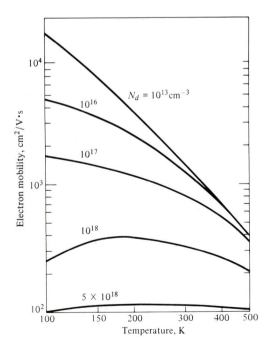

FIGURE 2-4
Mobility as a function of temperature in silicon.

heavily doped samples at low temperatures, where the lattice scattering can be ignored. The low temperature reduces the thermal velocity of carriers so that electrons and holes traveling past fixed charged ions will be deflected by the Coulomb force set up by the charged ions. As the temperature is increased, the fast-moving carriers become less likely to be deflected by the charged ions and the scattering is decreased. Therefore, the mobility increases with temperature, as seen most clearly for the electron mobility with doping of 10^{18} cm^{-3} in Fig. 2-4. This curve is shown schematically in Fig. 2-5 to illustrate the contribution from the two most important scattering mechanisms. The impurity or lattice scattering could each become the bottleneck for the electron movement so that

$$\frac{1}{\mu} = \frac{1}{\mu_L} + \frac{1}{\mu_I} \qquad (2\text{-}6)$$

where the subscripts denote the lattice and impurity scattering, respectively. The smaller of the two is the dominating process. Note also that the mobility decreases with increasing impurity concentration at a given temperature in Fig. 2-4.

Another important experimental result is shown in Fig. 2-6 for mobility as functions of donor and acceptor impurity concentrations. It is seen that the mobility is constant at low impurity concentrations. For impurity concentrations greater than 10^{16} cm^{-3} the mobility decreases as a result of impurity scattering. Note that the electron mobility is higher than that of holes, a situation found in most semiconductors. In silicon, the electron-to-hole mobility ratio is approximately 3, and the ratio for GaAs is about 20.

2-2   THE DRIFT CURRENT

The transport of carriers under the influence of an applied electric field produces a current called the *drift current*. Consider the semiconductor shown in Fig. 2-7. The current flow in the bar having n electrons per unit volume is given by

$$I_n = -qAnv_d = qAn\mu_n\mathscr{E} \qquad (2\text{-}7)$$

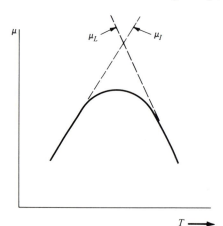

FIGURE 2-5
Mobility variation with temperature showing the effect of lattice and impurity scattering.

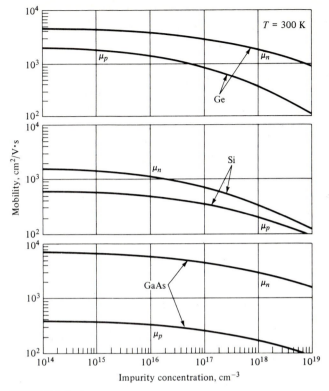

FIGURE 2-6
Electron and hole mobilities vs. impurity concentration for Ge, Si, and GaAs at 300 K. (*After Sze [1].*)

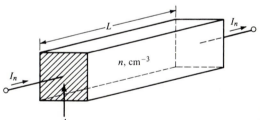

FIGURE 2-7
Current conducton in a semiconductor bar.

where q = magnitude of electronic charge

A = the cross-sectional area

L = length of bar

In addition Eq. (2-3) has been employed. If we now replace $\mathscr{E}$ by V/L in Eq. (2-7), and obtain the voltage-to-current ratio, we have

$$\frac{V}{I_n} = \frac{L}{qAn\mu_n} \qquad (2\text{-}8)$$

The resistance of the bar is defined by

$$R \equiv \rho \, \frac{L}{A} \equiv \frac{V}{I_n}$$

(2-9)

where ρ is the *resistivity*. Substituting Eq. (2-8) into Eq. (2-9) and solving for ρ, we obtain

$$\frac{1}{\rho} = \sigma = q\mu_n n$$

(2-10)

where σ is the *conductivity*. By analogy, the hole drift current can be written as

$$I_p = qA p \mu_p \mathscr{E}$$

(2-11)

The overall resistivity of a semiconductor including the effects of electrons and holes becomes

$$\frac{1}{\rho} = q\mu_n n + q\mu_p p$$

(2-12)

The resistivity of a semiconductor is an important parameter in device design. Figure 2-8 shows the relationship between the impurity concentration and resistivity for both *n*- and *p*-type silicon and GaAs at room temperature. The deviation from linearity in these curves is caused by the nonlinear mobility effect.

2-3 CARRIER DIFFUSION

Diffusion is a phenomenon that occurs frequently in our daily experience, for example, the propagation of flower fragrance or the tinting of a cup of hot water

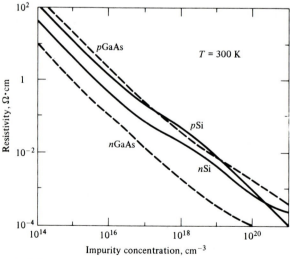

FIGURE 2-8
Resistivity vs. impurity concentration in Si and GaAs at 300 K. (*After Sze [1].*)

by adding a tea bag. In either case, although air convection or water motion could give rise to our senses, they are not necessary because perfectly stationary air or water can produce the same effects. They are the consequence of the diffusion of chemical molecules.

Let us consider the effect of dropping a tiny amount of highly concentrated red ink in a glass of water. Suppose that it is carefully done so that the water is not disturbed. We will observe that the ink spot is slowly enlarged. Physically, the ink droplet contains a large number of molecules of color pigments. The individual molecules move around because of thermal energy. Their motion is random, and they experience collisions just like electron scattering described in Sec. 2-1. In truly random movements, half the scattered color pigments will move toward the center of the ink droplet and the other half will move away from the center. Now, the number of molecules is fixed. The migration of the molecules reduces the density at the center of the color spot so that the spot becomes enlarged with a lighter color. This may be seen schematically in Fig. 2-9. Let the curve at t_1 represent the color-pigment concentration of the ink droplet shortly after it reaches the water. It has a peak density of unity and a small half-width. As time goes by, more molecules migrate away so that the density at the center goes down and the half-width increases. Eventually, the whole glass of water will be permeated with red ink to become a uniform light red liquid. Notice that the movement of color pigments is related to the concentration gradient so that the molecules have a net tendency to move from high concentration to low concentration. This is the basic physics of particle diffusion. When the concentration becomes uniform throughout the liquid, the molecules continue to move around. However, this random motion requires that movements in all directions are the same so that the concentration uniformity remains unchanged. Macroscopically, the density of the color pigments is the same throughout so that the diffusion stops. Therefore, the process of diffusion results from nonuniform distribution of the color pigments.

Similarly, in a semiconductor, the diffusion of electrons or holes results from their movement from high concentration to low concentration. Because electrons

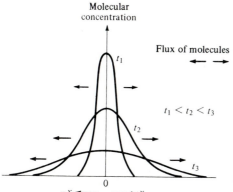

FIGURE 2-9
Spatial variation of molecules at different times.

and holes are charge carriers, their motion gives rise to a current flow known as the *diffusion current*. The diffusion flux obeys Fick's first law:

$$F = -D \frac{dN}{dx} \tag{2-13}$$

where $F = flux\ of\ carriers =$ numbers passing through $1\ \text{cm}^2/\text{s}$

$D =$ diffusion constant

$N =$ carrier density

With a cross-sectional area A, the current is given by qAF so that the diffusion currents of electrons and holes are

$$I_n = qAD_n \frac{dn}{dx} \tag{2-14}$$

$$I_p = -qAD_p \frac{dp}{dx} \tag{2-15}$$

where D_n and D_p are the diffusion constants for electrons and holes, respectively. The negative sign in Eq. (2-15) indicates that the hole current flows in the direction opposite the gradient of holes. This is shown in Fig. 2-10. The diffusion constants will be shown later to be related to the mobility by the Einstein relationship:

$$\frac{D_p}{\mu_p} = \frac{D_n}{\mu_n} = \frac{kT}{q} \tag{2-16}$$

For lightly doped silicon at room temperature $D_n = 38\ \text{cm}^2/\text{s}$ and $D_p = 13\ \text{cm}^2/\text{s}$.

The total electron and hole currents are obtained by adding the drift and diffusion components:

$$I_n = qA \left(\mu_n n \mathscr{E} + D_n \frac{dn}{dx} \right) \tag{2-17}$$

$$I_p = qA \left(\mu_p p \mathscr{E} - D_p \frac{dp}{dx} \right) \tag{2-18}$$

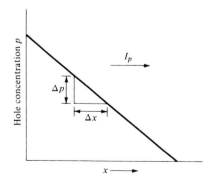

FIGURE 2-10
Steady-state hole concentration gradient producing diffusion current.

These equations will be used throughout the text to characterize semiconductor devices.

2-4 THE HALL EFFECT

If one places a semiconductor carrying a current I in a transverse magnetic field B, an electric field will be induced in the direction normal to both I and B. This phenomenon, known as the *Hall effect*, is frequently employed to distinguish an n-type from a p-type sample and to determine the carrier concentration. It can also be used to calculate the carrier mobility if the conductivity is measured separately.

Let us consider the silicon bar shown in Fig. 2-11 with a current $qnAv_x$ in the x direction and B_z in the z direction. A force $qv_x B_z$ will be exerted in the negative y direction on the current carriers. If the semiconductor is n-type, the majority carriers (electrons) will be deflected downward toward side 1. Thus, side 1 will become more negatively charged with respect to side 2, setting up an electric field $\mathscr{E}_y$ in the $-y$ direction. This electric field opposes the electron movement so that, eventually, an equilibrium is reached for zero deflection, given by the condition

$$q\mathscr{E}_y = qv_x B_z \qquad (2\text{-}19)$$

The induced voltage V_H, known as the *Hall voltage*, between sides 1 and 2 is

$$V_H = \mathscr{E}_y d = v_x B_z d = \frac{IB_z}{qwn} \qquad (2\text{-}20)$$

where Eqs. (2-7) and (2-19) have been used. The carrier concentration can be obtained from Eq. (2-20):

$$n = \frac{IB_z}{qwV_H} \equiv \frac{1}{qR_H} \qquad (2\text{-}21)$$

where the *Hall coefficient* R_H is defined. Equation (2-21) may be used to calculate the carrier concentration accurately since all parameters in the middle term are measurable. A carrier density as low as 10^{12} cm^{-3} is obtainable from this

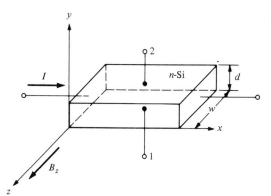

FIGURE 2-11
A semiconductor bar under a magnetic field. The cross-sectional area $A = w \cdot d$.

experiment. For a p-type sample, holes traveling in the opposite direction of the electrons will also be deflected toward side 1, setting up a Hall voltage with opposite polarity. For this reason, n-type and p-type specimens can be easily distinguished.

Let us now measure the conductance of the sample as

$$\sigma = q\mu_n n \tag{2-22}$$

which can be rewritten as

$$\mu_n = \frac{\sigma}{qn} = \sigma R_H \tag{2-23}$$

Thus, the mobility can be determined by using both the Hall and conductance measurements if conduction is due primarily to one type of carriers.

In the above discussion, it was assumed implicitly that all carriers travel with the same drift velocity v_x. When statistical distribution of velocity due to random thermal motions is taken into account, Eq. (2-23) must be modified to

$$\mu_n = r\sigma R_H \tag{2-24}$$

where r equals 1.18 for phonon scattering and 1.93 for ionized impurity scattering. In practice, a patterned semiconductor bar shown in Fig. 2-12 is employed, where the conductivity is obtained by measuring V across the outer pairs of contacts while the Hall voltage is measured as shown for a constant current I. Thus, carrier concentration and mobility of the sample are determined together accurately. A more convenient geometry using modern microelectronic fabrication technique is the van der Pauw method having four contacts for small samples of arbitrary shape. Interested readers are referred to the literature [2].

An obvious application of the Hall effect is the measurement of the magnetic field. Another interesting device is called the *Hall multiplier*, where the output signal is the product of the input current I and the magnetic field B_z. Of course, using the magnetic flux density as an input imposes a serious limitation, since it

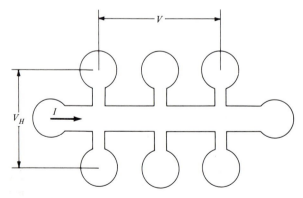

FIGURE 2-12
A semiconductor Hall sample with current flowing through the horizontal bar.

is very inconvenient to have a signal in the magnetic form. This analog multiplier was quite popular before integrated multipliers became available.

2-5   GENERATION, RECOMBINATION, AND INJECTION OF CARRIERS

Under thermal equilibrium, carriers in a semiconductor possess an average thermal energy corresponding to the ambient temperature. This thermal energy enables some valence electrons to reach the conduction band. The upward transition of an electron leaves a hole behind so that an electron-hole pair is produced. This process is called *carrier generation* and is represented by G_{th} in Fig. 2-13a. When an electron makes a transition from the conduction band to the valence band, an electron-hole pair is annihilated. This reverse process, called *recombination*, is represented by R_{th} in Fig. 2-13a. Under thermal equilibrium, the generation rate and the recombination rate must be equal so that the carrier concentrations remain constant. Thus, the condition $pn = n_i^2$ is maintained.

The equilibrium condition may be disturbed by the introduction of free carriers exceeding their thermal-equilibrium values. This process, called *carrier injection*, can be accomplished by either optical or electrical means. Optical injection involves an incident light having photon energy equal to or greater than the band-gap energy E_g. The photon energy is given by the product hv, where v is the frequency of light and h is Planck's constant. When the optical energy is absorbed

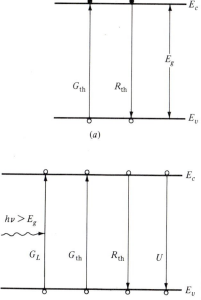

(a)

(b)

FIGURE 2-13
Band-to-band generation and recombination of electron-hole pairs (a) at thermal equilibrium and (b) under optical illumination.

by an electron in the valence band, the electron is excited to the conduction band and a hole is created in the valence band. The generaton rate of electron-hole pairs by light is shown in Fig. 2-13b as G_L. The injection of carriers increases the electron and hole densities such that $pn > n_i^2$. The additional carriers are called *excess carriers*. The excess electrons and holes are always in equal number so that space-charge neutrality is preserved.

Excess electrons in the conduction band may recombine with holes in the valence band, as represented by U in Fig. 2-13b. Thus, the optically generated electron-hole pairs may be annihilated and the thermal-equilibrium density may be reestablished. The energy released by the recombination is emitted as a photon or phonons, depending on the nature of the recombination mechanism. When a photon is emitted, the process is called *radiative recombination*. On the other hand, the lack of photon emission indicates a *nonradiative* recombination process, which emits phonons to the lattice in the form of heat dissipation.

The number of injected carriers in a semiconductor generally controls a device's behavior. Let us consider the case of an *n*-type silicon wafer uniformly illuminated by a light source under steady-state conditions. Before the light source is turned on, the silicon wafer is in equilibrium and there are no excess carriers. The majority-carrier density is equal to the donor concentration, and the minority-carrier density can be calculated from Eq. (1-32). After the light is turned on, two different conditions may exist. If the injected-carrier density is small compared with the donor concentration, the majority-carrier density remains essentially unchanged while the minority-carrier density is equal to the injected-carrier density. This condition is called *low-level injection*. If the injected-carrier density is comparable to or exceeds the donor concentration, it is called *high-level injection*. These conditions are illustrated by the example shown in Table 2-1. It should be pointed out that the total carrier density always equals the sum of the equilibrium and excess-carrier densities. High-level injection usually introduces additional complexity in the mathematical analyses. Since it provides little additional physical insight into device behavior, we shall ignore high-level injection effects whenever possible.

TABLE 2-1
n-type silicon with $N_d = 2.25 \times 10^{15}$ cm^{-3}

Carrier density, cm^{-3}	Injection condition		
	Equilibrium	Low level	High level
Excess Δn	0	10^{13}	10^{16}
Majority n_n	2.25×10^{15}	2.26×10^{15}	1.225×10^{16}
Minority p_n	10^5	10^{13}	10^{16}

2-6 CARRIER RECOMBINATION

While generation represents the creation of an electron-hole pair, recombination is the annihilation of the same pair of carriers which involves the capture of an electron by a hole. Since the electron and the hole are in two separate energy bands, different routes are possible to accomplish the same result. In Fig. 2-14, we have depicted three alternatives, the *direct*, *indirect* and *Auger* recombinations. When an electron makes a direct transition from the conduction band to the valence band as shown in (*a*), it is known as the *band-to-band* or *direct recombination*. This mechanism usually takes place in a direct-gap semiconductor such as GaAs, where the minimum of the conduction band aligns with the maximum of the valence band (Fig. 1-12*a*). Under this condition, an electron in the conduction band can simply give up its energy to move down and occupy the empty state (hole) in the valence band without a change in momentum. This is the most efficient way to annihilate an electron-hole pair in a direct-band semiconductor. When it happens, the energy given up by the electron will be emitted as a photon, the quantum of light. For this reason, we are able to make light-emitting diodes and lasers by using direct-band semiconductors.

The second type of recombination is the indirect transition through a recombination center (Fig. 2-14*b*). In silicon or germanium the minimum of the conduction band and the maximum of the valence band are separated in momentum space, as shown in Fig. 1-12*b*. For an electron to reach the valence band, it must experience a change of momentum as well as energy to satisfy the conservation principles. In the solid state, the change of momentum is accomplished by absorbing or emitting phonons, the quanta of lattice vibration. In the valence-bond model, this can be perceived as the scattering of an electron by lattice vibrations so that the electron will reach an empty site. The probability of such an event is quite low. However, there are electronic states deep in the otherwise forbidden gap that are created by defects or impurities. Here, the word *deep* indicates that the states are away from the band edges and near the center of the forbidden gap. These states are known as *recombination centers* or *traps*. A recombination center first captures

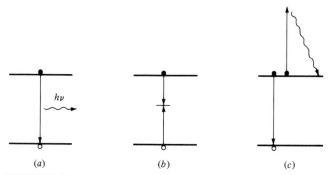

(*a*)   (*b*)   (*c*)

FIGURE 2-14
Three recombinaton processes: (*a*) direct and radiative, (*b*) indirect and nonradiative, and (*c*) Auger.

an electron and then a hole to complete the pair annihilation. The probability of such an indirect transition is much enhanced by the recombination center. It should be pointed out that the word *trap* is used interchangeably with the recombination center in this text. In the literature, there is the slight distinction that trapping implies the capturing of a carrier for a definite time, after which the captured carrier either is reemitted back to where it came from or recombines with the other type of carrier.

The third type of recombination is shown in Fig. 2-14c. The energy released by the direct recombination is absorbed by a second electron in the conduction band. This second electron, after being kicked upward, loses its energy to the lattice by scattering events. This is known as Auger recombination. There are different possibilities when traps are involved, and the one shown is the simplest example. Usually, Auger recombination is important when the carrier concentration is very high as a result of either high doping or high injection level.

The rate of recombination depends on the physical mechanism, which in turn relates to the specific material and doping concentration. In direct recombination, an electron in the conduction band makes a transition directly to the valence band to recombine with a hole. The rate of transition U is proportional to the available excess-electron density Δn in the conduction band and the number of holes in the valence band p_o.[1] Therefore

$$U = B \, \Delta n p_o \tag{2-25}$$

where B is a probability coefficient taking into account the statistical nature of recombination. The ratio of excess-carrier density to the recombination rate defines a characteristic time constant

$$\tau \equiv \frac{\Delta n}{U} \tag{2-26}$$

The quantity τ is called the *carrier lifetime* of the excess carriers. Thus, if the recombination rate is $10^{18} \, \text{cm}^{-3} \, \text{s}^{-1}$ and the minority-carrier lifetime is 100 μs, the excess-carrier density is $10^{14} \, \text{cm}^{-3}$, according to Eq. (2-26). Physically, τ represents the average time an electron remains free before it recombines with a hole.

In indirect recombination, the derivation of the recombination rate and carrier lifetime is more complicated. A schematic diagram showing the various steps involved in the recombination via an intermediate center is shown in Fig. 2-15, where the centers are located at E_t with a density of $N_t \, \text{cm}^{-3}$. There are four possible transition processes: (1) an electron is captured by an empty center, (2) an electron is emitted from an occupied center, (3) an occupied center captures a hole, and (4) an empty center emits a hole. By considering the transition processes in details, it is possible to derive the net recombination rate, which is given in Appendix B. The result is reproduced here for the case when the capture

---

[1] The more exact definition is $U = B(np - n_i^2)$, which is reduced to Eq. (2-25) in a p-type semiconductor under low-level injection; that is, $p_o \gg \Delta n$.

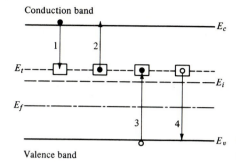

Conduction band

Valence band

FIGURE 2-15
Generation and recombination through intermediate states.

probabilities for electrons and holes are identical:

$$U = \frac{cN_t(pn - n_i^2)}{n + p + 2n_i \cosh[(E_t - E_i)/kT]}$$ (2-27)

where c is the capture probability coefficient.

Let us consider the recombination rate for an n-type silicon with a center located at E_i, the midgap position. At low-level injection where $n_n = n_{no}$, we have

$$U \approx cN_t(p_n - p_{no})$$ (2-28)

The last expression is obtained by using the mass-action law and $n_n \gg p_n + 2n_i$, a condition from the assumption of low-level injection. Using the definition of the minority-carrier lifetime, we find

$$\tau_p = \frac{1}{cN_t}$$ (2-29)

It is noted that the hole lifetime τ_p is inversely proportional to the number of recombination centers. In the past, some transistors and diodes were fabricated with a large trap density for high-speed operation. These traps were introduced by gold doping to achieve nanosecond switching speed. This practice is no longer common because of reliability problems associated with gold diffusion.

The carrier lifetime is an important device parameter because it indicates the quality of the semiconductor wafer. When the number of crystal defects is large, the lifetime is small. It has a direct influence on the performance of bipolar transistors and diodes. Even in a field-effect transistor, it will influence the leakage current, breakdown voltage, and device reliability. For this reason, starting materials for microelectronic structures should have a long carrier lifetime.

2-7 TRANSIENT RESPONSE AND PHOTOCONDUCTIVITY

Let us consider a slab of semiconductor illuminated by a light source whose photon energy is greater than the band-gap energy. Electron-hole pairs are generated, and their densities tend to increase with time. However, the increase of excess-carrier density is limited because the recombination process counterbalances the generation process. A steady state will be established when the recombination rate equals the generation rate. The behavior of excess carriers is therefore described by the *rate equation* which states that the time rate of change of the minority carriers is equal to the total generation rate minus the total recombination rate:

$$\frac{d\Delta p}{dt} = G_L - U \tag{2-30}$$

This equation can be rewritten for direct recombination under low-level injection as

$$\frac{d\Delta p}{dt} = G_L - \frac{\Delta p}{\tau_p} \tag{2-31}$$

The steady-state excess-carrier concentration is $\Delta p = G_L \tau_p$ at $t \leqslant 0$ because the derivative of the excess carrier is zero in Eq. (2-31). If we use this relation as the initial condition, Eq. (2-31) can be solved to yield

$$\Delta p = G_L \tau_p e^{-t/\tau_p} \tag{2-32}$$

Figure 2-16 shows the plot of Eq. (2-32). The analysis of the transient response for the indirect recombination in the general case is more involved. It does not necessarily lead to a solution with a simple exponential as expressed in Eq. (2-32). However, if we assume that the conditions leading to Eq. (2-29) prevail, Eq. (2-32) is a good approximation for the transient decay of carriers in indirect recombination as well.

The *photoconductivity* decay experiment is shown in Fig. 2-17. The conductivity is given by

$$\sigma = q[\mu_n n(t) + \mu_p p(t)] \tag{2-33}$$

Since $n(t) = n_o + \Delta n(t)$, $p(t) = p_o + \Delta p(t)$, and $\Delta n = \Delta p$, we have

$$\sigma(t) = \sigma_o + q(\mu_n + \mu_p)\, \Delta p(t) \tag{2-34}$$

where

$$\sigma_o = q(\mu_n n_o + \mu_p p_o) \tag{2-35}$$

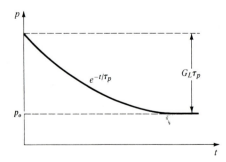

FIGURE 2-16
Transient decay of excess carriers after removal of light source.

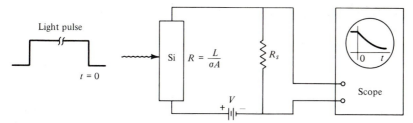

FIGURE 2-17
The photoconductivity decay experiment.

In the experimental setup, the voltage measured by the oscilloscope is given by

$$v = \frac{R_s V}{R + R_s} \approx \frac{R_s V}{R} \qquad \text{if } R \gg R_s \tag{2-36}$$

Using $R = L/\sigma A$, we rewrite Eq. (2-36) as

$$v = \frac{\sigma A R_s V}{L} \tag{2-37}$$

Since σ varies linearly with Δp in Eq. (2-34) and Δp varies exponentially with t/τ_p in Eq. (2-32), we can obtain the carrier lifetime from the oscilloscope waveform. The photoconductivity effect is useful for light detection, as will be described in Sec. 14-8.

2-8   SURFACE RECOMBINATION

The recombination processes described previously take place in the bulk of the semiconductor. It is conceivable that similar carrier activity occurs at the semiconductor surface. In fact, the presence of discontinuity in the lattice structure at the surface introduces a large number of energy states in the forbidden gap, as shown in Fig. 2-18a. These energy states, called *surface states*, greatly enhance the recombination rate at the surface region. In addition to the surface states, other imperfections exist resulting from adsorbed ions, molecules, or mechanical damage in the layer next to the surface. The adsorbed ions, for example, may be charged so that a space-charge layer is formed near the surface. Regardless of the origin of the surface imperfections, the recombination rate at the surface per unit area can be written in analogy to Eq. (2-28) as

$$U_s = cN_{ts}[p_n(0) - p_{no}] \tag{2-38}$$

where $p_n(0)$ is the average surface minority-carrier density and N_{ts} is the average recombination-center density *per unit area* in the surface layer. Since the product cN_{ts} has the dimension of centimeters per second, it is called the *surface recombination velocity* S. Thus Eq. (2-38) becomes

$$U_s = S[p_n(0) - p_{no}] \tag{2-39}$$

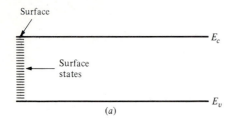

Surface

E_c

Surface
states

E_v

(a)

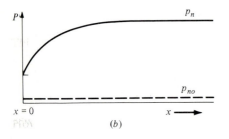

P

p_n

p_{no}

$x = 0$

$x \longrightarrow$

(b)

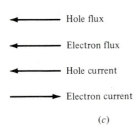

Hole flux

Electron flux

Hole current

Electron current

(c)

FIGURE 2-18
A semiconductor surface (a) energy band and surface states, (b) minority-carrier distribution, and (c) direction of fluxes and currents.

Equation (2-39) defines a hole recombination current density qU_s flowing into the surface from the bulk when excess carriers exist, as shown in Fig. 2-18b. The higher surface recombination rate leads to a lower carrier concentration at the surface. This gradient of holes yields a diffusion current density, Eq. (2-15), which is equal to the surface recombination current

$$-qD_p \left. \frac{dp_n}{dx} \right|_{x=0} = qU_s = qS[p_n(0) - p_{no}] \qquad (2\text{-}40)$$

However, an equal number of electrons is needed at the surface to accomplish recombination. Consequently, the electron and hole current cancel out, so that the net surface current is zero, as illustrated in Fig. 2-18c.

The numerical value of the surface recombination velocity may vary over a wide range depending on surface treatment received. It affects the surface leakage current and breakdown; both were serious problems in device performance. Modern planar silicon devices using silicon oxide passivation have overcome this difficulty.

2-9 ELECTROSTATIC FIELD AND POTENTIALS

The electric field $\mathscr{E}$ is defined as the negative gradient of the *electrostatic potential* ψ as expressed by

$$\mathscr{E} = -\frac{d\psi}{dx} \tag{2-41}$$

and the potential is related to the potential energy by

$$-q\psi = E \tag{2-42}$$

In a semiconductor, the potential energy for an electron in the conduction band is its lowest possible energy E_c. If an electron is located above E_c, the excess energy can only be in the form of kinetic energy. Similarly, the energy E_v denotes the potential energy of holes in the valence band, and a hole located below E_v has a kinetic-energy component. If there is no external applied field, Fig. 2-19a illustrates the relationship between energies and carrier positions in the energy-band diagram. The electrons situated at E_c (and holes at E_v) on the left-hand side of the diagram are carriers at rest, whereas the electron and hole on the right are carriers in motion.

When an external electric field is applied across the semiconductor, as shown in Fig. 2-19b, the band diagram is tilted to impart kinetic energy to electrons and holes. Here, the resting carriers are shown to be capable of moving by the electric field. Since E_c and E_v arc always in parallel with E_i and we are interested only in the potential gradient, we can express the electric field as

$$\mathscr{E} = \frac{1}{q}\frac{dE_i}{dx} = -\frac{d\psi}{dx} \tag{2-43}$$

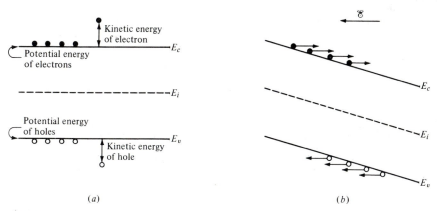

(a)

(b)

FIGURE 2-19
The energy-band diagram of a semiconductor under (a) zero electric field and (b) an electric field.

where

$$\psi \equiv -\frac{E_i}{q} \qquad (2\text{-}44)$$

Similarly, we define the *Fermi potential* as

$$\varphi \equiv -\frac{E_f}{q} \qquad (2\text{-}45)$$

Substitution of Eqs. (2-44) and (2-45) into Eqs. (1-25) and (1-26) yields

$$n = n_i e^{(\psi - \varphi)/\phi_T} \qquad (2\text{-}46)$$

$$p = n_i e^{(\varphi - \psi)/\phi_T} \qquad (2\text{-}47)$$

where

$$\phi_T \equiv \frac{kT}{q}$$

Since kT represents the thermal energy at temperature T, ϕ_T may be considered as the voltage equivalent of temperature. We will encounter ϕ_T frequently in later chapters. At 300 K, ϕ_T assumes a value of 26 mV.

Example. Show that the ratio of carrier mobility and diffusion constant is equal to ϕ_T.

Solution. Under thermal equilibrium, the Fermi potential is constant and can be taken as the zero reference. Therefore, Eq. (2-47) can be simplified to

$$p = n_i e^{-\psi/\phi_T} \qquad (2\text{-}48)$$

Differentiating the above equation, we have

$$\frac{dp}{dx} = -\frac{n_i}{\phi_T} e^{-\psi/\phi_T} \frac{d\psi}{dx} \qquad (2\text{-}49)$$

At equilibrium, the hole current must be zero so that

$$I_p = 0 = Aq\left(\mu_p p \mathscr{E} - D_p \frac{dp}{dx}\right) \qquad (2\text{-}50)$$

Substituting Eqs. (2-48) and (2-49) into the above equation and rearranging yields

$$\frac{D_p}{\mu_p} = \phi_T \qquad (2\text{-}51)$$

where Eq. (2-43) has been used. This is known as the *Einstein relationship*. A similar equation for electrons can be verified by the same procedure. It should be pointed out that Eq. (2-51) is a general relationship which is valid for the nonzero current condition.

2-10 INHOMOGENEOUS SEMICONDUCTOR AND BUILT-IN FIELD

During the growth of semiconductor crystals, special care is necessary to obtain uniform impurity distribution throughout the semiconductor. However, non-uniform distributions of impurities are sometimes introduced either accidentally or intentionally. An inhomogeneous impurity distribution results in a *built-in* electric field and is a useful technique for improving device performance.

The relationship between the built-in field and doping distribution can best be understood by means of the energy-band diagram. Let us consider an *n*-type silicon wafer with impurity distribution shown in Fig. 2-20*a*. The impurity concentration is limited to below 10^{18} cm^{-3} so that no part of the semiconductor is degenerate. Assume that all impurity atoms are ionized and that the electron density n is equal to $N_d(x)$ in Fig. 2-20*a*. Using this relation, we may rewrite Eq. (1-25) as

$$E_f - E_i = kT \ln \frac{N_d(x)}{n_i} \qquad (2\text{-}52)$$

The energy-band diagram is constructed by first taking the Fermi level E_f as the reference since E_f is constant at equilibrium. We may now obtain the position of

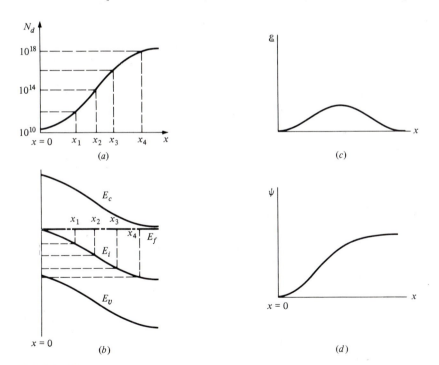

FIGURE 2-20
(*a*) A nonuniformly doped semiconductor donor distribution, (*b*) corresponding energy-band diagram, (*c*) electric field, and (*d*) electrostatic potential.

E_i from the above equation. Let us consider the donor concentraton at x_1 as 10^{12} cm^{-3}. The corresponding ΔE is

$$\Delta E = E_f - E_i = 26 \ln \frac{10^{12}}{1.5 \times 10^{10}} = 110 \text{ meV}$$

Similarly, we find ΔE at x_2, x_3, and x_4 for 10^{14}, 10^{16}, and 10^{18} cm^{-3} to be 230, 350, and 470 meV, respectively. Thus, for any value of N_d greater than n_i, the midgap energy is below the Fermi level, and the difference increases with increasing N_d. Plotting E or E_i with E_f as the reference, we obtain the diagram shown in Fig. 2-20b. Since the band-gap energy is a constant, the energy levels E_c and E_v are in parallel with E_i.

Let us differentiate Eq. (2-52):

$$\frac{dE_f}{dx} - \frac{dE_i}{dx} = \frac{kT}{N_d} \frac{dN_d}{dx} \tag{2-53}$$

Since the Fermi level E_f is constant,

$$\frac{dE_f}{dx} = 0 \tag{2-54}$$

If we now make use of Eq. (2-43) to relate the electric field and the midgap energy, we have

$$\mathscr{E} = \frac{1}{q} \frac{dE_i}{dx} = -\frac{\phi_T}{N_d} \frac{dN_d}{dx} \tag{2-55}$$

The electric field and the electrostatic potential are shown in Fig. 2-20c and d. Thus, we observe that a nonuniform spatial distribution of impurities produces a built-in electric field in the semiconductor. This built-in field is frequently introduced to improve device characteristics. When the semiconductor is doped homogeneously, the built-in electric field is zero.

2-11   QUASI-FERMI LEVELS

Under thermal equilibrium, the densities of electrons and holes are specified by the position of the Fermi level through Eqs. (1-25) and (1-26), but these equations do not apply in the nonequilibrium case because injected carriers render the Fermi level meaningless. Under nonequilibrium conditions, it is possible to define two quantities E_{fn} and E_{fp} to replace the Fermi level in Eqs. (1-25) and (1-26) such that

$$n = n_i e^{(E_{fn} - E_i)/kT} = n_i e^{(\psi - \varphi_n)/\phi_T} \tag{2-56}$$

$$p = n_i e^{(E_i - E_{fp})/kT} = n_i e^{(\varphi_p - \psi)/\phi_T} \tag{2-57}$$

where E_{fn} and E_{fp} are called the *quasi-Fermi levels* for electrons and holes, respectively, and φ_n and φ_p are the corresponding *quasi-Fermi potentials*. The quasi-Fermi level is simply a shorthand form of carrier concentration that is convenient in some situations. In most cases, the number of carriers is easier to identify and the quasi-Fermi level is more abstract. For this reason, this text will use carrier density as the basis of most analyses. But there are occasions when the probability of occupancy as implied by the use of a quasi-Fermi level is more useful, for example, in determining the activity of a trap in the forbidden gap. Its importance lies in giving a physical picture that is more than just numbers in some interface and optoelectronic problems.

Example. Calculate the quasi-Fermi levels at 300 K (room temperature) for a semiconductor with $N_a = 10^{16}$ cm^{-3}, $\tau_n = 10$ μs, $n_i = 10^{10}$ cm^{-3}, and $G_L = 10^{18}$ cm^{-3} s^{-1}.

Solution.
$$\Delta n = \Delta p = \tau_n G_L = 10^{13} \text{ cm}^{-3}$$

Therefore
$$p = p_o + \Delta p = N_a + \Delta p \approx 10^{16} \text{ cm}^{-3}$$

$$n = n_o + \Delta n = \frac{n_i^2}{N_a} + \Delta n = 10^4 + 10^{13} \approx 10^{13} \text{ cm}^{-3}$$

Rewrite Eq. (2-56) and use $kT = 26$ meV at 300 K to obtain

$$E_{fn} - E_i = kT \ln \frac{n}{n_i} = 0.026 \ln \frac{10^{13}}{10^{10}} = 0.18 \text{ eV}$$

Similarly, we obtain the quasi-Fermi level for holes by using Eq. (2-57):

$$E_i - E_{fp} = 0.026 \ln \frac{10^{16}}{10^{10}} = 0.36 \text{ eV}$$

Note that E_{fn} is above E_i and E_{fp} is below E_i.

The *pn* product under nonequilibrium is therefore, by multiplying n and p,

$$pn = n_i^2 \exp \frac{\varphi_p - \varphi_n}{\phi_T} \tag{2-58}$$

At equilibrium, $\varphi_n = \varphi_p = \varphi$ and $pn = n_i^2$. With the increase of injection, the difference of $E_{fn} - E_i$ in the first expression of Eq. (2-56) increases with n; thus E_{fn} moves away from E_i toward E_c. Similarly, from Eq. (2-57), E_{fp} moves away from E_i toward E_v with increasing injection. In other words, a higher concentration of holes brings E_{fp} closer to E_V, and a greater number of electrons moves E_{fn} closer to E_c.

2-12 BASIC GOVERNING EQUATIONS IN SEMICONDUCTORS

The mechanisms of carrier transport, generation, and recombination in a semiconductor have been discussed in the previous sections. These mechanisms are related to one another by the condition of current continuity. Figure 2-21 depicts the one-dimensional hole-current flow within a small increment Δx in a semiconductor *having unity cross-sectional area*. The continuity of carrier flow (the fact

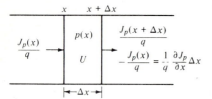

FIGURE 2-21
Current continuity in a semiconductor.

that charge can neither be created nor destroyed) requires that the rate of change of the number of holes in Δx be equal to the holes recombining plus holes leaving the increment. The net hole flow out of the increment is

$$\frac{J_p(x + \Delta x)}{q} - \frac{J_p(x)}{q} = \frac{1}{q} \frac{\partial J_p}{\partial x} \Delta x \tag{2-59}$$

The net recombination in the increment Δx, according to Eqs. (2-28) and (2-29), is

$$U \, \Delta x = \frac{p - p_o}{\tau_p} \Delta x \tag{2-60}$$

and the rate of change of holes in the increment is

$$-\frac{\partial p}{\partial t} \Delta x \tag{2-61}$$

The negative rate of change indicates a decrease of carriers. By adding Eqs. (2-59) and (2-60) and setting it equal to Eq. (2-61), we obtain the equation for the continuity of hole current as

$$\frac{1}{q} \frac{\partial J_p}{\partial x} + \frac{p - p_o}{\tau_p} = -\frac{\partial p}{\partial t} \tag{2-62}$$

Similarly, the continuity of electron current is

$$-\frac{1}{q} \frac{\partial J_n}{\partial x} + \frac{n - n_o}{\tau_n} = -\frac{\partial n}{\partial t} \tag{2-63}$$

The continuity equations are two of the most fundamental expressions for semiconductor device analyses.

In addition to the foregoing equations, we must consider the effect of space charge on the electric field. The semiconductor as a whole is charge-neutral. However, localized charged regions do exist. We can use Poisson's equation to describe the charged regions and their relationship to the electric field by

$$\frac{d\mathscr{E}}{dx} = \frac{\rho}{K_s \varepsilon_o} \tag{2-64}$$

where ρ = net space-charge density

K_s = dielectric constant

ε_o = permittivity of free space

The relationship between the charge distribution and the electric field is obtained by integrating Poisson's equation:

$$\mathscr{E} = \int \frac{\rho \, dx}{K_s \varepsilon_o} \tag{2-65}$$

Therefore, knowledge of the charge density leads to the magnitude of the electric field, as shown in Fig. 2-22, within a constant.

The net space charge in a semiconductor is the sum of the positive charges minus the sum of negative charges. Since an ionized donor atom has a fixed positive charge and an ionized acceptor atom has a fixed negative charge, the net space charge is given by

$$\rho = q[p + N_d - (n + N_a)] \tag{2-66}$$

Substituting Eqs. (2-43) and (2-66) into Eq. (2-64), and rearranging the result, we obtain

$$\frac{d\mathscr{E}}{dx} = -\frac{d^2\psi}{dx^2} = \frac{-q}{K_s \varepsilon_o}[(n - p) + (N_a - N_d)] \tag{2-67}$$

The current components I_n and I_p are related to the carrier concentrations by Eqs. (2-17) and (2-18), repeated here for convenient reference:

$$J_p = \frac{I_p}{A} = q\left(\mu_p p \mathscr{E} - D_p \frac{dp}{dx}\right) \tag{2-68}$$

$$J_n = \frac{I_n}{A} = q\left(\mu_n n \mathscr{E} + D_n \frac{dn}{dx}\right) \tag{2-69}$$

Equations (2-62) and (2-63) and Eqs. (2-67) to (2-69) constitute a complete set of equations which describe the carrier, current, and field distributions. They can be

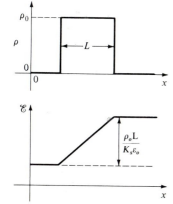

FIGURE 2-22
A box-type charge distribution and the corresponding electric field, which is obtained by integration of the charge distribution.

solved when appropriate boundary and initial conditions are available. In most cases, these equations will be simplified on the basis of physical approximations before a solution is attempted.

Example. Find the steady-state minority-carrier distribution in a semi-infinite homogeneous slab of n-type semiconductor if a density of excess carriers $p_n(0) - p_{no}$ is generated at $x = 0$.

Solution.   In the homogeneous semiconductor, $\mathscr{E} = 0$, and Eq. (2-68) becomes

$$J_p = -qD_p \frac{dp_n}{dx}$$

Substituting this expression into Eq. (2-62) for the time-indpendent case yields

$$D_p \frac{d^2 p_n}{dx^2} - \frac{p_n - p_{no}}{\tau_p} = 0$$

whose general solution is

$$p_n - p_{no} = A e^{x/L_p} + B e^{-x/L_p}$$

where $L_p = \sqrt{D_p \tau_p}$ and is known as the *diffusion length* for holes. Since no excess carriers are generated at $x = \infty$, we have

$$p_n - p_{no} = 0 \qquad \text{at } x = \infty \text{ or } A = 0$$

Using $p_n = p_n(0)$ at $x = 0$, we obtain $B = p_n(0) - p_{no}$. Thus, the solution is

$$p_n = p_{no} + [p_n(0) - p_{no}] e^{-x/L_p}$$

This equation is plotted in Fig. 2-23.

The minority-carrier diffusion length L_p in the above example is an important device parameter. Physically, L_p is the average length that an excess hole can diffuse before it recombines with an electron. It indicates how far the injected hole can reach as an effective agent of influence. The exponential relation shows that about 3 to 5 times L_p is the range of diffusion for minority carriers.

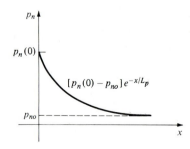

FIGURE 2-23
The steady-state minority-carrier distribution for a semiconductor slab with injection at $x = 0$.

REFERENCES

1. Sze, S. M.: "Physics of Semiconductor Devices," 2d ed, Wiley, New York, 1981.
2. Ghandi, S. K.: "VLSI Fabrication Principles," Wiley, New York, 1983.

ADDITIONAL READINGS

Grove, A. S.: "Physics and Technology of Semiconductor Devices," Wiley, New York, 1967. Chapter 5 on recombination theory.

Jonscher, A. K.: "Principles of Semiconductor Device Operation," Wiley, New York, 1960. Chapters 2 and 3 on recombination and transport of carriers.

Many, A., Y. Goldstein, and N. B. Grove: "Semiconductor Surfaces," Wiley, New York, 1965. Covers the topic of surface recombination in depth.

Pierret, R. F.: "Semiconductor Fundamentals," Addison-Wesley, Reading, Mass., 1983. On carrier transport.

Shockley, W., and W. T. Read: Statistics of the Recombination of Holes and Electrons, *Phys. Rev.,* **87**:835 (1952). The classic paper on recombination statistics.

PROBLEMS

2-1. The mobile carriers are in constant motion due to their thermal energy, i.e., due to the temperature of the lattice. The corresponding energy is $\frac{3}{2}kT$.

    (a) What is the velocity v_{th} of carriers having this average kinetic energy? Assume $m = 0.2m_o$, where m_o is the mass of a free electron, and $T = 300$ K.

    (b) What is the net average velocity $\bar{v}$ of carriers whose mobility is $\mu = 1000$ cm^2/V · s under the influence of an electric field $\mathscr{E} = 1000$ V/cm?

    (c) Compare (a) and (b) and explain why $v_{\text{th}} > \bar{v}$.

2-2. (a) Find the resistivity of intrinsic Ge and Si at 300 K.

    (b) If a shallow donor impurity is added to the extent of 1 atom per 10^8 Ge or Si, find the resistivity.

2-3. A semiconductor is doped with N_d and has a resistance R_1 ($N_d \gg n_i$). The same semiconductor is then doped with an unknown amount of acceptors N_a ($N_a \gg N_d$), yielding a resistance of $0.5R_1$. Find N_a in terms of N_d if $D_n/D_p = 50$.

2-4. (a) Use the data in Figs. 1-20 and 2-4 to find the carrier concentrations, mobility, and conductivity of the sample with $N_d = 10^{13}$ at 200 and 300 K.

    (b) Repeat (a) for $N_d = 10^{18}$ cm^{-3}.

    (c) Obtain the temperature coefficient of conductivity for the two samples. Which one would you select as the material for a temperature-sensitive device called the *thermistor*?

2-5. Using the result of Prob. 1-10, calculate the electron mobility in aluminum, for which the density is 2.7 g/cm^3, the resistivity is 3×10^{-6} Ω · cm, and the atomic weight is 27.

2-6. (a) A silicon wafer is doped with 2×10^{16} boron and 10^{16} phosphorus atoms. Calculate the electron and hole concentrations, E_f, and resistivity at room temperature.

    (b) Repeat (a) for 8×10^{15} boron atoms/cm^3.

2-7. The cross section of a Hall sample (Fig. 2-12) is 10×4 mils^2 and the spacing between contacts is 50 mils.

    (a) With $B = 10^{-4}$ Wb/cm^2 and $I = 1$ mA, we find $V = 500$ mV, $V_H = -10$ mV. Find the carrier density and mobility of the sample.

    (b) Design a multiplier $(B \times I)$ with the coefficient of unity for the sample shown in Fig. 2-11.

2-8. (a) From the definition of direct recombination, determine the average time an electron stays in the conduction band and the average time a hole stays in the valence band.

    (b) What is the relationship between the carrier lifetime τ and the average times obtained in (a)? Discuss this relationship for intrinsic and extrinsic semiconductors.

2-9. Derive an expression of carrier lifetime in direct recombination under steady-state high-level injection conditions.

2-10. Calculate the electron and hole concentration under steady-state illumination in an n-type silicon with $G_L = 10^{16}$ cm^{-3} s^{-1}, $N_d = 10^{15}$ cm^{-3}, and $\tau_n = \tau_p = 10\ \mu$s.

2-11. The energy level of electron traps can be measured by the thermally stimulated current (TSC) experiment, in which the semiconductor is first cooled to a very low temperature and then exposed to a strong light filling all the traps. After the steady state is reached, the light source is removed. The semiconductor is heated at a constant rate slowly in the dark, and its conductance is measured at different temperatures. The dark current must be subtracted in the measurement. A conductance peak σ_m is measured at temperature T_m, at which point the Fermi level aligns with E_t to release all the trapped electrons.

    (a) Show that the trap level is given by

$$E_c - E_t = kT_m \ln\left[q\, \frac{\mu_n N_c(T_m)}{\sigma_m} \right]$$

    (b) A TSC measurement in a GaAs crystal (with cross sectional area of 0.1 cm^2 and length of 1 cm) yields $T_m = -33°$C, and the thermally stimulated current peaks at 140 nA when biased with a voltage of 100 V. Calculate E_t.

2-12. (a) Show that the change of photoconductance of a semiconductor with illumination is

$$\Delta\sigma = q(\tau_n \mu_n + \tau_p \mu_p)G_L$$

    (b) Compare the sensitivity of devices using GaAs and Si. Assume $\tau_n = \tau_p = 1\ \mu$s for Si and $\tau_n = \tau_p = 10$ ns for GaAs.

2-13. Following the stated procedure, derive the Einstein relationship for both electrons and holes.

2-14. Given an n-type semiconductor doped with $N_d = 10^{17}$ cm^{-3}, $\tau_p = \infty$, $D_p = 15$ cm^2/s, $D_n = 37.5$ cm^2/s, $W = 2\ \mu$m, and $G_L(x) = 10^{15}$ cm^{-3}/s steady state. Also it is known that the surface recombination velocity is zero at $x = 0$ and infinity at $x = W$ (Fig. P2-14).

    (a) Determine $\Delta p(x)$.

    (b) From (a) find $J_p(x)$.

    (c) What is $J_n(x)$?

    (d) What is the electric field?

$S = 0$ $S = \infty$

$x = 0$ $x = w$ **FIGURE P2-14**

2-15. The excess hole concentration in an n-type silicon bar illuminated with steady-state radiation is given by

$$\Delta p(x) = 10^{13} \cos \frac{\pi x}{2W} + C \qquad \text{cm}^{-3}$$

for boundary conditions specified in Prob. 2-14 (Fig. P2-14). If $W = 10^{-2}$ cm, $\tau_p = \infty$, and $D_p = 15$ cm^2/s, calculate (a) the constant C, (b) $J_p(x)$, and (c) $G_L(x)$.

2-16. Assume that an n-type semiconductor has an infinite recombination lifetime. The semiconductor is illuminated by radiation, resulting in a carrier generation $G_L(t)$ given by

$$G_L(t) = \begin{cases} G_o t & \text{cm}^{-3}\,\text{s}^{-1} & \text{for } 0 < t < T_1 \\ 0 & & \text{otherwise} \end{cases}$$

Find $\Delta p(t)$ for all time.

2-17. The p-type semiconductor shown in Fig. P2-17 is illuminated by steady-state radiation that uniformly generates G_L electron-hole pairs/cm$^3 \cdot$s in the region $-L < x < L$. The minority-carrier lifetime τ_n is infinite. It is also known that $\Delta n(x = \pm 2L) = 0$. Find $\Delta n(x = 0)$ (G_L is a constant). *Hint:* $\Delta n(x)$ and $d\Delta n(x)/dx$ are continuous throughout the semiconductor.

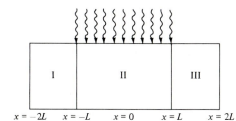

x = −2L   x = −L   x = 0   x = L   x = 2L   **FIGURE P2-17**

2-18. A semiconductor bar is uniformly illuminated with a steady-state light source that generates G_L electron-hole pairs/cm$^3 \cdot$s for $t < 0$. The intensity begins to decay exponentially at $t = 0$. Thus

$$G_L(t) = \begin{cases} G_L & t < 0 \\ G_L e^{-t/t_o} & t \geqslant 0 \end{cases}$$

Find the excess hole concentration as a function of time in the semiconductor. Express your answer in terms of G_L, t_o, and τ_p.

2-19. (a) A semi-infinite slab of n-type silicon is uniformly illuminated with a generation rate of G_L. Obtain the hole-continuity equation under these conditions.

(b) If the surface recombination velocity is S at $x = 0$, solve the new continuity equation to show that the steady-state hole distribution is given by

$$p_n(x) = p_{no} + \tau_p G_L \left(1 - \frac{\tau_p S e^{-x/L_p}}{L_p + S\tau_p} \right)$$

2-20. Derive the electric field in an n-type semiconductor if (a) $N_d = ax$, where a is a constant; (b) $N_d = N_o e^{-ax}$.

2-21. Plot the energy-band diagrams of Prob. 2-20*a* and *b*.

2-22. (*a*) Calculate the conductivity and quasi-Fermi levels for a silicon sample with
$N_d = 10^{15}$ cm^{-3}, $\tau_p = 1$ μs, and $G_L = 5 \times 10^{19}$ cm^{-3} s^{-1}.

(*b*) Find the value of G_L that produces 10^{15} holes/cm^3. What are the conductivity and quasi-Fermi levels?

2-23. The injection of carriers gives rise to the splitting of the electron and hole quasi-Fermi levels. Show that the nonequilibrium product (*pn*) of a semiconductor with an energy gap of E_g is the same as the equilibrum product $p_o n_o$ of a semiconductor with a band gap of $E_g - (E_{fn} - E_{fp})$.

2-24. Show that the electron current density in Eq. (2-17) can be expressed as $J_n = -q\mu_n n \, d\phi_n/dx$.

2-25. Sketch the electric field diagram for the charge distribution shown in Fig. P2-25.

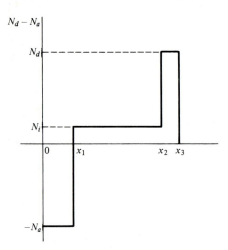

FIGURE P2-25

CHAPTER

3

pn JUNCTION ELECTROSTATICS

In this chapter, we shall briefly consider techniques for device fabrication. It is shown that impurities can be introduced so that both *p*-type and *n*-type regions may exist simultaneously in a semiconductor. The metallurgical boundary between a *p* region and an *n* region is called a *pn junction*. The primary concern in this chapter is the junction electrostatics. It includes the space-charge layer ,at the interface region, energy-band diagrams at equilibrium and under bias, capacitance effect, and junction breakdown.

3-1 JUNCTION FORMATION

When *n*-type and *p*-type silicon crystals are joined together, a *pn* junction is formed. However, joining two pieces of silicon mechanically creates all sorts of problems structurally, electrically, and chemically at the interface, making it unsuitable for device applications. In practice, a *pn* junction is formed by adding accepter impurities to an *n*-type wafer or donors to a *p*-type wafer. There are a variety of methods for junction formation. Of these, we shall describe briefly the techniques of alloying, epitaxy, diffusion, and ion implantation.

Alloyed Junction

The procedure of forming an alloyed junction diode is shown in Fig. 3-1. A thin aluminum film is evaporated onto a clean *n*-type silicon wafer, called a *substrate*, inside a vacuum chamber. The silicon is then placed in a furnace set at about 600 °C for 30 min, or alternatively, heated by a laser beam for a few seconds or less. It is important to point out that Al and Si constitute a *eutectic* system which has a melting point (known as the *eutectic*) lower than that of the individual components. If the heating temperature is above the eutectic (577 °C), the interface will be melted to form a liquid alloy. When the semiconductor is cooled down, the silicon from the liquid alloy will form a recrystallized layer which contains a significant amount of aluminium atoms. Since aluminum is a *p*-type impurity in silicon, the recrystallized region is a *p* region. Consequently, a *pn* junction is produced. The alloying process is simple and inexpensive in making a single *pn* junction, but it does not produce a uniform and smooth junction. The control of aluminum concentration is also difficult, and there are serious problems associated with the stability of the interface, which will be discussed in a later chapter. For these reasons, the alloyed junction is seldom used in practical devices.

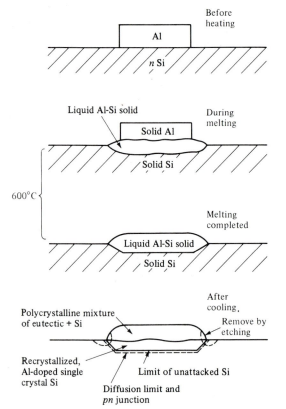

FIGURE 3-1
An alloyed silicon *pn* junction.

Epitaxy

The word *epitaxy* is derived from a Greek root meaning "arranged upon." It describes the growth technique of arranging atoms in single-crystal fashion upon a crystalline substrate so that the lattice structure of the newly grown film duplicates that of the substrate. An important reason for using this growth technique is the flexibility of impurity control in the epitaxial film. The dopant in the film may be *n* or *p* type and independent of the substrate doping. Therefore, epitaxial growth can be used to form a lightly doped layer on a heavily doped substrate, or a *pn* junction between the epitaxial film and the substrate. However, the *pn* junction formed by epitaxy is over the entire surface, and it is not suitable for localized *pn* junction formation. Three different methods are available to produce epitaxial films, and these are *vapor-phase epitaxy* (VPE), *liquid-phase epitaxy* (LPE), and *molecular-beam epitaxy* (MBE).

In VPE, hydrogen gas containing silicon tetrachloride is fed into a high-temperature reactor inside which silicon wafers are placed on a graphite susceptor (Fig. 3-2). The graphite is heated by a radio-frequency (rf) induction coil to a temperature above $1000\,^\circ C$. The rf energy heats the graphite but not the quartz tube so that there is no deposition of silicon on the tube. The high temperature is necessary for the deposited atoms to have high mobility so that they can find the proper position in the lattice to maintain a single-crystal layer. To grow a doped epitaxial layer, impurity atoms are introduced in the gas stream. For example, phosphine (PH_3) is used for *n*-type doping and diborane (B_2H_6) is used for *p*-type doping. The growth of a *p*-type silicon film where diborane has undergone decomposition is shown in Fig. 3-3. Notice that the interface is quite sharp and a *pn* junction is formed.

Experimentally, the transport of material to a heated substrate may be accomplished by direct contact with a liquid solution in LPE or by a molecular beam in an ultrahigh vacuum chamber in MBE. These two processes are capable of depositing multilayer films with a thickness in the submicron range. They are particularly useful for III-V compound semiconductors.

Diffusion

The most widely used technique in forming a *pn* junction is solid-state impurity diffusion. The diffusion of impurity is, in principle, the same as carrier diffusion

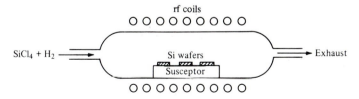

FIGURE 3-2
Schematic representation of an epitaxial growth system.

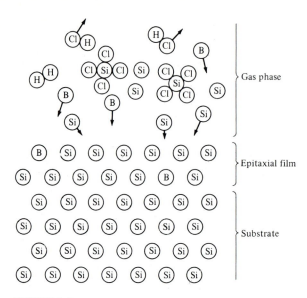

FIGURE 3-3
Incorporation of boron atoms in the epitaxial film during the vapor-phase epitaxial growth. (*After Warner and Fordemwalt [1].*)

described in Chap. 2 except that it occurs at very high temperature. In a typical diffusion system, a wafer is placed in a gaseous atmosphere containing impurity atoms inside a furnace. These impurity atoms may come from slices of boron nitride inserted between silicon wafers. The temperature of the furnace usually ranges between 900 and 1200°C. The number of impurity atoms taken in by the solid is limited by the *solid solubility*, which is the maximum impurity concentraton that the solid can accommodate at a given temperature. The values of solid solubility for B, P, and As in silicon in the normal range of diffusion temperatures are approximately 6×10^{20}, 10^{21}, and 2×10^{21} cm^{-3}, respectively. Under this condition, known as *predeposition*, the impurity concentration is specified by the following equation:

$$N(x, t) = N_o \text{ erfc } \frac{x}{2\sqrt{Dt}} \qquad (3\text{-}1)$$

where N_o is the surface impurity concentration. The complementary error function, erfc, is tabulated in Appendix C, together with some of its basic properties. The term $\sqrt{Dt}$ is called the diffusion length as explained in Chap. 2. The constant D is the *diffusivity* of the impurity and is given by

$$D = D_o e^{-E_o/kT} \qquad (3\text{-}2)$$

where E_o is an activation energy and D_o is a constant. As a first-order approximation, the values of E_o and D_o for B and P are the same, and they are 3.66 eV and 3.85 cm^2/s, respectively.

The total number of impurity atoms per unit surface is given by

$$Q(t) = \int_0^\infty N(x, t) \, dx = \frac{2}{\sqrt{\pi}} \sqrt{Dt} \, N_o \qquad (3\text{-}3)$$

If the solid is *n* type and the diffused impurity is *p* type, a *pn* junction is formed. The impurity distribution becomes

$$N(x, t) = N_o \, \text{erfc} \, \frac{x}{2\sqrt{Dt}} - N_{BC} \qquad (3\text{-}4)$$

where N_{BC} is the background doping density of the *n*-type solid. The junction depth from surface can be obtained by setting $N = 0$:

$$x_j = 2\sqrt{Dt} \, \text{erfc}^{-1} \frac{N_{BC}}{N_o} \qquad (3\text{-}5)$$

This junction depth is an important device parameter.

Example.   A *pn* junction is formed by diffusing boron into *n*-type silicon having 10^{16} phosphorus atoms/cm^3 at 1200°C. The surface concentration of boron is solubility-limited. Calculate the necessary diffusion time to realize a junction depth of 1 μm. What is the total number of boron atoms per square centimeter incorporated into silicon?

Solution.   At 1200°C, the solid solubility of boron in silicon is 6×10^{20}, and the diffusivity is 3×10^{-12} cm^2/s [Eq. (3-2)]. Setting $N(x, t) = 0$ and $x = x_j - 1$ μm in Eq. (3-4), we have

$$\text{erfc} \, \frac{x_j}{2\sqrt{Dt}} = \frac{10^{16}}{6 \times 10^{20}} = 1.66 \times 10^{-5}$$

From Fig. 3-4 or Appendix C we obtain

$$\frac{x_j}{2\sqrt{Dt}} = 3 \quad \text{or} \quad t = 92 \, \text{s} = 1.5 \, \text{min}$$

Using Eq. (3-3), we find

$$Q = \frac{2}{\sqrt{\pi}} \frac{10^{-4}}{6} (6 \times 10^{20}) = 1.13 \times 10^{16} \, \text{cm}^{-2}$$

After the predeposition step, the surface impurity concentration is nominally equal to the solid solubility at the diffusion temperature. For boron and phosphorus, the surface concentration is greater than 10^{20} cm^{-3}. Frequently, one would like to reduce the surface concentration and at the same time push the impurity atoms farther away from the surface into the bulk of the solid. This desirable impurity distribution can be achieved by a *drive-in* step after the predeposition. In a practical system, the surface of the wafer is sealed off by a thin oxide layer which prevents

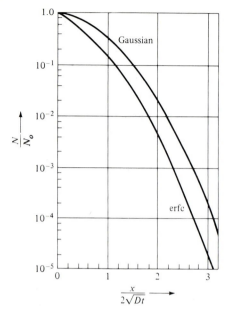

FIGURE 3-4
Normalized gaussian and erfc functions.

the escape of impurities through the surface. With the total impurity Q resulting from predepositon, the impurity distribution is found to be

$$N(x, t) = \frac{Q}{\sqrt{\pi D t}} e^{-x^2/4Dt} \tag{3-6}$$

which is a *gaussian distribution*. This is a good approximation for the dopant distribution of the two-step diffusion provided that the diffusion length of the drive-in step is at least 3 times larger than the diffusion length in predeposition. The normalized impurity profiles are shown in Fig. 3-4.

Ion Implantation

An alternative to high-temperature diffusion for introducing dopants into a semiconductor is *ion implantation*. A beam of dopant ions is accelerated through a desired energy potential ranging between 30 and 500 keV. The ion beam is aimed at the semiconductor target so that the high-energy ions penetrate the semiconductor surface. The energetic ions will lose their energy through collisions with the target nuclei and electrons so that the ions will finally come to rest. The distance traveled by the ions, i.e., the penetration depth, is called the *range*. The range is a function of the kinetic energy of the ions and the semiconductor's structural properties, e.g., lattice spacing and mass of atoms. A typical impurity-range distribution in an amorphous target is approximately gaussian, as depicted in Fig.

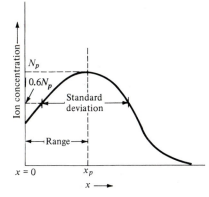

FIGURE 3-5
A typical ion-implanted impurity profile.

3-5, and is given by the empirical equation

$$N(x) = N_p \exp\left[\frac{-(x - x_p)^2}{2\sigma_R^2}\right] - N_{BC} \qquad (3\text{-}7)$$

where x_p = range

σ_R = standard deviaton

N_p = peak impurity concentration

N_{BC} = background doping concentration

Before an incident ion loses its kinetic energy, it collides with lattice atoms and the host atoms are dislodged. As a result, the crystalline region is turned into a disordered or amorphous layer. For this reason, it is necessary to anneal the semiconductor after implantation to reestablish the crystalline structure. In the annealing step, the semiconductor is placed in a high-temperature oven set at a temperature between 200 and 800°C. The annealing temperatures are typically well below those used in solid-state diffusion.

The ion-implantation process is attractive because it can be performed at low temperatures with negligible impurity diffusion. In addition, the impurity concentration introduced is better controlled than with standard diffusion techniques. Compared with other doping techniques, an ion-implanted layer is generally shallower, which is especially suited for submicron devices. In the last few years, ion implantation has emerged as the preferred technique for introducing dopants in silicon devices. It is used for practically every doping step in standard bipolar and MOS processes.

Oxidation

A high-quality natural oxide can be grown on a silicon surface in a high-temperature furnace. This oxide, SiO_2, is an excellent insulator, and it serves to protect and passivate the silicon–silicon oxide interface. In addition, the oxide is a good diffusion barrier against impurities such as B, P, and As. Therefore, it can be used as a diffusion mask (see next paragraph). Germanium and gallium arsenide are less fortunate since no useful native oxide can be formed.

Planar Technology

The fabrication steps of a planar *pn* diode are shown in Fig. 3-6. Using an n^+ silicon as the starting substrate,[1] a thin layer of *n* silicon is grown by the epitaxial process (Fig. 3-6a and b). Subsequently, a SiO_2 layer is formed by thermal oxidation (Fig. 3-6c). The oxidation step is followed by the photolithographic process. A thin film of organic polymer called *photoresist* is first put on in liquid form to cover the

---

[1] The superscript "+" represents heavily doped regions with an impurity concentration greater than $10^{19} \, cm^{-3}$. Superscript "−" is used for a doping level less than $10^{14} \, cm^{-3}$.

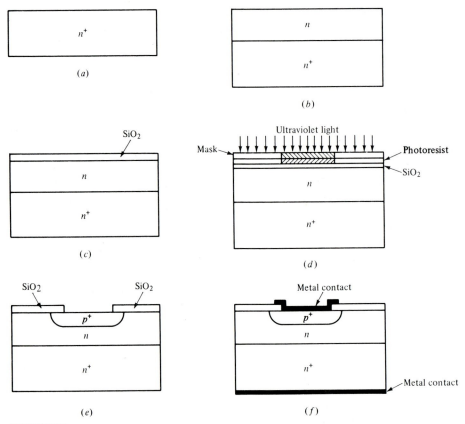

FIGURE 3-6
Planar technology: (*a*) the substrate, (*b*) with epitaxial layer, (*c*) after oxidation, (*d*) photolithography, (*e*) boron diffusion after oxide removal, and (*f*) the complete planar diode after metallization.

oxide and then dried in a baking oven. The photoresist material is soluble in a special solvent unless it is polymerized. A mask with transparent and opaque regions is now placed on the photoresist. When the entire structure is exposed to ultraviolet light, the exposed photoresist is polymerized but the unexposed area is removable (Fig. 3-6*d*). Etching of the oxide follows the removal of unexposed photoresist. Then, a *p*-type impurity, e.g., boron, is diffused through the oxide opening to make a *pn* junction (Fig. 3-6*e*). Finally a thin metal film is deposited and etched to obtain the structure shown in Fig. 3-6*f*. Using the basic steps described here, we can build a multijunction transistor or a complex integrated circuit.

3-2 PHYSICAL DESCRIPTION OF THE *pn* JUNCTION

As shown in Sec. 3-1, a *pn* junction diode may be fabricated by using the planar technology. The n^+ layer in Fig. 3-6*f* is important in forming a good ohmic contact at the back side, a subject to be discussed in a later chapter, but it is not crucial to the junction characteristics of the diode. The key section of the planar diode involves the p^+n junction, and the most remarkable property of the junction is that it *rectifies*. In other words, it allows the current to pass in one direction but not the opposite direction. If we connect the *p* side to the positive terminal of a battery and the *n* side to the negative terminal of the same, a large current flow is observed and its magnitude increases exponentially with the applied voltage. This is known as the *forward bias*, giving rise to a current-voltage curve as shown in Fig. 3-7*a* for a typical silicon *pn* junction diode. If we now interchange the two terminals of the power supply, the current will, for practical purposes, disappear. This is known as the *reverse-bias* condition. The current will remain negligible until the reverse voltage becomes very large; then at some critical voltage a large current surges through the diode under the condition of junction breakdown. The reverse *I-V* characteristic is shown in Fig. 3-7*b* in a larger horizontal scale.

A single *pn* junction can be used as a switch, rectifier, light detector, solar cell, microwave diode, or in compound semiconductors, as a light emitter or laser. More importantly, when two *pn* junctions are fabricated in close proximity·so that they interact with each other, the amplifying or switching operation of junction transistors and field-effect transistors can be realized. All microelectronic devices make use of the *pn* junction in some form. It is the most basic component that demands a thorough investigation of its properties.

As discussed in Sec. 3-1, the most common method of forming a *pn* junction is impurity diffusion or ion implantation. The resulting impurity distribution closely conforms to either the complementary error function or the gaussian function. In practice, a real *pn* junction is usually approximated by a step or linearly graded junction to avoid mathematical complexity in analyses. The impurity profile of the step junction assumes an abrupt transition between the *n*-type and *p*-type regions, as illustrated in Fig. 3-8*a*. The linearly graded junction has an impurity distribution changing gradually, as seen in Fig. 3-8*b*. Theoretical calculations using these two models are in excellent agreement with the first-order effects in practical

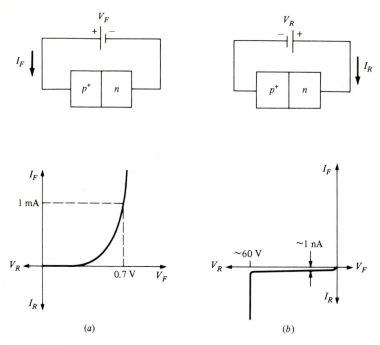

FIGURE 3-7
Current-voltage characteristic under (a) forward bias and (b) reverse bias.

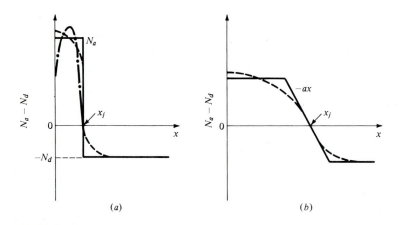

FIGURE 3-8
(a) A shallow diffused junction (dashed) and an implanted junction (dot-dashed) with the step-junction approximation (solid) and (b) a deep diffused junction (dashed) with the linearly graded junction approximation (solid).

junctions. Since the step junction is easier to describe analytically and most junctions belong to this group, we base our analyses on the step junction.

3-3 STEP *pn* JUNCTION AT EQUILIBRIUM

Let us redraw a *pn* junction in Fig. 3-9*a*, where the distribution of the donor and acceptor concentrations is shown, assuming uniform doping densities in both *n*- and *p*-type regions. At room temperature, the donor and acceptor ions are stationary positive- and negative-charge centers, respectively. Each donor contributes a free electron and each acceptor produces a free hole, and these carriers are free to roam around. It turns out that, away from the junction, the free electrons and holes will compensate the fixed ionic charges so that charge neutrality is maintained. These regions, away from the junction interface, are known as the *neutral* regions.

In the neighborhood of the junction, the situation is quite different. Since there are many more mobile electrons in the *n* side, they tend to diffuse away to reach the *p* side of the junction, leaving behind a positive fixed charge of ionized donors. Similarly, holes migrate from the *p* side to the *n* side, leaving behind negatively charged acceptor ions. Consequently, a space-charge region is formed, as shown in Fig. 3-9*b*. The two oppositely charged layers indicate that an internal electric field has been established. This field is in such a direction that further diffusion of electrons and holes is prevented. Therefore, the equilibrium condition at the junction is arrived at when the diffusion flux of carriers is neutralized by the drift flux resulting from the electric field. Making use of the graphical integration technique in Sec. 2-12, we may sketch the electric field and electrostatic potential as shown in Fig. 3-9*c* and *d*. Since the midgap potential $E_i = -q\psi$, we may turn the potential diagram upside down and obtain the spatial variation of E_i. The energy-band diagram may now be sketched as shown in Fig. 3-9*e* since both E_c and E_v are parallel to E_i.

Under thermal equilibrum, there is no net carrier movement, and this can happen only when the probability of occupancy at a given energy is the same everywhere. If the Fermi level is not the same, the different probability of occupancy at adjoining points will initiate carrier migration, which turns it into a non-equilibrium situation. In other words, E_f must be the same everywhere in a solid under thermal equilibrium even for a composite structure with layers of different materials. Thus, the energy-band diagram can also be constructed from the concept of the constancy of the Fermi level under thermal equilibrium. This approach is faster and less likely to introduce errors in more complicated devices than a *pn* junction. For this reason, we will show how to develop the energy-band diagram by using the Fermi level as the reference. Let us assume that the *n*-type and *p*-type materials are physically separate before the junction is formed. The Fermi level is near the conduction-band edge on the *n* side and near the valence-band edge on the *p* side, as shown in Fig. 3-10*a*. When the two sides are joined, the Fermi level at equilibrium must be the same throughout the structure. To reach this condition, the Fermi level in the *n* side must be lowered, which is realized when electrons are transferred to the *p* side. Similarly, the Fermi level in the *p* side is raised when holes migrate to the *n* side. The energy-band diagram is constructed by first drawing a horizontal line to represent the aligned Fermi level. Then, for the neutral regions far away from the junction, the band-edge energies E_c and E_v can be added for

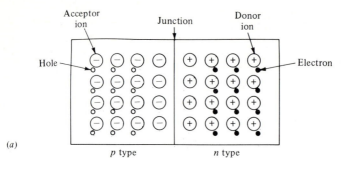

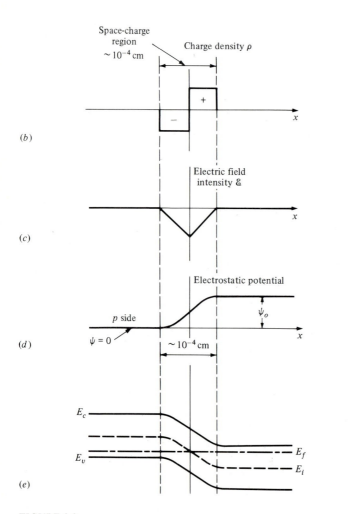

FIGURE 3-9
A schematic diagram of a *pn* junction, including the charge density, electric field intensity, potential, and energy band at the junction. (Not drawn to scale.)

both the p and n sides since they are undisturbed by the junction (Fig. 3-10b). The transition region may now be plotted by joining the respective band edges smoothly, as shown in Fig. 3-10c. The transfer of electrons and holes leaves behind uncompensated donor and acceptor ions N_d^+ and N_a^- in the n side and p side, respectively, creating two space-charge layers as shown in Fig. 3-10d.

The relationship between the charge distribution and the electric field is given by Poisson's equation (2-67), repeated here:

$$\frac{d\mathscr{E}}{dx} = \frac{\rho}{K_s \varepsilon_o} = \frac{-q}{K_s \varepsilon_o} [(n - p) - (N_d - N_a)] \tag{3-8}$$

The electron and hole densities are given by Eqs. (2-46) and (2-47) and are written in the following forms by taking the Fermi potential as the zero reference:

$$n = n_i e^{\psi/\phi_T} \tag{3-9}$$

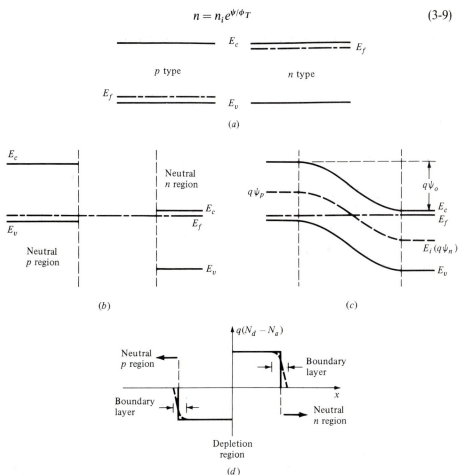

FIGURE 3-10
(a) Isolated p-type and n-type silicon before contact, (b) Fermi-level alignment, (c) the energy-band diagram after contact, and (d) the space-charge distribution of (c).

$$p = n_i e^{-\psi/\phi_T} \tag{3-10}$$

Equations (3-8) to (3-10) are applicable to the two regions of the *pn* junction: (1) the *neutral* regions away from the junction and (2) the *depletion* region where there are fixed charges but no free carriers.

In the neutral regions, the total space-charge density is zero. Thus, Eq. (3-8) becomes

$$\frac{d\mathscr{E}}{dx} = 0 \tag{3-11}$$

and

$$n - p - N_d + N_a = 0 \tag{3-12}$$

For an *n*-type neutral region, we may assume $N_a = 0$ *and* $p \ll n$. The potential of the *n*-type neutral region far from the junction, designated ψ_n in Fig. 3-10c, is derived by setting $N_a = p = 0$ in Eq. (3-12) and substituting the result into Eq. (3-9):

$$\psi_n = \phi_T \ln \frac{N_d}{n_i} \tag{3-13}$$

By using the same procedure in the *p*-type neutral region, we obtain the potential of the *p*-type neutral region as

$$\psi_p = -\phi_T \ln \frac{N_a}{n_i} \tag{3-14}$$

Therefore, the potential difference between the *n*-side and the *p*-side neutral region is

$$\psi_o = \psi_n - \psi_p = \phi_T \ln \frac{N_d N_a}{n_i^2} \tag{3-15}$$

where ψ_o is known as the *built-in* or *diffusion potential*. This potential difference exists in a *pn* junction under thermal equilibrium.

For the completely depleted region, free-carrier densities are zero ($n = p = 0$), and Eq. (3-8) becomes

$$\frac{d\mathscr{E}}{dx} = \frac{-q}{K_s \varepsilon_o}(N_a - N_d) \tag{3-16}$$

which will be solved in the next section.

Between the depletion and neutral regions, there is a boundary layer where carriers are partially depleted so that it does not belong to the neutral nor depletion region. Fortunately, this layer is usually very thin in comparison with the width of the depletion layer. In this text, we shall neglect its effects. From now on, we shall assume that the *pn* junction can be divided simply into the neutral and depletion regions. Since the depletion region will be the same as the space-charge region, the terms "depletion region" and "space-charge region" are frequently used interchangeably.

3-4   THE DEPLETION APPROXIMATION

The space-charge region of a step junction is well-represented by the box-type distribution shown in Fig. 3-11*a*. In this region, the free carriers are negligible so that Poisson's equation for the *n* side and *p* side can be simplified to

$$\frac{d\mathscr{E}}{dx} = \begin{cases} \dfrac{qN_d}{K_s\varepsilon_o} & \text{for} \quad 0 < x < x_n & (3\text{-}17a) \\[3ex] -\dfrac{qN_a}{K_s\varepsilon_o} & \text{for} \quad -x_p < x < 0 & (3\text{-}17b) \end{cases}$$

Since the electric field is zero in the neutral regions and at the edges of the depletion layer, we have

$$\mathscr{E}(x_n) = \mathscr{E}(-x_p) = 0$$

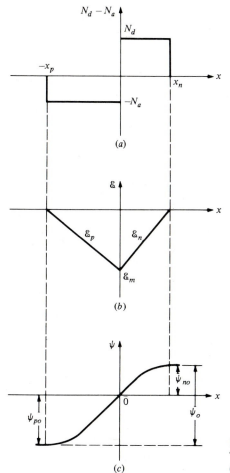

(a)

(b)

(c)

FIGURE 3-11
The step *pn* junction: (*a*) space-charge distribution, (*b*) electric field, and (*c*) potential diagrams.

as the boundary conditions. The electric field on the p-side of the depletion region is obtained by integrating Eq. (3-17b):

$$\int_{\mathscr{E}_p(-x_p)}^{\mathscr{E}_p(x)} d\mathscr{E} = \int_{-x_p}^{x} \frac{-qN_a}{K_s\varepsilon_o} \, dx$$

or

$$\mathscr{E}_p(x) = \frac{-qN_a}{K_s\varepsilon_o} (x_p + x) \tag{3-18}$$

where $\mathscr{E}_p(-x_p) = 0$ was used. This equation is plotted as the straight line on the p side in Fig. 3-11b. Similarly, the electric field on the n side of the depletion layer is given by integrating Eq. (3-17a):

$$\int_{\mathscr{E}_n(x)}^{\mathscr{E}_n(x_n)} d\mathscr{E} = \int_{x}^{x_n} \frac{qN_d}{K_s\varepsilon_o} \, dx$$

or

$$\mathscr{E}_n(x) = -\frac{qN_d}{K_s\varepsilon_o} (x_n - x) \tag{3-19}$$

where $\mathscr{E}_n(x_n) = 0$ was used. Equation (3-19) is plotted on the n side of Fig. 3-11b. Note that the $\mathscr{E}$ field is negative on both sides of the depletion layer, and the shape is the same as obtained by the graphical method shown in Fig. 3-9.

Since the electric field is continuous at the junction interface, Eqs. (3-18) and (3-19) must be equal at $x = 0$. This condition leads to

$$N_a x_p = N_d x_n \tag{3-20}$$

which states that the negative charge on the p side is exactly balanced by the positive charge on the n side in Fig. 3-11a. The semiconductor as a whole is therefore charge-neutral despite the existence of the localized positive and negative space-charge regions.

Let us now find the electrostatic potential by using Eq. (2-41), that is,

$$\mathscr{E} = -\frac{d\psi}{dx}$$

Therefore, Eq. (3-18) becomes

$$\frac{d\psi}{dx} = \frac{qN_a}{K_s\varepsilon_o} (x_p + x) \tag{3-21}$$

Integrating once,

$$\int_{\psi_p(-x_p)}^{\psi_p(x)} d\psi = \int_{-x_p}^{x} \frac{qN_a}{K_s\varepsilon_o} (x_p + x) \, dx$$

or

$$\psi_p(x) - \psi_{po} = \frac{qN_a}{2K_s\varepsilon_o} (x_p + x)^2 \tag{3-22}$$

where $\psi_p(-x_p) = \psi_{po}$ was used. Similarly for the n side of the depletion layer, we have

$$\int_{\psi(x)}^{\psi_n(x_n)} d\psi = \int_x^{x_n} \frac{qN_d}{K_s\varepsilon_o} (x_n - x)\, dx$$

or

$$\psi_{no} - \psi(x) = \frac{qN_d}{2K_s\varepsilon_o} (x_n - x)^2 \tag{3-23}$$

where $\psi_n(x_n) = \psi_{no}$ was employed. Equations (3-22) and (3-23) are plotted in Fig. 3-11c. Since the electrostatic potential must be the same at the interface, we set $\psi_n(0) = \psi_p(0)$ and obtain

$$\frac{qN_a}{2K_s\varepsilon_o} x_p^2 + \frac{qN_d}{2K_s\varepsilon_o} x_n^2 = \psi_o \tag{3-24}$$

where Eq. (3-15) was used. Solving Eqs. (3-20) and (3-24), we have the depletion-layer thickness

$$x_n = \left[\frac{2K_s\varepsilon_o\psi_o N_a}{qN_d(N_a + N_d)}\right]^{1/2} \tag{3-25}$$

and

$$x_p = \left[\frac{2K_s\varepsilon_o\psi_o N_d}{qN_a(N_a + N_d)}\right]^{1/2} \tag{3-26}$$

The total depletion-layer width is

$$x_d = x_n + x_p \tag{3-27}$$

If the impurity concentration on one side of the junction is much higher than that on the other side, the junction is called a *one-sided step junction*. The one-sided step junction is an excellent approximation for a diffused junction having a shallow junction depth. For example, if the doping densities of the two sides are of different magnitude such that $N_a \gg N_d$, then making use of Eqs. (3-20) and (3-27) leads to $x_n \gg x_p$ and $x_d = x_n$. Physically, this means that the width of the space-charge layer in the heavily doped side is negligible, and the depletion layer is essentially located in the n side alone.

Example. A silicon *pn* step-junction diode is doped with $N_d = 10^{16}\,\text{cm}^{-3}$ and $N_a = 4 \times 10^{18}$ on the n and p sides, respectively. Calculate the built-in potential, depletion-layer width, and maximum field at zero bias at room temperature.

Solution.

$$\psi_o = 0.026 \ln \frac{4 \times 10^{18} \times 10^{16}}{2.25 \times 10^{20}} = 0.826 \text{ V} \tag{3-15}$$

$$x_n \simeq \left(\frac{2K_s\varepsilon_o\psi_o}{qN_d}\right)^{1/2} = 3.28 \times 10^{-5} \text{ cm} \tag{3-25}$$

$$\mathscr{E}_n(0) = \mathscr{E}_m = \frac{-qN_dx_n}{K_s\varepsilon_o} = -5 \times 10^4 \text{ V/cm} \tag{3-19}$$

$$x_p = \frac{x_n N_d}{N_a} = 8.2 \times 10^{-8} \text{ cm} \tag{3-20}$$

Let $\psi(0) = 0$ as the potential reference, then

$$-\psi_{po} = \frac{qN_a}{2K_s\varepsilon_o} x_p^2 = 2 \text{ mV} \tag{3-22}$$

$$\psi_{no} = \frac{qN_a}{2K_s\varepsilon_o} x_n^2 = 824 \text{ mV} \tag{3.23}$$

It is clear from the above calculation that the depletion-layer width and the built-in potential in the heavily doped p side are very small and can be neglected. This example yields some useful order-of-magnitude values for a typical pn junction.

For a one-sided step junction with $N_a \gg N_d$, the electric field, electrostatic potential, built-in potential, and depletion-layer width can be derived (Prob. 3-6):

$$\mathscr{E} = \mathscr{E}_m\left(1 - \frac{x}{x_n}\right) \tag{3-28a}$$

$$\mathscr{E}_m = -\frac{qN_dx_n}{K_s\varepsilon_o} \tag{3-28b}$$

$$\psi = \psi_o\left[1 - \left(1 - \frac{x}{x_n}\right)^2\right] \tag{3-29}$$

$$\psi_o = \frac{qN_dx_n^2}{2K_s\varepsilon_o} \tag{3-30}$$

$$x_d = x_n = \left(\frac{2K_s\varepsilon_o\psi_o}{qN_d}\right)^{1/2} \tag{3-31}$$

These equations will be used in most of the analyses in the text since the two-sided step junction seldom occurs in real devices.

3-5 BIASING OF A pn JUNCTION

Under thermal equilibrium, the energy-band diagram of a p^+n step junction is shown in Fig. 3-12a along with the space-charge and electric-field distributions. The built-in potential ψ_o represents the potential barrier the carriers must overcome

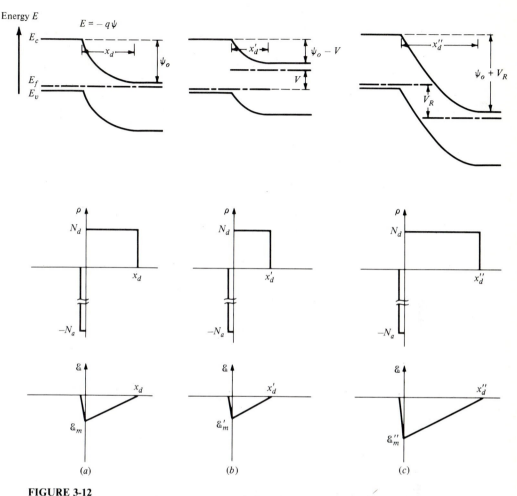

FIGURE 3-12
The energy-band, space-charge, and electric-field distribution of a one-sided step junction under (a) thermal equilibrium with a depletion-layer width x_d, (b) forward bias V with a depletion-layer width $x_d' < x_d$, and (c) reverse bias V_R with a depletion-layer width $x_d'' > x_d$.

in order to produce a current. As illustrated, essentially all the space-charge layer and the associated potential barrier are situated on the lightly doped side of the junction. When an external voltage source is connected across the *pn* junction, the thermal equilibrium is disturbed and a current will flow in the semiconductor. In general, the resistance of the space-charge region is so much higher than that of the neutral regions that the potential drops in the latter are negligible in comparison with the former. As a result, the external applied voltage is directly across the space-charge region. The magnitude of the conduction current depends strongly on the polarity of the applied voltage. If a positive voltage V is applied to the *p* side with respect to the *n* side, the potential-barrier height at the *pn* junction is lowered to $\psi_o - V$, as shown in Fig. 3-12b. The reduced potential-barrier height allows majority carriers to diffuse through the junction so that a large current is realized. This voltage polarity is in the *forward-bias* direction, which gives a low-resistance path for the *pn* junction. If a negative voltage $-V_R$ is applied to the *p* side with respect to the *n* side, the potential-barrier height is increased to $\psi_o + V_R$, as indicated in Fig. 3-12c. The increased potential barrier prevents carrier transport through the junction. Consequently, the current flow through the junction will be extremely small, and the impedance of the junction will be very high. This is the *reverse-bias* connection. It is observed that the depletion-layer width and the electric field expand with a reverse bias and contract with a forward bias.

Under reverse bias, Eq. (3-31) is valid provided that the built-in potential ψ_o is replaced by $\psi_o + V_R$. Thus the depletion-layer width becomes

$$x_d = \left[\frac{2K_s \varepsilon_o (\psi_o + V_R)}{qN_d} \right]^{1/2} \tag{3-32}$$

for the step junction, and the corresponding maximum electric field is given by

$$\mathscr{E}_m = \frac{qN_d x_d}{K_s \varepsilon_o} \tag{3-33}$$

where Eq. (3-28b) has been used with x_d replacing x_n. The depletion-layer width as a function of bias and doping concentration is shown in Fig. 3-13. Therefore, the maximum electric field increases with V_R through x_d. The biasing voltage will thus modulate the space charge and change the depletion-layer width, giving rise to a capacitance effect at the junction. At large reverse-bias voltage, the electric field may become very high so that the junction could reach a breakdown condition with large current. These two topics will be considered in the following sections.

Under the forward-bias condition, carriers are injected across the space-charge layer. For very small current, the injected-carrier densities do not seriously disturb the space-charge region, so that Eq. (3-32) can still be used if V_R is replaced by $-V$. However, as the current is increased, the carrier concentrations in the space-charge region will become comparable with the fixed impurity-ion density.

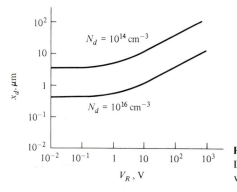

FIGURE 3-13
Depletion-layer width as a function of reverse-bias voltage in a p^+n junction.

Under this condition Eq. (3-32) no longer applies. In practice Eq. (3-32) cannot be used for most of the forward-current range.

3-6   TRANSITION CAPACITANCE AND VARACTOR

In Sec. 3-5, we have shown that the depletion-layer width is a function of the bias voltage. Since the charge stored in either half of the junction is directly proportional to the depletion-layer width, we have

$$Q = qAN_d x_d = A\sqrt{2qK_s\varepsilon_o(\psi_o + V_R)N_d} \qquad (3\text{-}34)$$

where Eq. (3-32) has been employed. The small-signal capacitance of the space-charge layer is defined by

$$C \equiv \frac{dQ}{dV_R} \qquad (3\text{-}35)$$

Substituting Eq. (3-34) into Eq. (3-35) leads to

$$C = A\left[\frac{qK_s\varepsilon_o N_d}{2(V_R + \psi_o)}\right]^{1/2} \qquad (3\text{-}36)$$

C is called the *transition* or *depletion-layer* capacitance. This equation can be rewritten

$$\frac{1}{C^2} = \frac{2}{qK_s\varepsilon_o N_d A^2}(V_R + \psi_o) \qquad (3\text{-}37)$$

An experimental plot of $1/C^2$ vs. V_R is shown in Fig. 3-14. From this figure we can calculate the donor density from the slope of the straight line. In addition, the built-in voltage can be obtained by extrapolating the line to the voltage axis. At the intercept, we have $1/C^2 = 0$ and $\psi_o = -V_R$ from Eq. (3-37).

The capacitance of a diffused junction can be calculated by solving Poisson's equation to obtain the depletion-layer width and then making use of Eq. (3-35). A computer calculation made by Lawrence and Warner gave the results shown in Fig. 3-15.

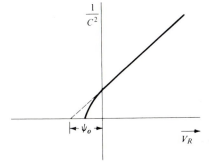

FIGURE 3-14
The capacitance-voltage (C–V) characteristic of a p^+n diode.

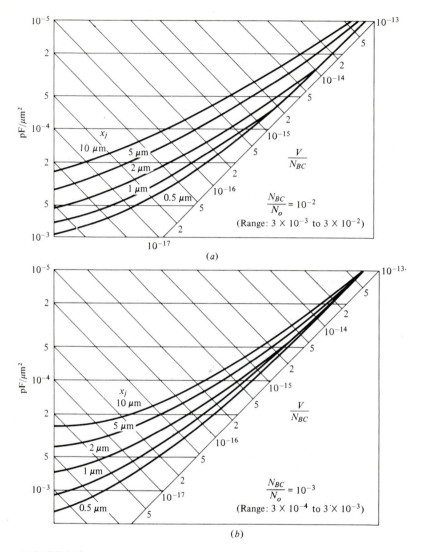

FIGURE 3-15
The junction capacitance of a diffused-silicon *pn* diode *C* per unit area vs. the total junction voltage
V divided by the background concentration N_{BC} for various junction depths x_j. (*After Lawrence and
Warner [2].*)

Example. Consider a *pn* junction fabricated by the two-step diffusion of boron into
a substrate with $N_d = 2 \times 10^{15}$ cm^{-3}. The surface concentration of boron is 10^{18} cm^{-3},
and the junction depth is 5 μm. Assume that the built-in potential is 0.8 V; obtain
the junction capacitance at a reverse bias of 5 V.

Solution. Since $N_{BC} = 2 \times 10^{15}$ cm^{-3} and $N_o = 10^{18}$ cm^{-3}, we have $N_{BC}/N_o = 2 \times 10^{-3}$. In addition

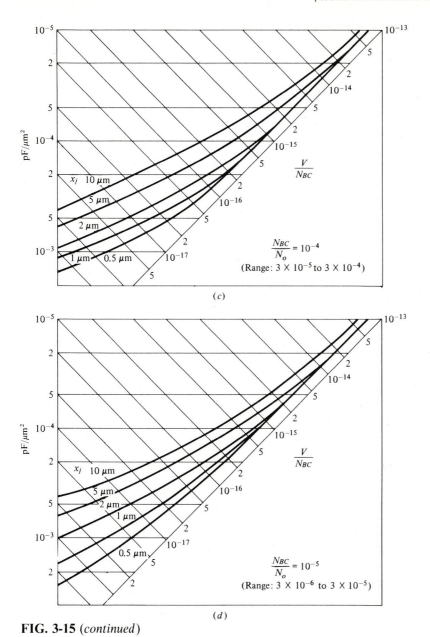

FIG. 3-15 (*continued*)

$$\frac{\psi_o + V_R}{N_{BC}} = \frac{5.8}{2 \times 10^{15}} = 2.9 \times 10^{-15}$$

Now, we use Fig. 3-15*b* to find $C = 4 \times 10^3$ pF/cm².

The capacitance of a *pn* junction biased in the reverse direction is useful in *LC* tuning circuits in which the resonant frequency is controlled by an external voltage. A diode made specially for this purpose is known as the *varactor* (from *vari*able re*actor*). The capacitance-voltage equation of a junction diode can be written

$$C = C_o(V_R + \psi_o)^{-n} \tag{3-38}$$

where V_R is the reverse voltage and $n = \frac{1}{2}$ for a one-sided step junction, as expressed in Eq. (3-36). The resonant frequency of an LC circuit including a pn junction capacitor is therefore given by

$$\omega_r = \frac{1}{\sqrt{LC}} = \frac{(V_R + \psi_o)^{n/2}}{\sqrt{LC_o}} \tag{3-39}$$

In circuit applications, it is frequently desirable to have a linear relationship between the resonant frequency and the control voltage, that is, $n = 2$. This special feature is obtained in the *hyperabrupt* varactor, and the detailed impurity profile necessary to achieve this function is derived by computer calculations. In designing varactors, we should minimize the series resistance and leakage current so that the quality factor is high. In addition, the range of capacitance variation should be large. A quality factor of 20 and a factor of 6 variation in capacitance are achievable design goals.

3-7   BREAKDOWN IN *pn* JUNCTIONS

One of the most important considerations in device design is the junction breakdown. A *pn* junction reaches the breakdown condition when a slight increase of the reverse-bias voltage produces a very large increment of current. The breakdown process is not inherently destructive, and it can be repeated as long as the current is limited. A very large current, however, can become destructive when it generates enough heat inside the junction. The high temperature could damage the device by short-circuiting the diode through interface melting or by lifting off the contact metal to become an opened circuit. For these reasons, an external resistance is needed to limit the maximum current flow. There are two mechanisms that may cause the junction breakdown, and these are known as *Zener* and *avalanche* breakdown.

Zener Breakdown

Consider a *pn* junction under reverse bias as shown in Fig. 3-16. The polarities of the bias are shown, and the splitting of the Fermi levels denotes the magnitude of the bias voltage. The electrons from the conduction band of the *p* side may slide down the "potential hill" to reach the *n* side, and the holes from the valence band of the *n* side can "float" up to the *p* side (like a bubble in water). The transport of these carriers produces a small current because both of them are minority carriers in minute amounts. The majority carriers, electrons in the *n* side and holes in the *p* side, are prevented from crossing the junction because of the increased

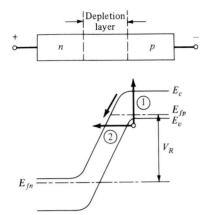

FIGURE 3-16
Zener breakdown.

potential barrier of the reverse bias. Detailed calculation of the reverse-bias current components will be discussed in the next chapter. Here, we would like to consider another source of current conduction, i.e., whether it is possible for valence electrons in the *p* side to pass through the depletion layer and reach the *n*-side conduction band. In the diagram shown, it appears that there are two possible routes. Path 1 requires the electron to first jump from the valence band to the conduction band and then to flow down the potential barrier. The energy necessary for this process is equal to or greater than the band-gap energy—a very unlikely possibility. The other option, path 2, is for the electron to penetrate the depletion layer by means of quantum-mechanical tunneling. This mechanism is possible if the depletion layer is very thin and the electric field is high. This process was proposed by Zener, whose theory is based on the breaking of covalent bonds in the depletion layer under high field. Experimentally, the Zener breakdown field is about 10^6 V/cm. If the doping densities on the two sides of the junction are greater than 10^{18} cm^{-3}, the depletion layer is thin and the critical field strength is reached with a reverse bias less than than 6 V.

Avalanche Breakdown

Let us consider the space-charge region of a reverse-biased junction as shown in Fig. 3-17a. A stray hole from the *n*-side region entering the space-charge layer acquires kinetic energy from the electric field as it travels toward the *p* region. With its high energy, the hole collides with the crystal lattice and ionizes an atom to generate an electron-hole pair at x_1. This is known as *impact ionization*. The ionization process is seen better with the energy-band diagram shown in Fig. 3-17b. At x_1, the electron-hole pair is produced by impact ionization initiated by the hole. Pictorially, we depict the electron as excited into the conduction band which then slides down the potential hill and gets back to the *n* side. Now, the extra hole remains in the valence band and travels along with the original hole. These holes continue to gain kinetic energy so that they will generate additional electrons by

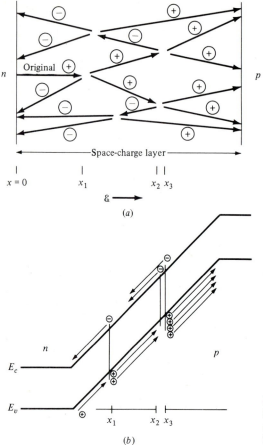

(a)

(b)

FIGURE 3-17
(a) Avalanche multiplication in the space-charge region initiated by a hole injection from the *n* side. *(b)* Energy-band diagram.

means of collision with the lattice. We have marked the additional collisions at x_2 and x_3. Thus, although we started with a single hole, four holes and three electrons cross the depletion layer to contribute to current conduction. Because the carrier generation is a multiplication process, it is known as *avalanche multiplication*. A multiplication factor is defined as

$$M \equiv \frac{I}{I_o + I_G} \tag{3-40}$$

where I is the total current, and I_o and I_G are the reverse saturation current and generation current, respectively. Since the collision probability depends on the kinetic energy of the carriers, it is a function of the electric field. In other words, a critical field must be reached before avalanche sets in. As a consequence, most carrier multiplication takes place at the location of maximum electric field. The electric-field distribution of a one-sided step junction has been shown in Fig. 3-12. There, the maximum field is shown to increase with the reverse bias, but the shape of the field distribution remain the same. Using Eqs. (3-32) and (3-33), the maximum field is given by

$$\mathscr{E}_{max} = \sqrt{\frac{2q(\psi_0 + V_R)N_d}{K_s\varepsilon_o}} \tag{3-41}$$

Let us set the maximum field equal to the critical avalanche breakdown field $\mathscr{E}_c$ at $V_R = BV$, where BV stands for the breakdown voltage. We obtain

$$BV = \frac{K_s\varepsilon_o\mathscr{E}_c^2}{2qN_d} \tag{3-42}$$

Equation (3-42) shows that the breakdown voltage is inversely proportional to impurity concentration at the lightly doped side. Using the experimentally measured critical field, the breakdown voltage has been calculated for Si, Ge, and GaAs, as shown in Fig. 3-18. It is clear that the breakdown voltage is inversely proportional to the doping density. In addition, the breakdown voltage is a function of the band-gap energy. Since the dielectric constant is about the same for all semiconductors, the change of the breakdown voltage has to come from the critical field. In a larger band-gap material, a higher kinetic energy is required to generate an electron-hole pair by collision. For this reason, the avalanche breakdown voltage in silicon is higher than that of germanium.

Figure 3-18 shows the ideal breakdown characteristics for junctions that are perfectly planar. In a real *pn* diode, as described in Sec. 3-1, the junction near the edge of the diffusion mask is curved and its curvature depends on the junction

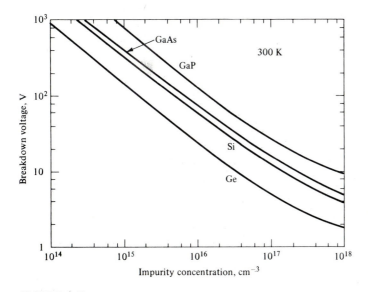

FIGURE 3-18
Avalanche breakdown voltage of a one-sided step junction as a function of impurity concentration in the lightly doped side. (*After Sze [3].*)

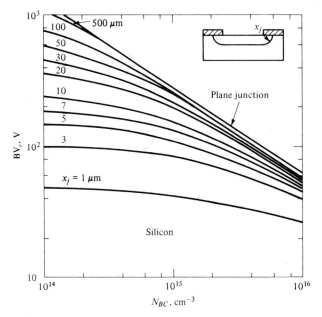

FIGURE 3-19
Effect of junction curvature on the breakdown voltage of a one-sided step junction. (*After Grove [4].*)

depth. Based on the solution of the two-dimensional Poisson equation, the curvature will create an electric-field pattern with higher field around the periphery. Using the junction depth x_j as a parameter, the breakdown voltage is calculated and shown in Fig. 3-19. It is seen that a shallower junction will have a lower breakdown voltage because a smaller x_j gives rise to a sharper curvature at the edges where the electric field is highest.

REFERENCES

1. Warner, R. M., Jr., and J. N. Fordemwalt: "Integrated Circuits," p. 275, McGraw-Hill, New York, 1965.
2. Lawrence, H., and R. M. Warner, Jr.: Diffused Junction Depletion Layer Calculations, *Bell Syst. Tech. J.*, **39**:389 (1960).
3. Sze, S. M.: "Physics of Semiconductor Devices," 2d ed., Wiley, New York, 1981.
4. Grove, A. S.: "Physics and Technology of Semiconductor Devices," Wiley, New York, 1967.

ADDITIONAL READINGS

Ghandi, S. K.: "VLSI Fabrication Principles," Wiley, New York, 1983.
Gray, P. E., D. DeWitt, A. R. Boothroyd, and J. F. Gibbons: "Physical Electronics and Circuit Models of Transistors," SEEC Series, vol. II, Wiley, NewYork, 1964.
Jonscher, A. K.: "Principles of Semiconductor Device Operation," Wiley, New York, 1960.
Neudeck, G. W.: "The *pn* Junction Diode," Addison-Wesley, Reading, Mass., 1983.

Shockley, W.: The Theory of *p-n* Junction in Semiconductors and *p-n* Junction Transistors, *Bell Syst. Tech. J.*, **28**:435 (1949).

Sze, S. M. (ed.): "VLSI Technology," McGraw-Hill, New York, 1983.

PROBLEMS

3-1. A uniformly doped *n*-type silicon epitaxial layer of $0.5\,\Omega \cdot$ cm resistivity is subjected to a boron diffusion with constant surface concentration of 5×10^{18} cm^{-3}. It is desired to form a *pn* junction at a depth of 2.7 μm. At what temperature should this diffusion be carried out if it is to be completed in 2 hours?

3-2. After the predeposition step, it is found that 5×10^{15} boron atoms/cm^2 are introduced into the silicon epitaxial layer with $N_d = 10^{16}$ atoms/cm^3. The sample is then subjected to a drive-in step. Assume the diffusion constant is 3×10^{-12} cm^2/s. Calculate the junction depth at the end of 60 min. Plot the impurity profile and indicate the junction depth.

3-3. An *n*-type silicon substrate of $0.5\,\Omega \cdot$ cm resistivity is subjected to a boron diffusion with constant surface concentration of 5×10^{18} cm^{-3}. The desired junction depth is 2.7 μm. Plot impurity concentrations (log scale) vs. distance from surface (linear scale) for the boron diffusion by calculation on this plot.

3-4. A Si step-junction diode is doped with $N_d = 10^{15}$ cm^{-3} on the *n* side and $N_a = 4 \times 10^{20}$ on the *p* side. At room temperature, calculate (*a*) the built-in potential and (*b*) the depletion-layer width and maximum field at zero bias.

3-5. An alternative method of deriving the built-in potential is to assume zero net electron or hole current at equilibrium. Derive Eq. (3-15) by this method.

3-6. From Poisson's equation, derive the expressions for the electric field, electrostatic potential, built-in potential, and depletion-layer width in a one-sided p^+n step junction.

3-7. Derive the expressions for the (*a*) electric field, (*b*) potential distribution, (*c*) depletion-layer width, and (*d*) built-in potential for a linearly graded *pn* junction with a doping gradient *a*.

3-8. The impurity profile of a *pn* junction is shown in Fig. P3-8.
 (*a*) Sketch the space charge and electric field distributions.
 (*b*) Derive an expression for the built-in potential ψ_o.

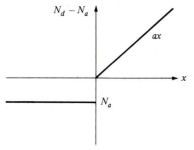

FIGURE P3-8

3-9. (*a*) For a *p-i-n* diode shown in Fig. P3-9, sketch the distribution of space charge, electric field, electrostatic potential, and energy diagram qualitatively.
 (*b*) Solve Poisson's equation to determine the depletion-layer width in the *p* and *n* regions, for $x_i = 0.5\,\mu$m.
 (*c*) Determine the electric field in the *i* region.

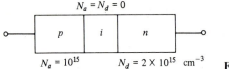

$N_a = N_d = 0$

| p | i | n |

$N_a = 10^{15}$ $N_d = 2 \times 10^{15}$ cm^{-3} **FIGURE P3-9**

3-10. The capacitance of a GaAs p^+n junction diode as a function of reverse bias is measured by using a capacitance meter at 1 MHz. The following capacitance data in picofarads are taken at 0.5-V increments from 0 to 5 V: 19.9, 17.3, 15.6, 14.3, 13.3, 12.4, 11.6, 11.1, 10.5, 10.1, 9.8. Calculate ψ_o and N_d. The diode area is 4×10^{-4} cm^2.

3-11. A p^+n silicon diode is used as a varactor. The doping concentrations on the two sides of the junction are $N_a = 10^{19}$ and $N_d = 10^{15}$, respectively. The diode area is 100 mil^2.
(a) Find the diode capacitance at $V_R = 1$ and 5 V.
(b) Calculate the resonant frequencies of a tank circuit using this varactor with $L = 2$ mH.

3-12. A capacitor in an integrated circuit is fabricated by the diffusion of boron into an n-type Si substrate with $N_{BC} = 5 \times 10^{16}$. Single-step diffusion is performed at $T = 1250°C$. The surface concentration is solubility-limited.
(a) What is the time required to obtain a junction depth of 10 μm?
(b) What is the corresponding capacitance per unit area at a reverse bias of 2 V?

3-13. Zener breakdown takes place when the maximum electric field approaches 10^6 V/cm in silicon. Assuming $N_a = 10^{20}$ cm^{-3} in the p side, what is the required donor concentration in the n side to obtain a Zener breakdown voltage of 2 V? Use one-sided step-junction approximation.

3-14. (a) Determine the critical avalanche breakdown field $\mathscr{E}_c$ from the plots in Fig. 3-18 at $N = 10^{16}$ cm^{-3} for Ge, Si, and GaAs.
(b) Using the $\mathscr{E}_c$ obtained in (a), calculate the breakdown voltage at $N = 10^{15}$, 10^{17}, and 10^{18} cm^{-3}. Compare your results with Fig. 3-18 and note the discrepancy at high doping.

3-15. To estimate the effect of junction curvature of a shallow junction, the electric field can be obtained by solving Poisson's equation in cylindrical coordinates (see Fig. P3-15):

$$\frac{1}{r}\frac{d}{dr}\left(r\frac{d\psi}{dr}\right) = -\frac{1}{r}\frac{d}{dr}(r\mathscr{E}) = -\frac{qN_d}{K_s\varepsilon_o}$$

(a) Show that the following electric field satisfies Poisson's equation:

$$\mathscr{E} = -\frac{qN_d}{2K_s\varepsilon_o}\frac{r_d^2 - r^2}{r}$$

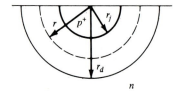

FIGURE P3-15

(b) Obtain an expression for the electrostatic potential by using the condition of zero potential at the metallurgical junction r_j.

(c) Using the critical field in silicon given in Prob. 3-14a, estimate the breakdown voltage for $N_d = 10^{15} \text{ cm}^{-3}$ and $r_j = 0.2, 0.5, 1$, and $5 \ \mu\text{m}$.

3-16. In a silicon linearly graded junction with a concentration gradient $a = 10^{20} \text{ cm}^{-4}$, estimate the breakdown voltage assuming the critical field of breakdown is the same as in Prob. 3-14a.

CHAPTER
4

THE
pn JUNCTION
UNDER
FORWARD
BIAS

In the last chapter, the potential barrier at a *pn* junction along with its space-charge effects were described. We were primarily concerned with the electrostatics, with emphasis on the equilibrium and reverse-biased characteristics. The current flow was briefly discussed without mentioning how it came about. In this chapter, we are interested in the details of current conduction under forward bias. We shall begin by finding out the relationship between the bias voltage and the injected-carrier density. The derivation of the current-voltage equation will be explained physically and then shown mathematically. The effect of recombination in the depletion layer will be analyzed, and the temperature dependence of the diode properties will be demonstrated. The last two sections will describe the junction diode under the operation of small-signal and transient conditions.

4-1 MINORITY-CARRIER INJECTION

Let us consider the energy-band diagram of the *pn* junction under thermal equilibrium shown in Fig. 4-1*a*. In this diagram, the directions of increasing potential energy for electrons and holes are noted. An electron in the conduction band of the *p* side has a higher potential energy so that it may slide down the "potential hill" of the depletion layer to reach the *n* side. The motion of the electron

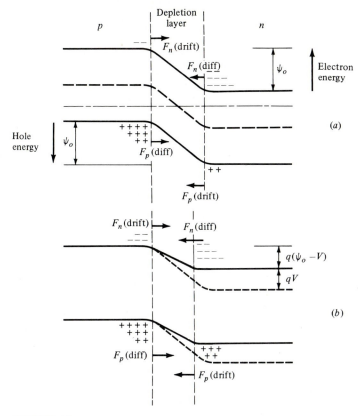

FIGURE 4-1
Schematic representation of the *pn* junction under (*a*) thermal equilibrium and (*b*) forward bias. Note that $qF_n = -I_n$ and $qF_p = I_p$.

to the right is caused by the potential difference or the built-in electric field, giving rise to a drift current. However, there are many more electrons in the *n* side, and the gradient of them produces a net diffusive electron motion to the left. Under equilibrium, the drift and diffusion components are equal in magnitude but opposite in direction so that the net electron current is zero. Using the same argument, we may say that the net hole current is also zero, and there is no net current flow across the diode. Under forward bias, the applied voltage reduces the potential barrier of the *pn* junction as depicted in Fig. 4-1*b*. The electric field at the junction is reduced, and the drift current components are decreased. At the same time, the barrier lowering enhances electron diffusion from the *n* side to the *p* side and hole diffusion from the *p* side to the *n* side. In other words, electrons are injected into the *p* side, and holes are injected into the *n* side. These injected carriers produce the current flow in a forward-bias *pn* junction.

Before we examine the problem of minority-carrier injection, we have to determine the relationship between the built-in potential and equilibrium carrier densities. Let us again use the subscripts n and p to denote the semiconductor type and the subscript o to specify the condition of thermal equilibrium. Thus, n_{no} and n_{po} are the equilibrium electron densities in the n side and p side, respectively. The expression of the built-in potential in Eq. (3-15) can be rewritten as

$$\psi_o = \phi_T \ln \frac{p_{po} n_{no}}{n_i^2} = \phi_T \ln \frac{n_{no}}{n_{po}} \tag{4-1}$$

where the majority-carrier densities p_{po} and n_{no} are used to replace N_a and N_d, respectively, and the mass-action law $p_{po} n_{po} = n_i^2$ has been employed. We can rearrange Eq. (4-1) to give

$$n_{no} = n_{po} e^{\psi_o/\phi_T} \tag{4-2}$$

Similarly, we have

$$p_{po} = p_{no} e^{\psi_o/\phi_T} \tag{4-3}$$

where p_{po} is the equilibrium hole density in the p side and p_{no} is the equilibrium hole density on the n side. From Eqs. (4-2) and (4-3) it is observed that the electron densities on the two edges of the space-charge layer of a junction barrier are related through the barrier height, in this case ψ_o. It is reasonable to assume that the same rule applies when the barrier height is changed by an external voltage.

By applying a forward bias V, the junction potential is reduced to $\psi_o - V$. Thus, Eq. (4-2) is modified to

$$n_n = n_p e^{(\psi_o - V)/\phi_T} \tag{4-4}$$

where n_n and n_p are the electron densities at the edge of the space-charge layer in the n side and p side, respectively. For low-level injection, the excess electron density in the n side is small compared with n_{no}, so that we can assume $n_n = n_{no}$. Substituting this condition and Eq. (4-2) into Eq. (4-4), we obtain

$$n_p = n_{po} e^{V/\phi_T} \tag{4-5}$$

Similarly, we have

$$p_n = p_{no} e^{V/\phi_T} \tag{4-6}$$

Equations (4-5) and (4-6) define the minority-carrier densities at the edges of the space-charge layer. These are the boundary conditions which specify the carrier densities uniquely for a given applied voltage.

4-2   SPACE-CHARGE EFFECT AND THE DIFFUSION APPROXIMATION

One of the fundamental contributions by Shockley was to show that *diffusion* of *minority carriers* controls the current conduction in a *pn* junction. This is an important issue that needs clarification. One may ask what has happened to the majority carriers, and why do they play a passive role? Is there an electric field induced by carrier injection? To answer these questions, let us consider the behavior of electrons and holes in the neutral n side after applying a forward bias. Under the assumption of low-level injection, the injected-hole density p_n is much smaller than the majority-carrier density n_{no} but greater than p_{no}. The excess holes, having the highest concentration at the edge of the depletion layer, will diffuse toward the lower concentration inside the neutral n region. The transport of holes forms a minority-carrier diffusion current. The excess holes will also recombine with the electrons so that the hole density and current decrease with distance from the space-charge layer edge, as illustrated in the lower part of Fig. 4-2.

The injected holes, each carrying a positive charge, will momentarily establish an electric field. This field acts on both majority and minority carriers to produce the drift currents $q\mu_n n_n \mathscr{E}$ and $q\mu_p p_n \mathscr{E}$. Since the density of the majority carriers n_n is much higher than that of the minority carriers p_n, the electric field gives rise to a large majority-carrier current but a negligible minority-carrier current. This field draws in excess electrons, as shown in the upper part of Fig. 4-2, which tend to neutralize the injected holes and reestablish the space-charge neutrality. For this reason, charge neutrality is preserved in the neutral n region and the hole current is essentially a pure diffusive current. The majority carriers respond to the minority-carrier injection but not the other way around. We may therefore neglect the majority-carrier effect in calculating the minority-carrier current.

Example. In a *pn* step junction, the doping concentrations of the two sides are $N_a = 10^{18}\ \text{cm}^{-3}$ and $N_d = 10^{15}\ \text{cm}^{-3}$, respectively.

(*a*) Calculate the minority-carrier concentrations at the depletion-layer edges for $V = 0.52$ V.

(*b*) Estimate the ratio of the drift currents inside the *n* region. Assume that the diode is at room temperature.

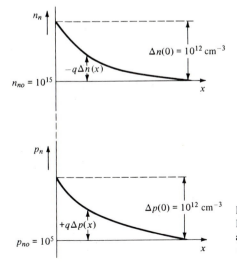

$$\Delta n(0) = 10^{12}\ \text{cm}^{-3}$$

$$\Delta p(0) = 10^{12}\ \text{cm}^{-3}$$

FIGURE 4-2
Hole injection into the *n* side of a p^+n junction and the resultant electron distribution. $+q$ and $-q$ are added to illustrate charge neutrality.

Solution

(a)

$$p_{no} = \frac{n_i^2}{N_d} = 2.25 \times 10^5 \text{ cm}^{-3}$$

$$n_{po} = \frac{n_i^2}{N_a} = 2.25 \times 10^2 \text{ cm}^{-3}$$

$$p_n(0) = p_{no} e^{V/\phi_T} = 10^{14} \text{ cm}^{-3}$$

$$n_p(0) = n_{po} e^{V/\phi_T} = 10^{11} \text{ cm}^{-3}$$

(b) Drift-current ratio in the n side is

$$\frac{q\mu_n n_{no} \mathscr{E}}{q\mu_p p_n \mathscr{E}} = \frac{\mu_n N_d}{\mu_p p_n(0)} = \frac{1350 \times 10^{18}}{450 \times 10^{14}} = 3 \times 10^4$$

The electron drift current is much larger than the hole current even though the maximum hole concentration is used.

In part (b) of the above example, we observe that the electric field effect on the majority carriers is much more evident than that of the minority carriers so that we can always neglect the minority-carrier drift current.

The minority-carrier density and current may be derived by assuming that the condition of space-charge neutrality prevails and that the minority carriers are the only type of carriers present. These minority carriers, being neutralized, are *uncharged*, and they are transported by diffusion in the neutral region. This is known as the *diffusion approximation*. Mathematically, the hole current is obtained by setting $\mathscr{E} = 0$ in Eq. (2-68). Thus, we have

$$I_p = -qAD_p \frac{dp_n}{dx} \tag{4-7}$$

and the hole continuity equation in Eq. (2-62) becomes

$$\frac{\partial p_n}{\partial t} = D_p \frac{\partial^2 p_n}{\partial x^2} - \frac{p_n - p_{no}}{\tau_p} \tag{4-8}$$

which is called the *diffusion equation* for holes. The injected-hole distribution and current magnitude are obtainable by solving Eqs. (4-7) and (4-8) with appropriate boundary conditions. The majority-carrier current can be obtained afterward by invoking current continuity, as will be shown later. Similarly, the electron current and diffusion equations on the p side of the junction are

$$I_n = qAD_n \frac{dn_p}{dx} \tag{4-9}$$

$$\frac{\partial n_p}{\partial t} = D_n \frac{\partial^2 n_p}{\partial x^2} - \frac{n_p - n_{po}}{\tau_n} \tag{4-10}$$

4-3 DC CURRENT-VOLTAGE CHARACTERISTICS

A *pn* junction diode refers to a packaged two-terminal rectifier in which current can flow in one direction only. In most cases a diode contains only one *pn* junction.

For this reason, the terms "junction" and "diode" are frequently used interchangeably. The current-voltage relationship in a *pn* junction can be obtained by solving the continuity and current equations (4-7) to (4-10). Let us consider the minority carriers in the *n* side of the junction for a *long* diode, where the external contact in the lightly doped side is very far from the junction space-charge region. In other words, all the injected carriers in a long diode recombine before they reach the contact. The impurity concentration in the *n* side is assumed to be uniform. The hole diffusion equation for the time-independent case is rewritten using $p_n - p_{no} = \Delta p_n$:

$$D_p \frac{d^2 \Delta p_n}{dx^2} - \frac{\Delta p_n}{\tau_p} = \frac{\partial p_n}{\partial t} = 0 \qquad (4\text{-}11)$$

where we have subtracted $d^2 p_{no}/dx^2$ to form the first term on the left-hand side since this added term is zero because of uniform doping. The general solution of Eq. (4-11) is

$$\Delta p = K_1 \exp\left(-\frac{x}{\sqrt{D_p \tau_p}}\right) + K_2 \exp \frac{x}{\sqrt{D_p \tau_p}} \qquad (4\text{-}12)$$

where K_1 and K_2 are constants to be determined. The boundary conditions are

$$p_n = \begin{cases} p_n(0) = p_{no} e^{V/\phi_T} & \text{at } x = 0 \qquad (4\text{-}13) \\ p_{no} & \text{at } x = \infty \qquad (4\text{-}14) \end{cases}$$

We have used the edge of the depletion layer as the origin $x = 0$. The first boundary condition is taken from Eq. (4-6), and the second boundary condition is obtained by assuming that the injected holes recombine before reaching the external contact. Substituting the boundary conditions into Eq. (4-12) leads to

$$\Delta p_n = p_{no}(e^{V/\phi_T} - 1)e^{-x/L_p} \qquad (4\text{-}15)$$

where

$$L_p = \sqrt{D_p \tau_p} \qquad (4\text{-}16)$$

The minority-carrier distribution is plotted in the right side of Fig. 4-3a. The symbol L_p is called the *hole diffusion length*. Physically, L_p is the mean distance traveled by an injected hole before recombination. The hole current is derived by substituting Eq. (4-15) into Eq. (4-7):

$$I_p = qA \frac{D_p p_{no}}{L_p} (e^{V/\phi_T} - 1)e^{-x/L_p} \qquad (4\text{-}17)$$

The hole current at the edge of the space-charge layer is evaluated at $x = 0$:

$$I_p(0) = qA \frac{D_p p_{no}}{L_p} (e^{V/\phi_T} - 1) \qquad (4\text{-}18)$$

Thus, the hole current distribution in Eq. (4-17) can be rewritten as

$$I_p = I_p(0)e^{-x/L_p} \qquad (4\text{-}19)$$

From this equation, we notice that the hole current in the n side decreases with distance away from the junction, as illustrated in Fig. 4-3b. Since the total current must be invariant with respect to x to satisfy current continuity, that is, $I_n + I_p = I$, the electron current must be increasing with x to compensate for the drop of the hole current. This situation is plotted in the right-hand side of Fig. 4-3c. In other words, the minority-carrier current is continuously transferred into the majority-carrier current via electron-hole pair recombination. Similarly, the electron distribution and electron current in the p side of the junction are

$$\Delta n_p = n_{po}(e^{V/\phi_T} - 1)e^{-x'/L_n} \tag{4-20}$$

$$I_n = I_n(0)e^{-x'/L_n} \tag{4-21}$$

where

$$I_n(0) = qA\frac{D_n n_{po}}{L_n}(e^{V/\phi_T} - 1) \tag{4-22}$$

and

$$L_n = \sqrt{D_n \tau_n} \tag{4-23}$$

L_n is the electron diffusion length, and x' is the spatial dimension in the p side with $x' = 0$ set at the edge of the depletion layer. Equations (4-20) and (4-21) are plotted in the left sides of Fig. 4-3a and b, respectively. Again, the transfer of electron current into hole current takes place on the p side of the junction. The complete current and minority-carrier distributions are illustrated in Fig. 4-3, where the exponential nature of current and carrier distribution are clearly observable. Equations (4-18) and (4-22) are applicable for the reverse bias provided a negative sign is assigned to the voltage V.

In Fig. 4-3, the total current of the junction can be obtained by adding the minority-carrier-current components at the respective space-charge-layer edge. Therefore

$$I = I_n(0) + I_p(0) = I_0(e^{V/\phi_T} - 1) \tag{4-24}$$

where

$$I_o = qA\left(\frac{D_p p_{no}}{L_p} + \frac{D_n n_{po}}{L_n}\right) \tag{4-25}$$

Equation (4-24) describes the ideal current-voltage characteristic of the pn junction. It turns out that Eq. (4-24) is applicable to both the forward- and reverse-bias voltages. The derivation of the reverse-bias $I-V$ equation is left as an exercise (Prob. 4-6). Under forward bias with $V/\phi_T \gg 1$, the current varies exponentially with the applied voltage. In reverse bias, the exponential term drops out and the current becomes independent of the bias. For this reason, I_o is called the *saturation current*, which is the ideal reverse-bias current of the junction. Among the reported current-voltage characteristics, the most ideal data reported recently are reproduced in Fig. 4-4. The normalized exponential current-voltage characteristic is striking, and the reverse current is essentially saturated in the temperature range of 300 to 400 K. It should be noted that the normalization factor I_o is itself strongly temperature-dependent.

Carrier concentrations

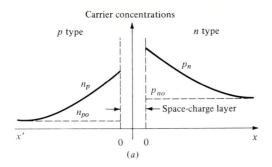

(a)

Minority-carrier
currents

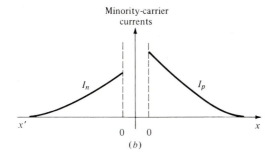

(b)

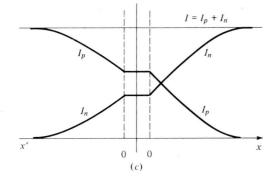

(c)

FIGURE 4-3
The forward-biased *pn* junction:
(a) minority-carrier distributions,
(b) minority-carrier currents, and
(c) electron and hole currents.

Example. A silicon *pn* junction diode has the following parameters: $N_d = 10^{16}$, $N_a = 5 \times 10^{18}$ cm^{-3}, $\tau_n = \tau_p = 1$ μs, $A = 0.01$ cm^2. Assume that the widths of the two sides of the junction are much greater than the respective minority-carrier diffusion length. Obtain the applied voltage at a forward current of 1 mA at 300 K.

Solution. From Fig. 2-6, we find the electron mobility in the *p* side ($N_a = 5 \times 10^{18}$ cm^{-3}) to be 200 cm^2/V·s and the hole mobility in the *n* side ($N_d = 10^{16}$ cm^{-3}) to be 500 cm^2/V·s. Since $\phi_T = 26$ mV at 300 K, we use Einstein's relationship to obtain

$$D_p = 0.026 \times 200 = 5.2 \text{ cm}^2/\text{s}$$
$$D_n = 0.026 \times 500 = 13 \text{ cm}^2/\text{s}$$

from Eq. (2-16)

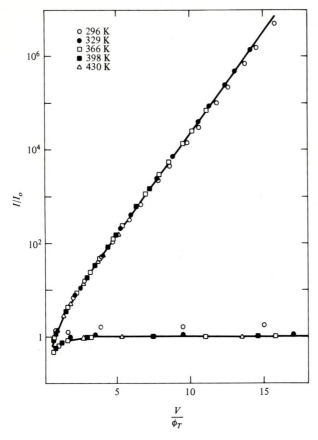

FIGURE 4-4
Semilog plot of I/I_o vs. V/ϕ_T. The upper curve applies to forward bias, the lower curve to reverse bias. All points lie on the same curve irrespective of the temperature. Solid line corresponds to Eq. (4-24). Diode area $= 6.25\ \text{mm}^2$. (*After Cappelletti [1].*)

Therefore $L_p = \sqrt{D_p \tau_p} = 2.3 \times 10^{-3}\ \text{cm}$ from Eq. (4-16)

$\qquad\qquad L_n = \sqrt{D_n \tau_n} = 3.6 \times 10^{-3}\ \text{cm}$ from Eq. (4-23)

In addition, $n_i = 1.5 \times 10^{10}\ \text{cm}^{-3}$ and

$$p_{no} = \frac{n_i^2}{N_d} = 2.25 \times 10^4\ \text{cm}^{-3} \qquad n_{po} = \frac{n_i^2}{N_a} = 45\ \text{cm}^{-3}$$

Thus

$$I_o = (1.6 \times 10^{-19})(0.01)\left[\frac{(5.2)(2.25 \times 10^4)}{2.3 \times 10^{-3}} + \frac{13 \times 45}{3.6 \times 10^{-3}}\right]$$

$$= 8.16 \times 10^{-14}\ \text{A} \qquad \text{from Eq. (4-25)}$$

Rewriting Eq. (4-24), we obtain

$$V = \phi_T \ln\left(\frac{I}{I_o} + 1\right) = 26 \ln \frac{10^{-3}}{8.16 \times 10^{-14}} = 604 \, \text{mV}$$

In most practical junctions, the current-voltage characteristic deviates significantly from Eq. (4-24) as a result of carrier recombination and generation inside the space-charge layer. This effect will be considered in a later section.

In general, the ratios of the diffusion constant to diffusion length for electrons and holes are of the same order of magnitude. For a one-sided step junction, the equilibrium minority-carrier density of the lightly doped side is much greater than that of the heavily doped side. Consequently, the $I-V$ characteristic is determined primarily by the lightly doped side.

In some diodes, the length of the lightly doped side W_n may be *much less* than the diffusion length. These are known as *short-* or *thin-base diodes*; a typical one is shown in Fig. 4-5 along with the minority-carrier distribution. Since the injected minority carriers have not decayed to the thermal-equilibrium density, the boundary condition given by Eq. (4-14) may have to be modified depending on the surface recombination velocity. There are two cases that are important. The first case is represented by the solid line in Fig. 4-5*b* where the minority-carrier concentration at the metal contact is zero. This corresponds to an infinite surface recombination velocity at the metal contact so that all arriving carriers will be removed immediately. The carrier distribution can be easily written as

$$\Delta p_n = p_{no}(e^{V/\phi_T} - 1)\left(1 - \frac{x}{W_n}\right) \tag{4-26}$$

The hole current I_p is spatially independent since recombination is negligible in

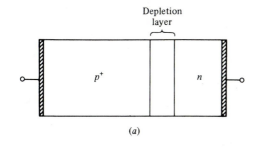

(a)

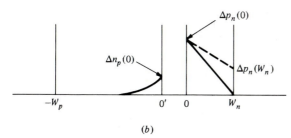

(b)

FIGURE 4-5
A p^+n diode with a short-base:
(a) structure and (b) minority-carrier distribution.

the short diode. Its value is given by

$$I_p(0) = I_p = -qAD_p \frac{dp_n}{dx} = qAD_p \frac{p_{no}}{W_n} (e^{V/\phi_T} - 1) \qquad (4\text{-}27)$$

Note that Eq. (4-27) is identical to Eq. (4-18) except that W_n has replaced L_p. In other words, the current is controlled by the base length rather than the diffusion length. This is exactly what happens in a bipolar transistor, a subject to be discussed in Chap. 5.

The second case involves a finite surface recombination velocity at the metal contact and is represented by the dashed line in Fig. 4-5b. The injected carriers are removed at a slower rate so that the carrier density is not zero at W_n. The boundary condition is defined by Eq. (2-40), where the carrier gradient is specified at the contact interface. Details are worked out in Prob. 4-9.

4-4   TEMPERATURE DEPENDENCE OF THE *V–I* CHARACTERISTICS

The temperature effect of the *V–I* characteristics of most diodes is contained implicitly in I_o and ϕ_T. We can derive the temperature dependence of both the current and voltage of the diode using Eq. (4-24). Neglecting the unity term in comparison with the exponential term in the diode equation, we derive the relations (Prob. 4-10)

and

$$\left.\frac{dV}{dT}\right|_{I = \text{const}} = \frac{V}{T} - \phi_T \left(\frac{1}{I_o} \frac{dI_o}{dT}\right) \qquad (4\text{-}28)$$

$$\left.\frac{dI}{dT}\right|_{V = \text{const}} = I \left(\frac{1}{I_o} \frac{dI_o}{dT} - \frac{V}{T\phi_T}\right) \qquad (4\text{-}29)$$

The temperature dependence of the current and voltage can be obtained if the saturation current as a function of temperature is known. Using the relations

$$p_{no} = \frac{n_i^2}{N_d}$$

$$n_{po} = \frac{n_i^2}{N_a}$$

the saturation current can be written as

$$I_o = qAn_i^2 \left(\frac{D_p}{L_p N_d} + \frac{D_n}{L_n N_a}\right) \qquad (4\text{-}30)$$

The parameters within the parentheses are relatively insensitive to temperature variation, and may be represented by a dependence of *T*. Using Eqs. (1-20) and (1-22) for the temperature dependence of n_i^2, we obtain

$$I_0 \propto T^4 e^{-E_g(0)/kT} \qquad (4\text{-}31)$$

Differentiating this equation and dividing the result by I_o leads to

$$\frac{I}{I_o}\frac{dI_o}{dT} = \frac{4}{T} + \frac{E_g(0)}{kT^2} \approx \frac{E_g(0)}{kT^2}$$

(4-32)

since in most cases, the term $4/T$ is negligible. Substitution of this expression into Eqs. (4-28) and (4-29) yields

$$\frac{dV}{dT} = \frac{qV - E_g(0)}{qT}$$

(4-33)

$$\frac{1}{I}\frac{dI}{dT} = \frac{E_g(0) - qV}{kT^2}$$

(4-34)

In a silicon diode where $E_g(0) = 1.17\,\text{eV}$ and the operating voltage is typically 0.6 V at room temperature (300 K), the current approximately doubles every 6°C, and the voltage decreases linearly with temperature with a coefficient of approximately $-2\,\text{mV}/°\text{C}$. The temperature dependence of a silicon diode under both forward and reverse bias is plotted in Figs. 4-6 and 4-7. The ideal exponential dependence of current on voltage should be a straight line in Fig. 4-6. This is not seen because it is compromised by the space-charge recombination current (see next section) in

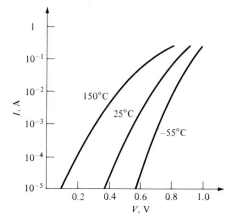

FIGURE 4-6
Temperature effect on the current-voltage characteristic of a silicon planar diode.

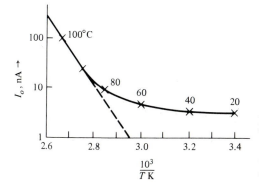

FIGURE 4-7
The reverse saturation current as a function of temperature in a silicon *pn* diode. The dashed line represents the exponential dependence of I_o on $1/T$.

the low-bias region and by the ohmic drop across the series resistance at high current. In general, the current decreases as the temperature is lowered. For reverse leakage current which is not sensitive to temperature. In most diodes, the leakage current is negligible.

4-5 RECOMBINATION AND GENERATION CURRENT IN THE SPACE-CHARGE REGION

In a preceding section on the forward characteristics of a *pn* junction diode, we have implicitly assumed that there is no carrier recombination within the space-charge region. However, the forward bias increases the carrier concentration at the edges of the space-charge layer, so that $pn > n_i^2$. These excess carriers are crossing the space-charge layer, so that the carrier concentrations may exceed the equilibrium values. Therefore, recombination is expected in the junction space-charge layer. The recombination current is defined by

$$I_{rec} = qA \int_0^{x_d} U \, dx \qquad (4\text{-}35)$$

where U is given in Eq. (2-27). It can be shown that the maximum recombination rate occurs when $n = p$ (Prob. 4-12), and we have

$$n = p = n_i e^{V/2\phi_T} \qquad (4\text{-}36)$$

and
$$U_{max} = \frac{n_i(e^{V/\phi_T} - 1)}{2\tau_o(e^{V/2\phi_T} + 1)} \qquad \text{if} \quad E_t = E_i \qquad (4\text{-}37)$$

where $\tau_o = 1/CN_t$ is the effective minority-carrier lifetime in the depletion region. For $V \gg \phi_T$, the maximum recombination rate becomes

$$U_{max} = \frac{n_i}{2\tau_o} e^{V/2\phi_T} \qquad (4\text{-}38)$$

Substitution of Eq. (4-38) into Eq. (4-35) yields

$$I_{rec} = I_R e^{V/2\phi_T} \qquad (4\text{-}39)$$

where
$$I_R = \frac{qAn_i x_d}{2\tau_o} \qquad (4\text{-}40)$$

I_{rec} is derived on the basis of the worst-case condition. Strictly speaking, x_d in Eq. (4-40) should be replaced by a width within which the maximum recombination rate occurs. This width is significantly smaller than x_d. In both silicon and gallium arsenide diodes, the recombination-current component dominates at low-current levels, as shown in Fig. 4-8 for a typical silicon diode. The semilog plot shows the change of slope from $1/2\phi_T$ to $1/\phi_T$ for increasing current, indicating the increase of the diffusion current. At high-current levels, the series resistance provides a large ohmic drop to dominate the V–I characteristics.

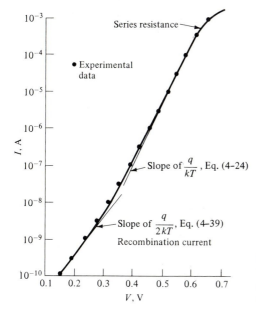

FIGURE 4-8
The current-voltage characteristic of a silicon-diffused junction with a substrate doping of 10^{16} cm^{-3}.

In a forward-biased junction, we have the condition $pn > n_i^2$. It is therefore reasonable to expect $pn < n_i^2$ in the space-charge layer of a reverse-biased junction. Substitution of this condition into Eq. (2-27) leads to a negative recombination rate. From the physical picture shown in Chap. 2, a negative recombination rate means a positive generation rate, and the resulting current is a generation current instead of a recombination current. By using the approximation $|V/\phi_T| \gg 1$, the generation current is derived from the total recombination within the space-charge layer. Thus,

$$I_G = qA\,|\,U\,|\,x_d = \frac{qn_i A}{2\tau_o}\,x_d \qquad (4\text{-}41)$$

where we have used the relation

$$U = -\frac{n_i}{2\tau_o} \qquad (4\text{-}42)$$

Since the space-charge-layer width increases with reverse bias, the generation current increases with the reverse-bias voltage. Introduction of gold will reduce the carrier lifetime and thus increase the generation rate. In a practical silicon junction diode, the generation current is usually greater than the saturation current expressed in Eq. (4-25), so that the reverse current is not saturated.

Using a technique called *gettering*, defects and impurities in silicon can be reduced significantly. Gettering involves a heat treatment of the silicon wafer at high temperature for 2 to 4 hours so that oxygen within 5 to 10 μm below the surface will be removed. Other gettering methods include the use of chlorine and controlled damage to the back side and annealing. In well-gettered materials, the recombination and generation current may be negligible, as demonstrated in Fig. 4-4.

4-6   TUNNELING CURRENT

In our previous discussion of the *pn* junction, we considered the diffusion current resulting from injected carriers that overcame the junction potential barrier. When both the *p* side and *n* side are heavily doped with impurities, some carriers may penetrate (instead of going over) the potential barrier to produce an additional current. This mechanism, called *quantum-mechanical tunneling*, can take place when (1) the Fermi level is located within the conduction and valence band, (2) the width of the space-charge layer is so narrow that a high tunneling probability exists, and (3) at the same energy level, electrons are available in the *n*-type conduction band and empty states are available in the *p*-type valence band. These conditions are satisfied when both sides of the junction are heavily doped so that they become degenerate semiconductors. Figure 4-9*a* shows such a junction at 0 K without external bias. The absolute zero temperature is chosen so that all states below the Fermi level will be occupied and all states above the Fermi level will be empty. This assumption simplifies the graphical description without losing the essence of the physical reality at room temperature. When a forward bias is applied, the band diagram is changed to that of Fig. 4-9*b*. Note that the energy of some electrons in the conduction band of the *n* side is now raised to a level corresponding to the level of empty states in the *p* side of the junction. As a result, electrons may tunnel through the junction potential barrier to produce a current. The magnitude of this tunneling current is limited either by the available electrons in the *n* side for tunneling or by the available empty states at the same energy level to which electrons can tunnel. Consequently, the maximum current is reached at the bias condition shown in Fig. 4-9*c*. Further increases of the forward bias would reduce the current since there are fewer empty states on the *p* side to accommodate the tunneling electrons. The condition of zero tunneling current is realized in the band diagram in Fig. 4-9*d*. Under reverse bias, the band diagram depicted in Fig. 4-19*e* leads to increasing reverse tunneling current with the reverse-bias voltage.

The analysis of the tunneling mechanism requires a background in quantum mechanics and is beyond the scope of this text. However, it is instructive to examine a simple model so as to gain physical insight into this important effect. Let us consider the potential barrier of the tunneling junction under forward bias, illustrated in Fig. 4-10, where χ_B is the potential-barrier height and is roughly equal to the band-gap energy E_g. The thickness of the barrier x_d is the space-charge-layer width, and *n* is the number of electrons available for tunneling. In this situation, the simplified tunneling probability (given here without proof) is [2]

$$T_t = \exp\left(-\frac{8\pi\sqrt{2qm_e}\,\chi_B^{3/2}}{3h\mathscr{E}}\right) \tag{4-43}$$

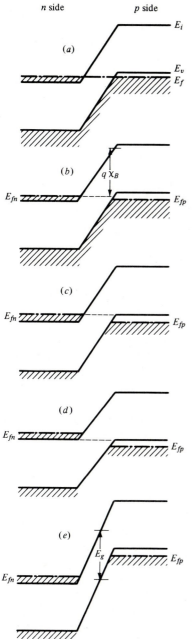

FIGURE 4-9
The energy-band diagram of a tunneling junction under various biasing conditions.

In Fig. 4-10 we have

$$\mathscr{E} = \frac{\chi_B}{x_d} \qquad (4\text{-}44)$$

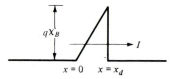

FIGURE 4-10
The potential barrier of the forward-biased tunneling junction corresponding to Fig. 4-9*b*.

Substitution of Eq. (4-44) into (4-43) leads to

$$T_t = \exp\left[-\frac{8\pi x_d}{3h}(2qm_e\chi_B)^{1/2} \right] \tag{4-45}$$

Assuming that the number of empty states on the other side of the barrier, that is, $x > x_d$, is very large, the tunneling current is given by

$$I_t = qAv_{\text{th}}nT_t \tag{4-46}$$

where v_{th} is the velocity of the tunneling electrons. As an example, let a silicon diode be doped heavily on both sides so that the space-charge-layer width is 5 nm. The barrier height is 1.1 eV, the diode area is 10^{-4} cm^2, $m_e = 0.2$ m, and $n = 2 \times 10^{20}$ cm^{-3}. Then the tunneling probability is 10^{-7}, and the tunneling current is approximately 2 mA. Note that the tunneling probability is quite small at a significant current level.

In addition to the tunneling-current component, the diffusion current becomes important at high forward-bias voltages. The current-voltage characteristics of a junction including the tunneling and diffusion components are summarized in Fig. 4-11*a*. A diode having this *I–V* curve is called the *tunnel diode* or *Esaki diode*. The impurity densities are typically 5×10^{19} cm^{-3}, and the depletion-layer thickness is in the order of 5 to 10 nm. If the doping densities are slightly reduced so that the forward tunneling current is negligible, the current-voltage curve will be modified to that shown in Fig. 4-11*b*. This is called a *backward diode*.

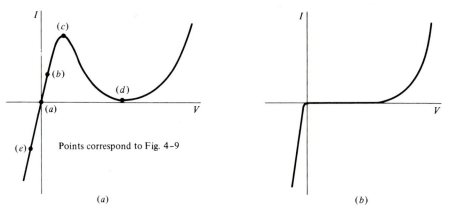

FIGURE 4-11
Current-voltage characteristics of (*a*) the Esaki diode and (*b*) the backward diode.

The tunneling current is relatively insensitive to temperature variation, and the negative resistance region of a silicon Esaki diode remains the same even at a temperature of 150°C. Because the time required for tunneling is very small, the switching speed of an Esaki diode is very high. Applications of this unique device include oscillators, bistable and monostable multivibrators, high-speed logic circuits, and low-noise microwave amplifiers. However, problems in using a two-terminal active device and difficulties in fabricating them in integrated-circuit form have limited the use of tunnel diodes.

4-7 SMALL-SIGNAL AC ANALYSIS

The small-signal characteristics of a *pn* junction are important for circuit applications. In this section, we present the derivation of the diode impedance by first solving for the small-signal carrier distribution. With a small-signal voltage superimposed on the dc voltage V, the total applied voltage is expressed as

$$v = V + v_a e^{j\omega t} \tag{4-47}$$

The small-signal condition is satisfied if the applied ac signal v_a is small compared with ϕ_T. By substituting Eq. (4-47) into Eq. (4-6) and using the approximation

$$\exp \frac{v_a e^{j\omega t}}{\phi_T} = 1 + \frac{v_a}{\phi_T} e^{j\omega t} \quad \text{for } \frac{v_a}{\phi_T} \ll 1 \tag{4-48}$$

we have

$$p_n(0) = P_n(0) + p_{a1} e^{j\omega t} \tag{4-49}$$

where $P_n(0)$ is given by Eq. (4-13) and

$$p_{a1} = \frac{p_{no} v_a}{\phi_T} e^{V/\phi_T} \tag{4-50}$$

The boundary condition can be separated into an ac component and a dc component. In addition, we can separate the hole diffusion equation into dc and ac equations by substituting

$$p_n = P_n + p_a e^{j\omega t} \tag{4-51}.$$

into Eq. (4-8). The dc equation is identical with Eq. (4-11), and the ac equaton is

$$D_p \frac{d^2 p_a}{dx^2} - \frac{p_a}{\tau_p} = j\omega p_a \tag{4-52}$$

for which the ac boundary conditions are

$$p_a = \begin{cases} p_{a1} & \text{at } x = 0 \\ 0 & \text{at } x = \infty \end{cases}$$

Using these boundary conditions, we derive the expression for the ac hole density:

$$p_a = p_{a1} \exp\left(\frac{-x}{L_p} \sqrt{1 + j\omega\tau_p}\right) \tag{4-53}$$

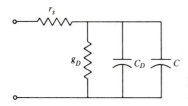

FIGURE 4-12
An equivalent circuit for a *pn* junction diode.

Differentiating this hole density and substituting the result into Eq. (4-7), we obtain the hole current at $x = 0$:

$$i_p(0) = \frac{v_a}{\phi_T} I_p(0)\sqrt{1 + j\omega\tau_p} \qquad (4\text{-}54)$$

where $I_p(0)$ is given in Eq. (4-18). Similarly, we find the electron current in the *p* side as

$$i_n(0) = \frac{v_a}{\phi_T} I_n(0)\sqrt{1 + j\omega\tau_n} \qquad (4\text{-}55)$$

The total ac current of the *pn* junction is therefore

$$i = i_p(0) + i_n(0) \qquad (4\text{-}56)$$

For a p^+n junction, $i_p(0) = i$ and $I_p(0) = I$. Thus, the *diode admittance* is reduced to

$$y = \frac{i_p(0)}{v_a} = \frac{I}{\phi_T}\sqrt{1 + j\omega\tau_p} \approx \frac{I}{\phi_T} + j\frac{\omega\tau_p I}{2\phi_T} \qquad (4\text{-}57)$$

The last equation is obtained by using $1 \gg \omega\tau_p$ for a low-frequency approximation. Note that the admittance has a conductive and a capacitive component. The capacitance in Eq. (4-57) is called the *diffusion capacitance*:

$$C_D = \frac{\tau_p I}{2\phi_T} \qquad (4\text{-}58)$$

For the diode of the example in Sec. 4-3, we obtain the small-signal conductance $g_D = I/\phi_T = 0.04\,\text{S}$ and $C_D = 2 \times 10^{-8}\,\text{F}$. The small-signal equivalent circuit of a *pn* junction diode is shown in Fig. 4-12. In addition to the diode resistance and diffusion capacitance, we have included a series resistance and a transition capacitance. The series resistance is caused by the ohmic drop across the neutral semiconductor regions and the contacts. The transition capacitance arises from the junction space-charge layer as shown in Sec. 3-6.

4-8 CHARGE STORAGE AND REVERSE TRANSIENT: THE STEP-RECOVERY DIODE

With a constant forward bias, carriers are injected and maintained in a *pn* junction diode. The carriers so maintained in the steady-state condition cannot be removed instantaneously when the forward bias is abruptly switched to a reverse bias. Figure 4-13 shows the injected-carrier distributions at different times after the application

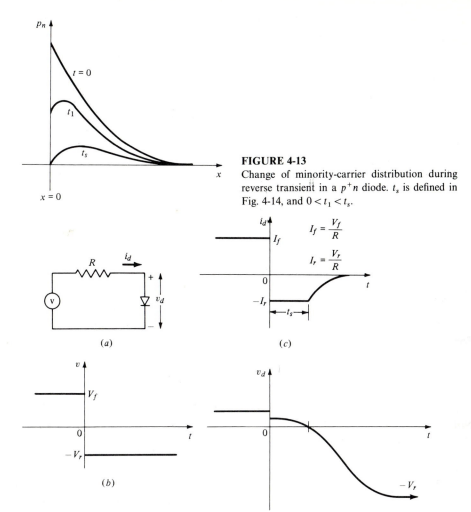

FIGURE 4-13
Change of minority-carrier distribution during reverse transient in a p^+n diode. t_s is defined in Fig. 4-14, and $0 < t_1 < t_s$.

FIGURE 4-14
The reverse transient of a long p^+n diode: (*a*) the biasing circuit, (*b*) the driving voltage waveform, and (*c*) the current and (*d*) voltage waveforms across the diode.

of a step-function reverse current. The temporary charge-storage effect of minority carriers is called the *reverse recovery transient*. The current and voltage waveforms are shown in Fig. 4-14. We consider the charge-storage effect by using a long p^+n diode.

Let us define the total stored charge in the *n* side as

$$Q_s = qA \int_0^{W_n} \Delta p_n \, dx \tag{4-59}$$

where W_n is the *n*-layer thickness. By integrating the continuity equation Eq. (2-62) once from 0 to W_n and using Eq. (4-59), we obtain

$$I_p(0) - I_p(W_n) = \frac{dQ_s}{dt} + \frac{Q_s}{\tau_p} \tag{4-60}$$

This is called the *charge-control* equation. In a long diode, we can assume $I_p(W_n)$ is zero. Therefore, the steady-state forward current of the diode is obtained by setting $dQ_s/dt = 0$:

$$I_f = I_p(0) = \frac{Q_{sf}}{\tau_p} \tag{4-61}$$

where Q_{sf} is the stored charge under steady-state, forward-bias condition. Thus,

$$Q_{sf} = I_f \tau_p \tag{4-62}$$

Let us now assume that a negative current I_r is applied at $t = 0$ by reversing the bias voltage as shown in Fig. 4-14. The charge-control equation becomes

$$-I_r = \frac{dQ_s}{dt} + \frac{Q_s}{\tau_p} \tag{4-63}$$

Solving the foregoing equation with Eq. (4-62) as the initial condition, we derive

$$Q_s(t) = \tau_p[-I_r + (I_f + I_r)e^{-t/\tau_p}] \tag{4-64}$$

Let us now define the storage time t_s at which point when all the stored charge has been removed, that is, $Q_s \approx 0$. Consequently

$$t_s = \tau_p \ln\left(1 + \frac{I_f}{I_r}\right) \tag{4-65}$$

An exact analysis by solving the time-dependent diffusion equation yields

$$\text{erf}\sqrt{\frac{t_s}{\tau_p}} = \frac{I_f}{I_r + I_f} \tag{4-66}$$

Both results are plotted in Fig. 4-15, which shows that the charge-control analysis is a reasonable approximation. However, the tail end of the reverse recovery transient shown in Fig. 4-14 can be obtained only through the exact solution, which is not given here [3].

In general, we can assume that the time t_s defines the end of the reverse transient. In other words, the low-impedance state of a forward-biased junction is returned to the high-impedance state of a reverse-biased junction after t_s. From Eq. (4-65), we find that the minority-carrier lifetime should be small to achieve high switching speed in a *pn* diode. Typical storage time of a switching diode ranges from $\frac{1}{2}$ to 10 ns. For power rectifiers the storage time ranges from 1 μs to 10 ms.

The reverse transient waveforms can be modified by introducing a built-in electric field in the diode. Such an electric field is obtained by doping the diode nonuniformly. For example, if the impurity concentration in the lightly doped side of the p^+n diode is

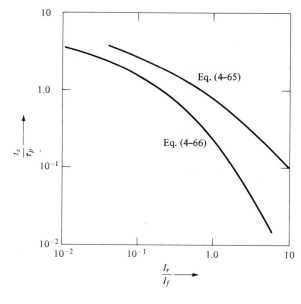

FIGURE 4-15
Normalized storage time vs. diode-current ratio.

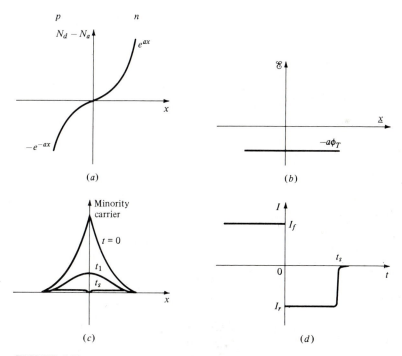

FIGURE 4-16
The step-recovery or snap-action diode: (*a*) impurity profile near the junction, (*b*) electric field distribution, (*c*) injected minority-carrier distribution, and (*d*) current waveform.

$$N_d = N_d(0)e^{-ax} \tag{4-67}$$

where $N_d(0)$ is the impurity value at the junction and a is a constant, then, according to Eq. (2-35), the electric field is

$$\mathscr{E} = -a\phi_T \tag{4-68}$$

Substituting Eq. (4-68) into (2-68), we find that both drift- and diffusion-current components have the same sign and the currents are cumulative. In other words, the direction of the field expressed in Eq. (4-68) helps hole transport and is known as the *aiding field*. If the donor-impurity distribution is described by a positive exponent in Eq. (4-67), the electric field is positive and is a *retarding* field for holes. In either case, the field-free approximation used in previous sections must be modified to incorporate the drift-current term.

An unusual application of the built-in field is demonstrated by the *step-recovery* diode shown in Fig. 4-16. The impurity distribution in the junction region provides a retarding field for both injected electrons and holes, so that carriers are confined closely to the junction region. Upon the application of a step reverse current from its forward-bias condition, the stored charge is reduced, just as in a regular diode, but the injected carriers at the junction do not reach zero until all stored charge has been removed. For this reason, the diode abruptly returns to its high-impedance state, and the current waveform shows a snap-action switching in less than 1 ns. The tail end of the current recovery in a regular diode is absent here.

REFERENCES

1. Cappelletti, P., G. F. Cerfolinni, and M. L. Polignano: Current-Voltage Characteristics of Ideal Silicon Diodes, *J. Appl. Phys.*, **57**:646 (1985).
2. Sze, S. M.: "Physics of Semiconductor Devices," 2d ed., Wiley, New York, 1981.
3. Lax, B., and S. F. Neustadter: Transient Response of a *p–n* Junction, *J. Appl. Phys.*, **25**:1148 (1954).

ADDITIONAL READINGS

Ghandi, S. K.: "Semiconductor Power Devices," Wiley, New York, 1977.
Gray, P. E., D. DeWitt, A. R. Boothroyd, and J. F. Gibbons: "Physical Electronics and Circuit Models of Transistors," SEEC Series, vol. II, Wiley, New York, 1964.
Jonscher, A. K.: "Principles of Semiconductor Device Operation," Wiley, New York, 1960.
Neudeck, G. W.: "The *p-n* Junction Diode," Addison-Wesley, Reading, Mass., 1983.
Shockley, W.: The Theory of *p-n* Junction in Semiconductors and *p-n* Junction Transistors, *Bell Syst. Tech. J.*, **28**:435 (1949).

PROBLEMS

4-1. (*a*) Consider the one-sided step junction shown in Fig. 3-12*b*. Obtain the boundary conditions of carriers, p_n and n_p, under the conditions that the injection into the *p* side is low-level but injection into the *n* side is no longer low-level.

(*b*) Simplify the expressions in (*a*) if high-level injection is approached in the *n* side of the junction.

4-2. A silicon *pn* junction long diode has the following parameters at 300 K: $N_d = 2 \times 10^{15}$, $N_a = 10^{15}$ cm^{-3}, $\tau_n = \tau_p = 1$ μs, $A = 0.01$ cm^2.
   (a) Calculate the reverse-saturation hole-current component.
   (b) Calculate the reverse-saturation electron-current component.
   (c) Calculate $n_p(0)$ and $p_n(0)$ at $V_A = 0.6$ V ($V_A \equiv$ forward-bias voltage).
   (d) Determine the electron and hole currents in (c).
   (e) Repeat (c) and (d) for $V_A = 0.7$ V.

4-3. Repeat Prob. 4-2 for $N_d = 2 \times 10^{17}$ cm^{-3} and $N_a = 10^{14}$ cm^{-3}.

4-4. Determine the applied-bias voltage at the point where the injected carriers have reached one-tenth the doping concentration in Probs. 4-2 and 4-3. This is the condition that the assumption of low-level injection is no longer valid.

4-5. (a) Obtain the expressions of the hole density and current distribution if the boundary condition in Eq. (4-14) is changed to $p_n = p_{no}$ at $x = W_n$, where W_n has the same order of magnitude as L_p.
   (b) Plot the hole density and current distribution obtained in (a) for $W_n/L_p = 0.1$, 1, and 10.

4-6. Derive and plot the carrier and current distributions for a reverse-biased *pn* junction similar to those depicted in Fig. 4-3a and c.

4-7. (a) The hole-injection efficiency of a *pn* junction is defined as I_p/I at $x = 0$. Show that this efficiency can be written

$$\gamma = \frac{I_p}{I} = \frac{1}{1 + \sigma_n L_p / \sigma_p L_n}$$

   (b) What should you do to make γ approach unity in a practical diode?

4-8. Calculate the injection efficiency for the diodes in Probs. 4-2 and 4-3, using the formula given in Prob. 4-7. Note the significant difference of the two diodes.

4-9. In a p^+n junction diode, the width of the n region W_n is much smaller than L_p. Using $I_p|_{x=W_n} = qAS\Delta p_n$ as one of the boundary conditions, derive the carrier and current distributions. Sketch the shape of minority carriers in the n side for $S = 0$ and $S = \infty$.

4-10. Derive Eqs. (4-28) and (4-29).

4-11. A silicon diode operates at a forward voltage of 0.5 V. Calculate the factor by which the current will be multiplied when the temperature is increased from 25 to 150°C. Assume that $I \approx I_o e^{V/2\phi_T}$ and that I_o doubles every 6°C.

4-12. Show that the maximum recombination rate occurs at $p = n$ in the junction space-charge layer and derive Eq. (4-36).

4-13. Derive the small-signal ac hole distribution and diode admittance of a p^+n diode if the width of the n region W_n is of the same order as the diffusion length. Assume infinite surface recombination velocity at $x = W_n$.

4-14. The hole lifetime of a p^+n diode is measured by the diode-recovery method.
   (a) For $I_f = 1$ mA and $I_r = 2$ mA, t_s is found to be 3 ns in an oscilloscope with a 0.1-ns rise time. Find τ_p.
   (b) If the fast scope in (a) is not available and you have to use a slower scope with a 10-ns rise time, how can you make an accurate measurement? Describe your result.

CHAPTER
5

BIPOLAR TRANSISTOR FUNDAMENTALS

The *bipolar junction transistor* (BJT), an active three-terminal device that can be used as an amplifier or switch, was the most widely used electronic component before 1975. Although its importance is now overshadowed by the metal-oxide-semiconductor (MOS) field-effect transistor, there are still areas of applications in which the bipolar transistor is superior, for example, in high-power devices and in high-speed logics for high-performance computers. The situation is expected to remain in the foreseeable future.

In this chapter, the amplifying principle is first described qualitatively. Then, the carrier distribution is derived by solving the diffusion equation of minority carriers in the base. On the basis of carrier distribution, we obtain the current components, current gains, and current-voltage characteristics. The Ebers-Moll and hybrid-pi models are shown to be valuable in describing the transistor behavior. In addition, the frequency response and switching properties are examined, and limitations due to the base resistance and junction breakdown are discussed.

If an additional *pn* junction is fabricated on a bipolar transistor, we have a four-layer *pnpn* structure. The *pnpn* devices are bistable switches whose operation depends on an internal feedback mechanism giving rise to high- and low-impedance stable states. They are available in a wide range of voltage and current ratings. The low-power devices are designed for switching circuitry, while the high-power devices find applications in industry as ac switches, phase-control devices, power inverters, and dc choppers.

116

5-1 THE TRANSISTOR ACTION

A bipolar transistor consists of a layer of p-type silicon sandwiched between two layers of n-type silicon. Alternatively, it may have a layer of n-type between two layers of p-type material. In the former case, the transistor is referred to as an npn transistor and in the latter case as a pnp transistor. The standard fabrication technique is the planar technology described in Chap. 3. An npn transistor can be made by introducing an additional n-type impurity diffusion subsequent to the step shown in Fig. 3-6e. First, an oxide layer is grown to cover the entire surface. Then a window is opened by means of photoresist-masked etching, and a phosphorus diffusion is made to form a second junction. The final structure after contact metallization and lead attachment, shown in Fig. 5-1a, is known as the silicon *double-diffused planar* transistor.

To study the basic characteristics of the transistor, we take a one-dimensional section (shown between the dashed lines in Fig. 5-1a) to represent the transistor and redraw it in Fig. 5-1b. The three layers are known as *emitter*, *base*, and *collector*. In the following discussion, our attention is directed toward this simplified model because it demonstrates all the important features of transistor action. The circuit representation of the npn transistor is shown in Fig. 5-1c, where the arrow on the emitter lead specifies the direction of current flow when the emitter-base junction is forward-biased. It should be noted that the external circuit currents I_E, I_B, and I_C are assumed to be positive when currents flow *into* the transistor. The symbols V_E (for V_{BE}) and V_C (for V_{BC}) are the base-to-emitter and base-to-collector voltages, respectively.

Under normal operation, the emitter junction is forward-biased and the collector junction is reverse-biased. The energy-band diagram is shown in Fig. 5-2a for unbiased (dashed) and normal bias (solid) conditions. Forward biasing of the emitter junction lowers the emitter-base potential barrier by V_E, whereas reverse biasing of the collector junction increases the collector-base potential barrier by V_C. The lowering of the emitter-base barrier results in an injection of electrons into the base and holes into the emitter. The injected holes constitute the emitter hole current I_{pE}. The injected electrons, which are the minority carriers in the base, diffuse across the base layer and give rise to a current I_{nE}. Some injected electrons recombine with holes in the base, but the majority of the injected electrons survive recombination and reach the collector junction. The electrons which reach the collector junction fall down the potential barrier and are *collected* by the collector as I_{nC}. Consequently, a large current may flow in a *reverse-biased* (collector) junction resulting from carrier injection of a nearby (emitter) junction. This is the basis of the transistor action, and it is realized only when the two junctions are close enough physically to interact in the manner described. The difference between I_{nE} and I_{nC} represents the holes supplied by the base to recombine with electrons in the base layer. These current components are illustrated in Fig. 5-2b where x_{dE} and x_{dC} are the depletion-layer widths for the emitter and collector junctions respectively. Note that the arrows in this figure represent the direction of carrier motion. Therefore, the direction of electron current is in the direction opposite the arrows, whereas

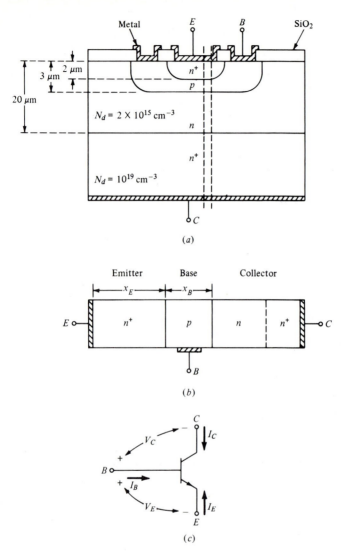

FIGURE 5-1
The *npn* bipolar transistor: (*a*) planar structure, (*b*) simplified model, and (*c*) circuit symbol.

the hole current is in the direction shown. In addition, the emitter space-charge-layer recombination current I_{rg}, the collector generation current I_G, and reverse-saturation currents I_{nCO} and I_{pCO} are included. In practice, the components I_G, I_{nCO}, and I_{pCO} are very small and are usually combined and designated as I_{CO}, the *collector reverse-saturation current*.

The emitter, base, and collector currents are obtained by examining Fig. 5-2*b*, noting that the dashed lines represent the negative direction of electron-current

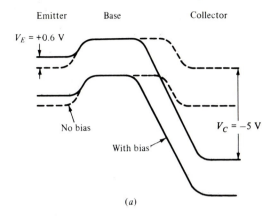

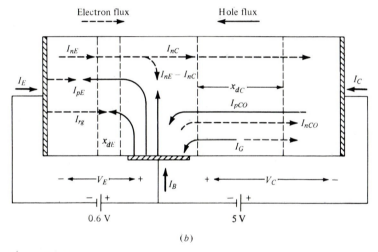

FIGURE 5-2
(a) Energy-band diagram and (b) current components.

components:

$$-I_E = I_{pE} + I_{nE} + I_{rg} \tag{5-1}$$

$$I_B = I_{pE} + I_{rg} + (I_{nE} - I_{nC}) - I_{CO} \tag{5-2}$$

$$I_C = I_{nC} + I_{CO} \tag{5-3}$$

Furthermore, the sum of all currents entering the transistor must be zero, so that

$$I_E + I_B + I_C = 0 \tag{5-4}$$

These equations will be used in later sections in deriving the current gains and the current-voltage characteristics.

5-2 CURRENT GAINS AND CURRENT-VOLTAGE CHARACTERISTICS

The ratio of the output current to the input current is known as the *current gain*. In Fig. 5-2b, the emitter and collector are the input and output terminals, respectively, and the base is the common terminal. This arrangement is known as the *common-base configuration*. The common-base current gain α is defined by the current components crossing the emitter and collector junctions:

$$\alpha \equiv -\frac{I_C - I_{co}}{I_E} \tag{5-5}$$

The current I_{co} is subtracted from the collector current since it is independent of the input current. Making use of Eqs. (5-1) and (5-3), we obtain

$$\alpha = \frac{I_{nC}}{I_{nE} + I_{pE} + I_{rg}} = \gamma \beta_T \tag{5-6}$$

where

$$\gamma \equiv \frac{I_{nE}}{I_{nE} + I_{pE} + I_{rg}} \tag{5-7}$$

$$\beta_T \equiv \frac{I_{nC}}{I_{nE}} \tag{5-8}$$

In the above equations, we define the emitter injection efficiency γ and the base transport factor β_T.

The injection efficiency specifies the portion of minority carriers injected into the base. Only these carriers may be collected by the collector junction to become a useful output. The other components, I_{rg} and I_{nE}, do not contribute to the output current, and they should be reduced as much as possible. The space-charge recombination current is determined by the quality of the starting wafer and is not under the control of the device designer. Its magnitude is quite small in well-gettered silicon, and it could be neglected in most recently made devices. In this case, the emitter injection efficiency is specified by the ratio of I_{pE}/I_{pE}, which is controllable by doping concentrations, to be described in the next section.

As the injected electrons traverse the base, some will recombine with the majority carriers (holes) but most will survive to reach the collector. Thus, the base transport factor designates the portion of the surviving electrons. It is determined by the electron lifetime in the base and the base width. To obtain a high transport factor, i.e., approaching unity, the carrier lifetime in the base must be long so that the diffusion length is much larger than the base width.

Since both the emitter injection efficiency and the base transport factor are smaller than 1, α is always less than unity, although it approaches unity for well-fabricated transistors. Using I_E as the input independent variable and I_C as the output dependent variable, we may rewrite Eq. (5-5) as

$$I_C = -\alpha I_E + I_{co} \tag{5-9}$$

Equation (5-9) is valid if the emitter is forward-biased and the collector is reverse-biased. Under these conditions, the transistor is said to be in its *active* or *normal region* of operation, where the collector current is essentially independent of collector voltage. If both forward- and reverse-bias conditions for the collector junction are considered, however, I_{co} in Eq. (5-9) has to be replaced by the diode equation (4-24) with I_o replaced by $-I_{co}$ and V by V_C. The complete expression for I_C becomes

$$I_C = -\alpha I_E + I_{co}(1 - e^{V_C/\phi_T}) \qquad (5\text{-}10)$$

Note that if the collector junction is reverse-biased, V_C is negative (that is, V_{CB} is positive in Fig. 5-1c), and Eq. (5-10) becomes Eq. (5-9) provided that the bias voltage is greater than ϕ_T. Therefore, the output current increases linearly with the input current as long as the collector junction is reverse-biased. For each given I_E, I_C is a constant giving rise to a set of horizontal straight lines in the I–V characteristics for positive V_{CB} (Fig. 5-3a). When V_C is positive and V_{CB} is negative, the output current becomes

$$I_C = -\alpha I_E - I_{co}e^{V_C/\phi_T} \qquad (5\text{-}11)$$

The second term on the right of Eq. (5-11) represents a diode I–V characteristic that is turned upside down. When it is superimposed on the constant current αI_E, the current bends downward for a negative V_{CB}, as shown in Fig. 5-3a. This is known as the *common-base output I–V characteristics*.

In most circuit applications, the emitter is used as the common terminal, with the base and collector as the input and output terminals, respectively. The relationship between the collector and base currents can be obtained by substituting I_E from Eq. (5-4) into Eq. (5-9):

$$I_C = \frac{\alpha}{1-\alpha} I_B + \frac{I_{co}}{1-\alpha} = h_{FE}I_B + I_{CEO} \qquad (5\text{-}12)$$

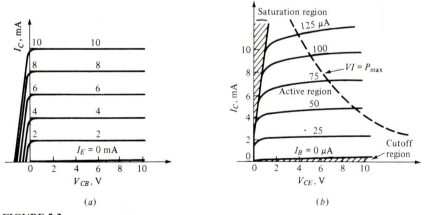

(a)   (b)

FIGURE 5-3
Collector current-vs.-voltage characteristics for (a) the common-base configuration and (b) the common-emitter configuration.

where we define

$$h_{FE} \equiv \frac{\alpha}{1-\alpha} \tag{5-13}$$

$$I_{CEO} \equiv \frac{I_{CO}}{1-\alpha} \tag{5-14}$$

The symbol h_{FE}, also known as β, is called the *common-emitter current gain*. Experimental current-voltage characteristics of the common-emitter configuration are shown in Fig. 5-3b, where three regions of operation are specified. The *cutoff* region corresponds to the condition that both the emitter and collector junctions are reverse-biased. The *normal active* region indicates that the emitter junction is forward-biased and the collector junction is reverse-biased. In the *saturation* region, both junctions are forward-biased. As an amplifier, the transistor is operated within the normal active region. However, a transistor switch usually traverses between the saturation (ON) and cutoff (OFF) regions.

The maximum power P_{max} that can be dissipated in a transistor without overheating is given in transistor specifications. The constant product $VI = P_{max}$ is plotted in Fig. 5-3b, and the transistor should not be operated beyond this curve.

5-3 DERIVATION OF CURRENT COMPONENTS AND GAIN EXPRESSIONS

The injected-current components shown in Fig. 5-2b can be derived by solving the diffusion equations in Sec. 4-2. To simplify our analysis, we assume that the doping concentration is uniform in the base and emitter and that the injection level is low. The electron current and diffusion equations (4-9) and (4-10) are rewritten here for the dc case:

$$I_n = qAD_n \frac{dn_p}{dx} \tag{5-15}$$

$$D_n \frac{d^2 n_p}{dx^2} - \frac{n_p - n_{po}}{\tau_n} = 0 \tag{5-16}$$

Equation (5-16) has a general solution

$$n_p - n_{po} = K_1 e^{-x/L_n} + K_2 e^{x/L_n} \tag{5-17}$$

where K_1 and K_2 are constants to be determined by the boundary conditions. With a forward-biased emitter junction, the boundary condition at the edge of the emitter depletion layer in the base side is, according to Eq. (4-5),

$$n_p(0) = n_{po} e^{V_E/\phi_T} \tag{5-18}$$

The boundary condition at the depletion-layer edge of the reverse-biased collector junction is usually assumed to be[1]

$$n_p(x_B) = 0 \qquad (5\text{-}19)$$

This is a reasonable assumption since the electric field of the collector junction sweeps carriers into the collector so that the collector is nearly a perfect sink.

In almost all modern transistors, the base width is made very narrow such that the minority-carrier recombination in the base is negligible. Under this condition, the boundary conditions specify the two endpoints of the base carrier concentration with a straight line between them, as shown in Fig. 5-4. There is no need of solving Eq. (5-16) since we can write the base carrier concentration as

$$n_p = n_{po} e^{V_E/\phi_T} \left(1 - \frac{x}{x_B} \right) \qquad (5\text{-}20)$$

For those who are mathematically inclined, Prob. 5-3 is given as an exercise. Now, we may make use of Eq. (5-15) to obtain the current I_{nE} at $x = 0$ as

$$I_{nE} = -qAD_n \frac{n_i^2}{N_a x_B} e^{V_E/\phi_T} \qquad (5\text{-}21)$$

where the mass-action law has been employed, and N_a is the base-doping density.

The injected hole current I_{pE} can be obtained in a similar manner provided that the emitter junction depth (x_E) is much shorter than the hole diffusion length in the emitter. This is indeed an excellent approximation, as depicted in Fig. 5-4 for the hole distribution, which leads to

$$I_{pE} = -qAD_p \frac{n_i^2}{N_{dE} x_E} e^{V_E/\phi_T} \qquad (5\text{-}22)$$

where N_{dE} is the donor concentration in the emitter. The emitter injection efficiency

---

[1] Strictly speaking, x_B is a function of the collector voltage which controls the extension of the collector depletion-layer width into the base. Known as the *Early effect*, this represents a feedback from the collector to the base. The Early effect is not important in double-diffused transistors in which the collector doping is lower than the base doping (see Fig. 6-3).

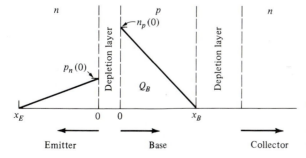

FIGURE 5-4
Minority-carrier concentrations under normal active bias with negligible recombination.

may now be written as

$$\gamma = \frac{I_{nE}}{I_{nE} + I_{pE}} = \left(1 + \frac{D_p N_a x_B}{D_n N_{dE} x_E}\right)^{-1} \tag{5-23}$$

Since the emitter and base widths are of the same order of magnitude, the control of the injection efficiency comes from the selection of the ratio of the doping concentrations in the emitter and the base. To achieve a high injection efficiency, the doping concentration in the emitter should be much larger than that of the base. A typical ratio is a thousand.

Previously, we have assumed that the base recombination is zero, which is clearly an exaggeration. We may however estimate the recombination current since the total minority carriers in the base Q_B is given by the triangle shown in Fig. 5-4:

$$Q_B = \frac{q n_p(0) x_B}{2} \tag{5-24}$$

The base recombination current for a carrier lifetime of τ_n is

$$-I_{nB} = \frac{A Q_B}{\tau_n} = \frac{q A x_B n_{po}}{2\tau_n} e^{V_E/\phi_T} \tag{5-25}$$

This base current represents the portion of minority carriers that fail to reach the collector junction. Therefore, the collector current I_{nC} is given by

$$I_{nC} = I_{nE} - I_{nB} = I_{nE}\left(1 - \frac{x_B^2}{2L_n^2}\right) \tag{5-26}$$

where $L_n = \sqrt{D_n \tau_n}$, and the base transport factor is given by

$$\beta_T = \frac{I_{nC}}{I_{nE}} = 1 - \frac{x_B^2}{2L_n^2} \tag{5-27}$$

Example. In a bipolar junction transistor, the doping concentrations for the emitter, base, and collector are 10^{19}, 5×10^{17}, and 10^{15} cm^{-3}, respectively. The emitter and base widths are 1 and 0.7 μm, respectively, and the diffusion length is 20 μm. Calculate the emitter efficiency, base transport factor, and current gains.

Solution. For $N_{dE} = 10^{19}$ cm^{-3} and $N_a = 5 \times 10^{17}$ cm^{-3}, we find from Fig. 2-6 that $\mu_p = 90$ and $\mu_n = 450$ cm^2/V $\cdot$ s, respectively.

Using $\mu_n/\mu_p = D_n/D_p$ and Eq. (5-23), we obtain

$$\gamma = \left(1 + \frac{90 \times 5 \times 10^{17} \times 0.7}{450 \times 10^{19} \times 1}\right)^{-1} = 0.993$$

The base transport factor is

$$\beta_T = 1 - \frac{0.7^2}{2 \times 20^2} = 0.99938$$

Therefore

$$\alpha = \gamma \beta_T = 0.992 \quad \text{and} \quad h_{FE} = 124$$

In the foregoing analysis, we have assumed that the recombination current in the space-charge layer is negligible. This is true only in well-prepared transistors where the crystal defects and metallic impurity have been reduced by gettering. For most silicon bipolar transistors, the space-charge recombination current must be taken into account. Using the result given in Sec. 4-5, we have

$$I_{rg} = \frac{qAn_i x_{dE}}{2\tau_o} e^{V_E/2\phi_T} \tag{5-28}$$

where x_{dE} is the emitter depletion-layer width. The base current according to Eq. (5-2) is sometimes written, after adding Eqs. (5-22), (5-25), and (5-28) and simplifying,

$$I_B \propto e^{V_E/\eta\phi_T} \tag{5-29}$$

where η is a factor with a value between 1 and 2 in a silicon transistor. An experimental current-voltage plot on semilog paper is shown in Fig. 5-5 for both I_B and I_C. The effect of I_{rg} is clearly observed in the base-current data in the

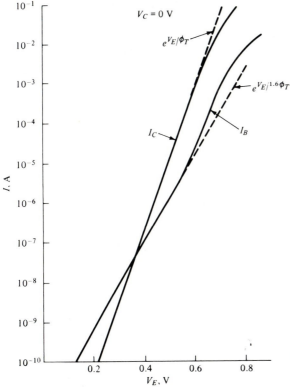

FIGURE 5-5
Static current-voltage characteristics of an *npn* transistor.

low-current region with $\eta = 1.6$. On the other hand, I_C follows the exponential with a slope of $1/\phi_T$ for more than 6 decades of current, according to Eq. (5-26) with I_{nE} given by Eq. (5-21).

In a typical transistor, β_T approaches unity, and $I_C = I_{nC} \approx I_{nE}$. Therefore, the reciprocal of the common-emitter current gain expressed in Eq. (5-12) with $I_{CEO} \approx 0$ can be written as

$$\frac{1}{h_{FE}} = \frac{I_B}{I_C} \approx \frac{I_{pE}}{I_{nE}} + \frac{I_{nE} - I_{nC}}{I_{nE}} + \frac{I_{rg}}{I_{nE}} \tag{5-30}$$

Using Eqs. (5-21), (5-22), (5-26), and (5-28), we obtain

$$\frac{1}{h_{FE}} = \frac{N_a x_B D_p}{N_{dE} x_E D_n} + \frac{x_B^2}{2L_n^2} + \frac{N_a x_B x_{dE}}{2D_n n_i \tau_0} e^{-V_E/2\phi_T} \tag{5-31}$$

The contribution of the last term on the right-hand side of Eq. (5-31) leads to a nonlinear relationship between the applied emitter voltage and current gain. Thus, the current gain is not constant and is a function of the current in the transistor. In general, the larger the I_{rg} the lower the h_{FE}. At a sufficiently high current level, however, the effect of I_{rg} becomes negligible due to the factor $1/2\phi_T$ in the exponent.

Using the data given in Fig. 5-5, we plot the current-gain variation with collector current in Fig. 5-6. The solid curve is in good agreement with Eq. (5-31) except in the high-current region. When the collector current is greater than 10 mA, h_{FE} starts to level off and then decreases with further increase of current. The drop of gain at high currents is due to the fact that the injected minority-carrier density becomes comparable to the majority-carrier density in the base. As a result, the fraction of total emitter current carried by holes injected from base to emitter increases, leading to a decrease in emitter efficiency and a decline in h_{FE}.

Example. A silicon bipolar transistor's base doping is 10^{17} cm^{-3}. Find the condition where the high-level injection is reached and use it to estimate the bias voltage at the onset of decrease in current gain.

Solution. For high-level injection, we may neglect the space-charge recombinaton current. Since $N_B = 10^{17}$ cm^{-3}, the equilibrium minority-carrier density is

$$n_{po} = \frac{n_i^2}{N_B} = 2.25 \times 10^3 \text{ cm}^{-3}$$

and the forward bias for reaching a minority carrier of 10^{17} cm^{-3} is, using Eq. (5-18),

$$V_F = 0.026 \ln \frac{10^{17}}{2.25 \times 10^3} = 0.81 \text{ V}$$

When the minority-carrier density reaches $0.1N_B$, however, the injection ratio starts to be modified and the current gain is affected. The corresponding bias voltage is

$$V_F = 0.026 \ln \frac{10^{16}}{2.25 \times 10^3} = 0.75 \text{ V}$$

At this bias voltage, the current gain starts to decrease with increasing current.

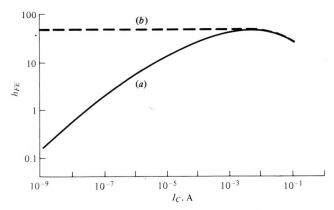

FIGURE 5-6
Dependence of current gain on the collector current: (*a*) conventional transistor and (*b*) well-gettered modern transistor.

In new devices fabricated on well-gettered wafers, the space-charge recombination current is quite small. The gain dependence on current is shown in the dashed line of Fig. 5-6 where the current gain is relatively flat for the low-current regime.

5-4 THE EBERS-MOLL MODEL

A bipolar transistor has two junctions, either one of which may be forward- or reverse-biased. Therefore, we have four modes of operation, and these are known as the normal, cutoff, saturation, and inverse regions. The bias conditions for each of these modes are listed in Table 5-1.

When the emitter junction is forward-biased and the collector is reverse-biased, the transistor is operated in the *normal* mode. Most amplifiers are operated in this region, and its common-base current gain is known as the *normal alpha*, α_N. If both junctions are reverse-biased, no current will flow through the transistor. The transistor is operated in its *cutoff* mode, and it acts as an open circuit. On the

TABLE 5-1
Transistor operating modes

Emitter bias	Collector bias	Operating mode
Forward	Reverse	Normal
Reverse	Reverse	Cutoff
Forward	Forward	Saturation
Reverse	Forward	Inverse

other hand, when both junctions are forward-biased, the transistor is in its *saturation* mode, having very low resistance. As a switch, the saturation and cutoff regions correspond to the ON and OFF states. When the emitter is reverse-biased and the collector is forward-biased, the transistor is operated in the *inverse* mode, and the current gain is known as the *inverse alpha*, α_I. For a symmetrical transistor with identical emitter and collector, the normal and inverse alphas are the same. However, most transistors are optimized in such a way that α_N is nearly unity and α_I is small. In fact, a transistor is seldom used in its inverse region; a notable exception is the multiple-emitter transistor in a TTL logic gate.

From a practical standpoint, it is desirable to have a circuit model with a single set of equations to describe all four operating regions. Such an approach is available in the Ebers-Moll model. To appreciate the work of Ebers and Moll, let us consider the minority-carrier distribution in the base under different bias conditions. When both junctions are reverse-biased, there is no carrier injection into the base, and the result is depicted in Fig. 5-7a. The electron concentration is given by its thermal equilibrium value with zero concentration at the junction boundaries. Under normal bias, electrons are injected from the emitter, but the zero carrier concentration remains at the collector junction, as shown in Fig. 5-7b.

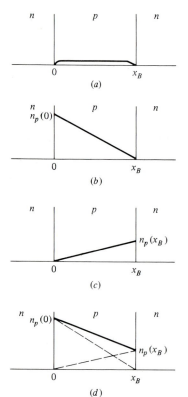

FIGURE 5-7
Minority-carrier density in the base at (a) cutoff, (b) normal, (c) inverse, and (d) saturation.

In the inverse mode of bias, carriers are injected from the collector side but not from the emitter side, as seen in Fig. 5-7c. When both junctions are forward-biased, the minority-carrier concentration increases with the bias voltages, as depicted in Fig. 5-7d. This carrier distribution may be decomposed into two components, as shown by the dashed lines. By comparing this diagram with Fig. 5-7b and c, it is clear that a transistor under saturation can be represented by two transistors connected together with one operated in the normal mode and the other in the inverse mode. Recalling the discussion in Sec. 5-2, we may rewrite the current equation for the normal mode as

$$I_C = -\alpha_N I_E + I_{CO}(1 - e^{V_C/\phi_T}) \tag{5-32}$$

Note that α_N has replaced α in Eq. (5-10). We may use the similar procedure to develop an equation for the inverse mode of operation:

$$I_E = -\alpha_I I_C + I_{EO}(1 - e^{V_E/\phi_T}) \tag{5-33}$$

Notice that Eq. (5-33) is the same as Eq. (5-32) except that α_N is replaced by α_I and all subscripts C and E are interchanged. These two equations are known as the *Ebers-Moll equations*. In each equation, the first term on the right represents a current source and the second term represents a *pn* junction diode. Schematically, these can be incorporated into an equivalent circuit, as shown in Fig. 5-8. The four parameters α_N, α_I, I_{CO}, and I_{EO} are not independent but are related by

$$\alpha_N I_{EO} = \alpha_I I_{CO} \tag{5-34}$$

This condition can be verified by solving the continuity equations with appropriate boundary conditions (Prob. 5-9).

The Ebers-Moll equations and equivalent circuit are adequate to describe the current-voltage characteristics of the four different regions of operation. For example, one may want to find out the ON voltage at the output of a common-emitter switch. When the transistor is ON, its output voltage is V_{CE}(sat), representing the collector-to-emitter voltage at saturation. For a given collector

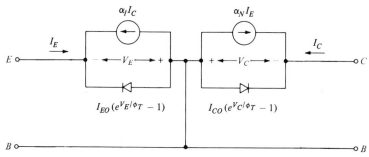

FIGURE 5-8
The Ebers-Moll model.

and base currents, the result is (Prob. 5-10)

$$V_{CE}(\text{sat}) = \phi_T \ln \frac{(I_B + I_C - \alpha_I I_C)\alpha_N}{(\alpha_N I_B - I_C + \alpha_N I_C)\alpha_I} \tag{5-35}$$

An interesting observation of Eq. (5-35) is that $V_{CE}(\text{sat})$ is insensitive to the currents as long as the saturation conditions are satisfied.

Example. In the bipolar junction transistor example in Sec. 5-3, additional device parameters are $A = 10^{-4}$ cm^{-2} and collector width $x_C = 50\,\mu$m. Calculate (a) α_I, (b) I_{EO} and I_{CO}, and (c) $V_{CE}(\text{sat})$ at $I_C = 1$ mA and $I_B = 50\,\mu$A.

Solution
(a) To determine α_I, let us first calculate the injection efficiency of the collector junction making use of Eq. (5-23) with appropriate modifications:

$$\gamma_I = \left(1 + \frac{450 \times 5 \times 10^{17} \times 0.7}{450 \times 10^{15} \times 10}\right)^{-1} = 0.222$$

where we have determined $\mu_n = \mu_p = 450$ cm^2/V·s from Fig. 2-6. In addition, the collector diffusion length is used to replace its width since the hole current is established in a long diode, i.e., using Eq. (4-18) instead of Eq. (4-27) for the hole current. The inverse alpha is

$$\alpha_I = \beta_T \gamma_I = 0.22$$

(b) We may use the preexponential terms in Eqs. (5-21) and (5-22) for the electron and hole current components:

$$I_{EO} = qA\left(\frac{D_n n_i^2}{N_a x_B} + \frac{D_p n_i^2}{N_{dE} x_E}\right)$$

Since $D_n = \mu_n \phi_T = 11.7$ cm^2/V·s and $D_p = \mu_p \phi_T = 2.34$ cm^2/V·s, we have

$$I_{EO} = 1.2 \times 10^{-15}\,\text{A}$$

Therefore,

$$I_{CO} = \frac{\alpha_N I_{EO}}{\alpha_I} = 5.45 \times 10^{-15}\,\text{A}$$

(c) Using Eqs. (5-32) and (5-33), we have

$$V_E = \phi_T \ln\left(1 - \frac{I_E + \alpha_I I_C}{I_{EO}}\right) = 0.71\,\text{V}$$

$$V_C = \phi_T \ln\left(1 - \frac{I_C + \alpha_N I_E}{I_{CO}}\right) = 0.59\,\text{V}$$

Therefore,

$$V_{CE}(\text{sat}) = V_E - V_C = 0.12\,\text{V}$$

5-5 HYBRID-PI EQUIVALENT CIRCUIT

It is frequently necessary to represent the transistor by a simple equivalent circuit in order to calculate the small-signal circuit response. The most widely used equivalent circuit is the *hybrid-pi model*, shown in Fig. 5-9. It represents a transistor operated in the normal active mode in the common-emitter configuration. A very important feature of the hybrid-pi model is that all the parameters are related to the physical processes in the transistor.

Basically, the hybrid-pi model represents the dynamic incremental variation of the stored charge in the base upon the incremental change of the applied emitter voltage. As illustrated in Fig. 5-10, a change of $+\Delta V_E$ gives rise to an increase of ΔQ_B and thus ΔI_C. The *transconductance* g_m is defined by

$$g_m \equiv \frac{dI_C}{dV_E} \tag{5-36}$$

From the discussion in Sec. 5-3, we have

$$n_p(0) = n_{po} e^{V_E/\phi_T} \tag{5-37}$$

and

$$|I_C| = \frac{qAD_n n_p(0)}{x_B} \tag{5-38}$$

Thus, we derive

$$g_m = \frac{|I_C|}{\phi_T} \tag{5-39}$$

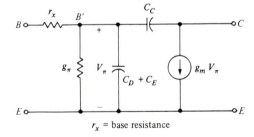

r_x = base resistance

FIGURE 5-9
The hybrid-pi equivalent circuit.
$V_\pi = V_E$ if $r_x = 0$.

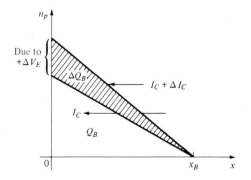

FIGURE 5-10
Schematic representation of the relationship between V_E, Q_B, and I_C.

The transconductance represents the incremental change of the (output) collector current induced by the (input) emitter voltage increment; therefore, it is the basic parameter that characterizes the amplifying function. The higher transconductance a transistor has, the higher gain it could achieve. At room temperature (300 K) where $\phi_T = 25.6 \, \text{mV}$, the transconductance is 0.04 S at 1 mA of collector current and is the same for all bipolar transistors regardless of material used. If the operating temperature is reduced to 100 K, we obtain $g_m = 0.12 \, \text{S}$. The realizable gain is improved by a factor of 3 at this low temperature.

The parameter g_π represents the small-signal *input conductance* of the emitter junction looking at the input of the base, and it is defined by

$$g_\pi \equiv \frac{dI_B}{dV_E} \tag{5-40}$$

If the space-charge current is ignored in Eq. (5-29), I_B is given by

$$I_B = q A n_i^2 \left(\frac{D_p}{x_E N_{dE}} + \frac{D_n x_B}{2 N_a L_n^2} \right) \left(e^{V_E/\phi_T} - 1 \right) \tag{5-41}$$

Therefore, we find

$$g_\pi = \frac{I_B}{\phi_T} = \frac{I_C}{h_{FE} \phi_T} = \frac{g_m}{h_{FE}} \tag{5-42}$$

Since h_{FE} is usually known at a given collector current and g_m is specified by the collector current, g_π is readily calculated.

The *base resistance* r_x describes the ohmic drop between the base contact and the active base region where carriers are injected. Although it is small and can be neglected in a first-order approximation, it could be significant at high frequencies. Its value could be reduced if the doping in the base is higher and the lateral dimension is made smaller (Fig. 5-1a).

Another important parameter in this equivalent circuit is the *diffusion capacitance* which represents the change in the base charge storage as affected by the emitter bias voltage. It is defined by

$$C_D \equiv \frac{dQ_B}{dV_E} \tag{5-43}$$

Since the total charge stored in the base is

$$Q_B = \frac{q}{2} n_p(0) x_B A \tag{5-44}$$

we obtain the following expression by using Eqs. (5-37) to (5-39):

$$C_D = \frac{q A x_B n_p(0)}{2 \phi_T} = g_m \frac{x_B^2}{2 D_n} \tag{5-45}$$

The hybrid-pi equivalent circuit is completed when the junction capacitances are added. These capacitances C_E and C_C represent the incremental change of charge in the emitter and collector junction depletion layers as described in Chap. 3.

With the above equivalent circuit, we are now in a position to analyze the transistor behavior at high frequencies. Because the frequency response is a function of the load resistance, it is desirable to eliminate this effect by short-circuiting the output terminals, as shown in Fig. 5-11. Note that the grounding of the collector allows the three capacitances to be combined into one. With an input current I_b, the voltage V_π is given by

$$V_\pi = \frac{I_b}{g_\pi + j\omega C} \tag{5-46}$$

where $C = C_D + C_E + C_C$. Since $I_c = g_m V_\pi$, we find

$$\frac{I_c}{I_b} = \frac{g_m}{g_\pi + j\omega C} = \frac{h_{FE}}{1 + j\omega C h_{FE}/g_m} \tag{5-47}$$

The characteristic 3-dB frequency, known as ω_β, is given by

$$\omega_\beta = \frac{g_m}{(C_D + C_E + C_C)h_{FE}} \tag{5-48}$$

and the gain-bandwidth product for $C_D \gg C_E + C_C$ is

$$\omega_T = h_{FE}\omega_\beta = \frac{2D_n}{x_B^2} \tag{5-49}$$

ω_T is a frequently quoted figure of merit for a transistor. It represents the maximum frequency the transistor is capable of to achieve a gain of unity.

Example. For the bipolar junction transistor example in Sec. 5-3, determine the hybrid-pi equivalent circuit parameters at $I_C = 5$ mA and $V_C = 5$ V. The cross-sectional area of the transistor is 10^{-4} cm^{-2}.

Solution

$$g_m = \frac{I_C}{\phi_T} = 0.192 \text{ S}$$

$$g_\pi = \frac{g_m}{h_{FE}} = 1.5 \times 10^{-3} \text{ S}$$

$$D_n = \mu_n \phi_T = 11.7 \text{ cm}^2/\text{s}$$

$$C_D = 0.192 \frac{(0.7 \times 10^{-4})^2}{2 \times 11.7} = 40 \text{ pF}$$

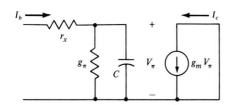

FIGURE 5-11
The hybrid-π circuit with output terminals short-circuited.

$x_{dC} = 2$ μm and $x_{dE} = 1$ μm are estimated values of the depletion-layer widths obtained from Fig. 3-13. Thus

$$C_C = \frac{AK_s\varepsilon_o}{x_{dC}} = 0.5 \text{ pF}$$

$$C_E = \frac{AK_s\varepsilon_o}{x_{dE}} = 1 \text{ pF}$$

Therefore, we have

$$C \approx C_D = 40 \text{ pF}$$

$$\omega_\beta = 3.75 \times 10^7 \text{ s}^{-1}$$

$$f_T = \frac{\omega_T}{2\pi} = 740 \text{ MHz}$$

5-6   BASE SPREADING RESISTANCE

The base current is a majority-carrier current which flows in the direction perpendicular to the minority-carrier injected from the emitter. As illustrated in Fig. 5-12, this current produces a lateral potential drop in both the intrinsic and extrinsic base regions. The effective extrinsic base resistance can be obtained from the geometry of the structure, which is

$$R_1 = \frac{\rho_{BX}L}{A} = \frac{\rho_{BX}d}{(x_E + x_B)W} \tag{5-50}$$

The intrinsic base resistance is

$$R_2 = \frac{\rho_{BI}h}{3x_B W} \tag{5-51}$$

This formula is derived from considering the two-dimensional current flow and is given here without proof. The last term that contributes to the base resistance comes from the base contact R_C, which is given in Chap. 7.

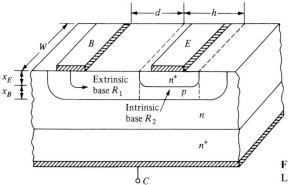

FIGURE 5-12
Lateral base current and ohmic drop.

The base spreading resistance is, therefore,

$$R_b = R_C + R_1 + R_2 = r_x \qquad (5\text{-}52)$$

5-7 FREQUENCY RESPONSE OF TRANSISTORS

In this section, we consider the transistor small-signal characteristics in the normal active mode of operation. The small-signal common-base and common-emitter current gains are defined as

$$\alpha \equiv \left. \frac{dI_C}{dI_E} \right|_{V_{CB} = \text{const}} \qquad h_{fe} \equiv \left. \frac{dI_C}{dI_B} \right|_{V_{CE} = \text{const}}$$

At a low frequency, the current gains are independent of the operating frequency. However, the magnitude of the gains decreases after a certain critical frequency is reached. A sketch of the typical frequency response of the current gains is given in Fig. 5-13. The various frequencies shown in this figure are defined as (1) the *common-base cutoff frequency* ω_α, (2) the *common-emitter cutoff frequency* ω_β, and (3) the *gain-bandwidth product* ω_T, that is, the frequency at which h_{fe} becomes unity. Since the slope of α vs. frequency is 20 dB/decade, it can be described by

$$\alpha = \frac{\alpha_o}{1 + j\omega/\omega_\alpha} \qquad (5\text{-}53)$$

where α_o is the current gain at low frequency. At $\omega = \omega_\alpha$, the magnitude of α is $0.707\alpha_o$ (3 dB down), so that ω_α is also known as the 3-dB frequency. Various factors limit the cutoff frequency, and the four most important ones are described below. Each of these factors introduces a time delay when the signal propagates from the emitter to collector.

Base Transit Time

The most severe limitation on the transistor frequency response is the carrier transport through the base layer. It is possible to derive the equations describing

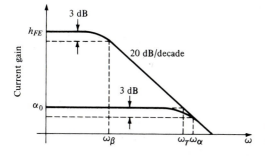

FIGURE 5-13
Current gains as a function of frequency.

the transistor frequency characteristic by solving the ac diffusion equation for minority carriers in the base following the procedures outlined in Sec. 4-7. Instead of carrying out the rigorous derivation, however, we present here a simple technique known as the *transit-time analysis*, which yields the same information as the solution of the diffusion equation. Let us define the effective minority-carrier velocity $v(x)$ in the base by the current equation

$$I_n = qAn_p(x)v(x) \tag{5-54}$$

Since $dx = v(x) \, dt$, the time required for an electron to traverse the base is obtained by using Eq. (5-54), and, by integration,

$$\tau_B = \int_0^{x_B} \frac{dx}{v(x)} = \int_0^{x_B} \frac{qAn_p(x)}{I_n} \, dx \tag{5-55}$$

For most transistors, we have $x_B/L_n \ll 1$, and the electron distribution is given by Eq. (5-20). Substituting Eq. (5-20) into Eq. (5-55) and integrating leads to (Prob. 5-20):

$$\tau_B = \frac{x_B^2}{2D_n} \tag{5-56}$$

A small τ_B means a short delay of signal or high frequency of operation. Therefore, transistors are designed with a small base width to achieve better frequency response. In a transistor with a built-in field such as the double-diffused transistor, the base transit time is found to be one-half that given by Eq. (5-56), so that the maximum frequency is roughly double that of a transistor with uniform doping in the base.

Emitter Transition-Capacitance Charging Time

The forward-biased emitter-junction transition capacitance C_E is a function of the bias voltage, and it is difficult to measure because of the shunting of the diffusion capacitance (see Sec. 5-5). A rough estimate of this capacitance is 4 times the zero-bias capacitance given by Eq. (3-36) with $V_R = 0$. This capacitance is in parallel with a junction resistance r_e, and the charging-time constant is

$$\tau_E = r_e C_E = \frac{4\phi_T}{I_E} C_E(0) \tag{5-57}$$

The resistance r_e is derived by taking dI_E/dV_E from the diode equation (4-24) with I replaced by I_E and V by V_E.

Collector Depletion-Layer Transit Time

With a high reverse bias across the collector junction, the depletion-layer width is significant, and it takes time for carriers to get through. Since the electric field is very high there, the carriers may be assumed to have reached the saturation velocity v_{th}. Therefore, the transit time for carriers to traverse the depletion layer is

$$\tau_d = \frac{x_m}{v_{th}} \tag{5-58}$$

where x_m is the total thickness of the collector depletion layer.

Collector Capacitance Charging Time

The collector junction is reverse-biased, so that the shunting resistance across the junction capacitance is very large. As a result, the charging-time constant is determined by the capacitance C_C and the collector series resistance r_{sc}:

$$\tau_C = r_{sc} C_C \tag{5-59}$$

The collector resistance of the planar epitaxial transistor in Fig. 5-1 is small because of the heavily doped epitaxial substrate, and τ_C is negligible. However, it should be included in the calculation for integrated transistors.

The cutoff frequency ω_α is equal to the reciprocal of the total time delay incurred in signal propagation from the emitter to collector. Therefore, we have

$$\frac{1}{\omega_\alpha} = \tau_E + \tau_B + \tau_d + \tau_C \tag{5-60}$$

By using the relationship between α and h_{fe}, we find

$$h_{fe} = \frac{\alpha}{1 - \alpha} = \frac{h_{FE}}{1 + j\omega/\omega_\beta} \tag{5-61}$$

where

$$\omega_\beta = \omega_\alpha(1 - \alpha_o) \quad \text{and} \quad h_{FE} = \frac{\alpha_o}{1 - \alpha_o} \tag{5-62}$$

From these equations we find that the common-emitter cutoff frequency is much lower than ω_α. But the gain-bandwidth product is

$$\omega_T = h_{FE}\omega_\beta \tag{5-63}$$

which is equal to ω_α. In practical transistors, however, ω_T is always smaller than ω_α because an excess phase shift exists in the transport of carriers from the emitter to the collector.

An important time constant that has not been accounted for is the charging time of capacitances through the base resistance r_x. Because the diffusion capacitance C_D is usually the largest capacitance in the transistor, the time constant $r_x C_D$ is in most cases a major factor in limiting the cutoff frequency. A smaller r_x is always desirable, and it can be obtained by a heavily doped base. Unfortunately, a high base doping reduces the injection efficiency and current gain. A better approach is to use self-aligned structures to reduce the extrinsic base resistance, a process to be described in Chap. 6.

5-8 THE TRANSISTOR AS A SWITCH

In addition to its application as an amplifier, the transistor is used as an ON-OFF switch. We consider the transistor circuit shown in Fig. 5-14a. Driven by a current pulse waveform shown in Fig. 5-14b, the transistor is made to operate in the cutoff region and in saturaton as depicted in Fig. 5-14c. The transistor is considered to be ON in saturation because the collector current is large and its resistance is low. It is considered to be OFF in the cutoff region since there is no current flow under this condition. In saturation, the collector current is limited by the load resistance so that

$$I_{CS} = I_C(\text{sat}) = \frac{V_{CC} - V_{CE}(\text{sat})}{R_L} \tag{5-64}$$

where $V_{CE}(\text{sat})$ is the collector-to-emitter voltage in saturation (typically 0.2 V for silicon transistors). The minimum base current I_{B1} needed to drive the transistor into saturation is

$$I_{B1} \geqslant \frac{I_C(\text{sat})}{h_{FE}} \approx \frac{V_{CC}}{h_{FE}R_L} \tag{5-65}$$

where $V_{CE}(\text{sat})$ is neglected.

Physically, the saturation condition is satisfied if both emitter and collector junctions are forward-biased, and the cutoff condition is satisfied if both junctions

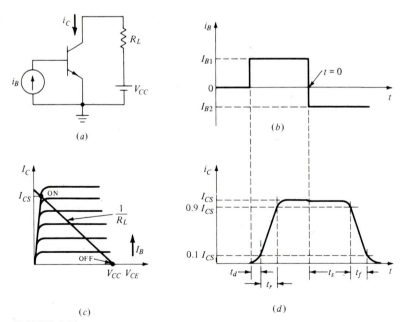

FIGURE 5-14
Switching operation of a bipolar transistor: (a) circuit diagram, (b) base-current drive, (c) output V-I characteristics, and (d) the output-current waveform.

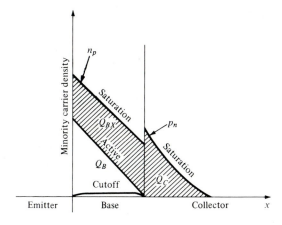

FIGURE 5-15
Charge storage in the base and collector at saturation. Also shown are the base charge at cutoff and active region.

are reverse-biased. The corresponding minority carriers in the base and collector for the planar epitaxial transistor are shown in Fig. 5-15. The minority-carrier storage in the collector is usually small compared with the total charge storage in the base and will be neglected in our consideration. Switching between the ON and OFF state is accomplished by a change of the carrier distribution; as with the *pn* junction diode, these carrier densities cannot be changed instantaneously. The transition time from one state to the other, known as the *switching time*, corresponds to the establishment or removal of the minority carriers involved. A typical switching waveform of the collector current is shown in Fig. 5-14*d*. The definition of the switching times is given in the following paragraphs.

Delay Time

The turn-on delay time t_d is the time lapse between the application of the input step pulse to the time the output current reaches 10 percent of its final value $0.1I_C(\text{sat})$. It is limited by (1) the charging time of the junction transition capacitances from reverse bias to the new voltage levels and (2) the carrier transit times through the base and the collector depletion layer.

Rise Time and Fall Time

The rise time t_r is the time required for the current to rise from 10 to 90 percent of $I_C(\text{sat})$. It corresponds to the establishment of minority carriers in the base to reach 90 percent of the collector saturation current and is affected by the output time constant $C_C R_L$. The turn-off fall time t_f specifies the interval during which the collector current falls from 90 to 10 percent of its maximum value. This is the reverse process of the rise time and is limited by the same factors.

Storage Time

The storage time t_s is measured between the negative step in the base current and the time when the collector current reaches $0.9I_C(\text{sat})$. This is the most important parameter in limiting the switching speed of a switching transistor. When the transistor is in saturation, both the emitter and collector junctions are forward-biased. As a result, excess carriers above and beyond those necessary to maintain the active mode of operation, are stored in the base and collector, as depicted in the shaded area in Fig. 5-15. The transistor output current cannot change until the excess carriers are removed. Thus, the storage time corresponds to the time required to remove the excess minority carriers.

The complete switching-time response can be calculated with the Ebers-Moll model given in Fig. 5-8. The procedure involves (1) replacing α_N and α_I by their frequency-dependent form, (2) analyzing the Ebers-Moll equations in the frequency domain with the input base current as a step function, and (3) using the Laplace transform to obtain the time response. Although the procedure is straightforward, the details of such an analysis are quite complicated. For this reason, we consider the charge-control analysis first introduced in Sec. 4-8. Since the storage time is the dominant parameter, we shall limit our discussion to its derivation.

The charge-control equation for the base region in the normal active mode is given by Eq. (4-60), with $I_p(0)$ replaced by i_B, Q_s by Q_B, and τ_p by τ_n:

$$i_B = \frac{dQ_B}{dt} + \frac{Q_B}{\tau_n} \tag{5-66}$$

where
$$Q_B = qA\,\frac{n_p(0)x_B}{2} \tag{5-67}$$

It should be pointed out that the collector current is related to the charge storage by Eqs. (5-38) and (5-67), so that the collector current is specified if Q_B is known. Under steady-state conditions, the time-dependent term in Eq. (5-66) is zero, and the base current is given by

$$I_B = \frac{Q_B}{\tau_n} \tag{5-68}$$

Let us designate the base current that brings the transistor to the onset of saturation as I_{BA}. Using Eq. (5-65), we find

$$I_{BA} = \frac{V_{CC}}{h_{FE}R_L} \tag{5-69}$$

When the saturation sets in, the total charge is given by $Q_B + Q_{BX}$, and the charge-control equation becomes (assuming $Q_C = 0$ in Fig. 5-15)

$$i_B = \frac{Q_B}{\tau_n} + \frac{Q_{BX}}{\tau_s} + \frac{dQ_B}{dt} + \frac{dQ_{BX}}{dt} \tag{5-70}$$

where Q_{BX} is the excess amount of charge storage and τ_s is the time constant associated with the removal of Q_{BX}. Let us now abruptly change the base current from I_{B1} to $-I_{B2}$ (Fig. 5-14b); the excess charge starts to decrease, but the active charge Q_B remains the same between $t = 0+$ and t_s. During this time interval, we may set

$$\frac{dQ_B}{dt} = 0 \tag{5-71}$$

and

$$\frac{Q_B}{\tau_n} = I_{BA} \tag{5-72}$$

The second equation is obtained since the ratio Q_B/τ_n corresponds to the onset of saturation. At $t = 0-$, the excess charge is obtained by setting the time-dependent terms in Eq. (5-70) to zero and using Eq. (5-72):

$$Q_{BX}(0) = \tau_s(I_{B1} - I_{BA}) \tag{5-73}$$

This is the initial condition for Eq. (5-70), and the solution for Q_{BX} is given by

$$Q_{BX} = \tau_s(I_{B1} - I_{B2})e^{-t/\tau_s} + \tau_s(I_{B2} - I_{BA}) \tag{5-74}$$

At $t = t_s$, all the excess minority carriers are removed, and $Q_{BX} = 0$. Thus, we find

$$t_s = \tau_s \ln \frac{I_{B1} - I_{B2}}{I_{BA} - I_{B2}} \tag{5-75}$$

τ_s is related to the minority-carrier lifetime in the base and is usually given in the specifications of a switching transistor.

5-9   BREAKDOWN VOLTAGES

The basic limitation of the maximum voltage in a transistor is the same as that in a *pn* junction diode, i.e., the avalanche or Zener breakdown. However, the voltage breakdown depends not only on the nature of the junction involved but also on the external circuit arrangement. For example, the breakdown voltage for the common-emitter circuit differs from that of the common-base circuit, and the external base impedance influences the maximum operating voltage before breakdown. In the following paragraphs, we consider the breakdown voltages under different operating conditions. The mechanism of *punch-through* breakdown, unique to transistors, is also discussed.

Common-Base Configuration

The maximum allowable reverse-bias voltage between the collector and the base terminals of the transistor with the emitter lead open-circuited is designated by BV_{CBO}. This voltage is determined by the avalanche breakdown voltage of the collector-base junction. The avalanche multiplication factor given in Eq. (3-40) can be expressed empirically by

$$M = \frac{1}{1 - (V_{CB}/BV_{CBO})^n} \tag{5-76}$$

for the common-base circuit where n is a constant. The collector current-voltage characteristic for $I_E = 0$ in the breakdown region is sketched in Fig. 5-16. The abrupt increase of I_C at BV_{CBO} is the distinct feature of the breakdown phenomenon. The collector current is related to the emitter current by

$$\alpha^* = M\alpha \tag{5-77}$$

indicating that the effective current gain is increased by the M factor. The breakdown is satisfied when M approaches infinity.

Common-Emitter Configuration

Since $h_{FE} = \alpha/(1 - \alpha)$, it follows that the common-emitter current gain incorporating the avalanche effect h_{FE}^* is

$$h_{FE}^* = \frac{\alpha^*}{1 - \alpha^*} = \frac{M\alpha}{1 - M\alpha} \tag{5-78}$$

Accordingly, the new current gain becomes infinite when the condition $M\alpha = 1$ is reached. Since α is very close to unity, the common-emitter breakdown condition is satisfied when M is not much greater than unity. The breakdown voltage is denoted BV_{CEO} when the base is open-circuited. By setting $V_{CB} = BV_{CEO}$ in Eq. (5-76) and equating M to $1/\alpha$, we can solve for BV_{CEO} and obtain

$$BV_{CEO} = BV_{CBO}\sqrt[n]{1 - \alpha} \approx BV_{CBO}(h_{FE})^{-1/n} \tag{5-79}$$

With an n value between 2 and 4 in silicon and a large h_{FE}, the common-emitter breakdown voltage BV_{CEO} could be much lower than the common-base breakdown voltage.

The breakdown voltage between the collector and emitter can be made greater than BV_{CEO} by returning the base to the emitter through a base resistor R_B. If R_B is very large, the base is essentially an open circuit and the breakdown voltage approaches BV_{CEO}. If R_B is small and close to zero, the base is essentially short-circuited to emitter and the breakdown voltage approaches BV_{CBO}. For a finite R_B, the breakdown voltage is somewhere in between. The common-emitter breakdown characteristic with a finite R_B is also shown in Fig. 5-16.

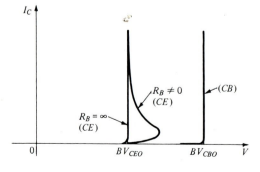

FIGURE 5-16
Breakdown voltage for common-emitter and common-base circuits.

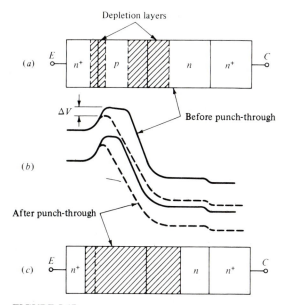

FIGURE 5-17
Punch-through in an *npn* transistor: (*a*) space-charge regions before punch-through, (*b*) energy-band diagrams, and (*c*) space-charge regions after punch-through.

Punch-Through Breakdown

A transistor is *punched through* if the space-charge region of the collector junction reaches the emitter junction before the avalanche can take place. The space charge and energy-band distributions of an n^+pn transistor are shown in Fig. 5-17. Under this condition, the emitter and collector regions are joined as a continuous space-charge region so that the potential barrier at the emitter is lowered by ΔV by the punch-through collector voltage. As a result, a large emitter current can flow in the structure and the breakdown occurs. Usually, the *V-I* curve for the punch-through breakdown is not as abrupt as the avalanche breakdown. The punch-through voltage is obtained by using the space-charge-layer width given by Eq. (3-32).

5-10 The *pnpn* DIODE

A *pnpn* device consists of four layers of semiconductor doped alternately with *p*- and *n*-type impurities. A *pnpn* device having external connections only to the end regions is called a *pnpn diode*. Figure 5-18 shows such a diode and its voltage-current characteristic. The terminal connected to the p_1 region is called the *anode*, and the terminal connected to the n_2 region is called the *cathode*. A typical *V-I* characteristic for a silicon *pnpn* diode is shown in Fig. 5-18*b*. When an external voltage is applied to make the anode positive with respect to the cathode, the *V-I* characteristic

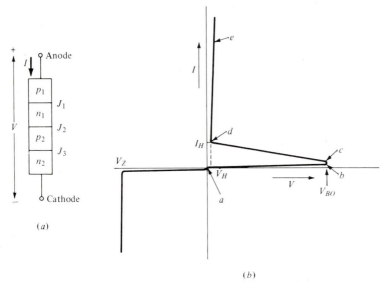

FIGURE 5-18
A *pnpn* diode and its characteristic.

exhibits four distinct regions:

a-b At low voltages, junctions J_1 and J_3 are forward-biased and junction J_2 is reverse-biased. The external impressed voltage appears principally across the reverse-biased junction. The device behaves essentially like a reverse-biased *pn* junction. Therefore, this is called the OFF, or *high-impedance*, region.

b-c As the voltage V is increased, the current increases slowly up to voltage called the *breakover voltage* V_{BO}, where the current increases abruptly.

c-d The device then traverses a region of differential negative resistance; i.e., the current increases as the voltage decreases sharply.

d-e The ON, or *low-impedance*, region. In this region J_2 is forward-biased so that the voltage across the device is essentially that of a single forward-biased *pn* junction, i.e., about 0.7 V. If the current flowing through the diode is reduced, the diode will remain in its ON state until $I = I_H$. This current and the corresponding voltage V_H are called the *holding current* and *voltage*, respectively. When the current is reduced below this value, the diode switches back to its high-impedance state.

If the diode is biased negatively, the current remains very small because junctions J_1 *and* J_3 are reverse-biased. When the device is reverse-biased to the point that both junctions J_1 and J_3 break down, it behaves like an avalanche or Zener diode with $V_Z = V_{Z_1} + V_{Z_3}$. This region is of no particular interest.

In Fig. 5-19a, the *pnpn* diode has been split into two parts to demonstrate

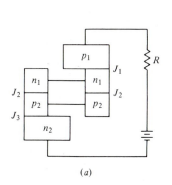
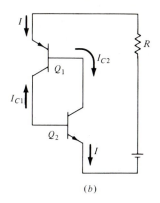

FIGURE 5-19
Equivalent circuit of a *pnpn* diode.

that the device can be considered as two transistors connected back to back. One transistor is a *pnp* type, whereas the second is an *npn* type. The n_1 region is the base of the *pnp* transistor and the collector of the *npn* transistor. The p_2 region is the base of the *npn* transistor and the collector of the *pnp* transistor. The junction J_2 is the collector junction shared by both transistors. In Fig. 5-19*b* the arrangement in Fig. 5-19*a* has been redrawn using transistor-circuit symbols, and a voltage source has been applied through a resistor across the switch, giving rise to a current I. Collector currents I_{C1} and I_{C2} for transistors Q_1 and Q_2 are indicated. In the active region the collector current of a transistor is given by Eq. (5-9). When we apply Eq. (5-9) to Q_1 and Q_2,

$$I_{C1} = -\alpha_1 I + I_{CO1} \tag{5-80}$$

$$I_{C2} = \alpha_2 I + I_{CO2} \tag{5-81}$$

Setting equal to zero the sum of the currents into transistor Q_1, we have

$$I + I_{C1} - I_{C2} = 0 \tag{5-82}$$

Combining Eqs. (5-80) to (5-82), we find

$$I = \frac{I_{CO2} - I_{CO1}}{1 - \alpha_1 - \alpha_2} = \frac{I_{CO}}{1 - \alpha_1 - \alpha_2} \tag{5-83}$$

where $I_{CO} = I_{CO2} - I_{CO1}$ is the total reverse saturation current of junction J_2. We observe that as the sum $\alpha_1 + \alpha_2$ approaches unity, the current I increases without limit; i.e., the device breaks over.

In a transistor, the magnitude of α increases with collector voltage and with collector current at low current levels, the latter being shown in Fig. 5-6. The effect of collector voltage on α is particularly pronounced as the voltage is increased to the avalanche voltage.

Accordingly, as the voltage across the four-layer device is increased, the collector current and the magnitude of alphas in the two equivalent transistors increase. When the sum of α_1 and α_2 approaches unity, the current I increases sharply. The increase of current I, in turn, further increases the alphas. When the sum of the avalanche-enhanced alphas equals unity, $\alpha_1 + \alpha_2 = 1$, breakover takes place. At this point, the current is large, and α_1 and α_2 might be expected individually to attain values in the neighborhood of unity. If such were the case, Eq. (5-83) indicates that the current might be expected to reverse. What provides stability to the ON state of the switch is that in the ON state the center junction becomes forward-biased. Now both transistors are in saturation, and the current gain α is again small. Thus, stability is attained by virtue of the fact that the transistors enter saturation to the extent necessary to maintain the condition $\alpha_1 + \alpha_2 = 1$. The current I increases to the point at which it is limited primarily by the external circuitry. This region is the ON region. In the ON region, all three junctions in the diode are forward-biased, and normal transistor action is no longer effective. The voltage across the device is very nearly equal to the algebraic sum of these three saturation junction voltages. The magnitude of the ON voltage is of the order of 1.0 V because the voltage drop across the center junction J_2 is in the direction opposite that of the voltages across junctions J_1 and J_3.

To keep the diode in its low-impedance state, the condition $\alpha_1 + \alpha_2 = 1$ must be satisfied. The holding current corresponds to the minimum current at which $\alpha_1 + \alpha_2 = 1$ is satisfied. Further reduction of current violates the foregoing condition, and the device switches back to its high-impedance region.

In practical operation, when a positive voltage is applied to the anode to turn the device from the OFF to ON state, the junction capacitance across the junction J_2 is charged. The charging current flows through the emitter junctions of the two transistors. If the rate of change of the applied voltage with time is high, the charging current may be large enough to raise the alphas of the two transistors sufficiently to turn on the diode. This is called the *rate effect*. It may reduce the breakover voltage to half or less of its static value.

5-11 SILICON-CONTROLLED RECTIFIERS

In the *pnpn*-diode operation, it is necessary to increase the external applied voltage so that the junction J_2 is in the avalanche multiplication region. The breakover voltage of the diode is fixed in fabrication. The shape of the voltage-current characteristic can be controlled if a third terminal, called the *gate*, is provided. The three-terminal device is commonly known as a *silicon-controlled rectifier* (SCR) or *thyristor*, and one of its possible structures is shown in Fig. 5-20. Using the standard planar technology, the p_1 and p_2 regions are diffused simultaneously, and the subsequent n_2 diffusion completes the four-layer device. The $p_1 n_1 p_2$ transistor is known as the *lateral transistor*, and its operation will be described in the next chapter. The current flowing through the gate terminal can now increase α_2 independently of V and I. The effect of gate current on the V-I characteristic is shown in Fig. 5-21. A family of V-I characteristics is obtained with I_G maintained constant at various values.

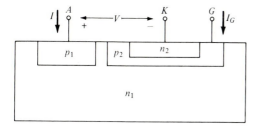

FIGURE 5-20
A low-power SCR structure.

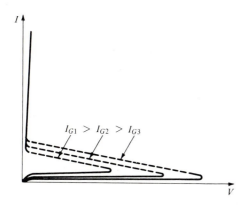

FIGURE 5-21
Current-voltage characteristics of a silicon-controlled rectifier.

The most obvious effects of increasing I_G are to increase the OFF current and to decrease both the breakover voltage and the holding current. These changes can be explained qualitatively in terms of the two-transistor equivalent circuit. In the OFF state the device behaves essentially like a normal *npn* transistor with the *pnp* transistor acting like an emitter-follower with a very small $h_{fe} \approx 0.1$. Increasing the gate current increases the collector current of the *npn* transistor in the usual way. The increase of the collector current, in turn, increases the anode current. This larger current causes an increase in the alphas (Fig. 5-6). The condition $\alpha_1 + \alpha_2 = 1$ is therefore reached at lower values of the avalanche multiplication factor, and the breakover voltage is reduced. In the ON state, the flow of gate current again increases the alphas. Thus, the current I can fall to lower values at which the holding-point condition is still satisfied.

The voltage at which the device goes from an OFF to an ON state is controlled by a small gate signal. In high-power SCRs once the device is in the ON state, the gate circuit has little effect on the operation. However, in low-power SCRs the gate circuit can be used to turn the device both on and off. A negative current is usually required to turn the device off.

ADDITIONAL READINGS

Gentry, F. E., et al.: "Semiconductor Controlled Rectifiers," Prentice-Hall, Englewood Cliffs, N.J., 1965.
Ghandi, S. K.: "Semiconductor Power Devices," Wiley, New York, 1977.

Gray, P. E., D. DeWitt, A. R. Boothroyd, and J. F. Gibbons: "Physical Electronics and Circuit Models of Transistors," SEEC Series, vol. II, Wiley, New York, 1964.

Muller, R. S., and T. I. Kamins: "Device Electronics for Integrated Circuits," 2d ed., Wiley, New York, 1986.

Neudeck, G. W.: "The Bipolar Junction Transistor," Addison-Wesley, Reading, Mass., 1983.

PROBLEMS

5-1. (*a*) Sketch the energy-band diagram for a *pnp* transistor at equilibrium and under the normal active mode of operation.

   (*b*) Sketch a schematic diagram to represent the transistor and indicate all current components.

   (*c*) Rewrite Eqs. (5-1) to (5-3) for this transistor.

5-2. Consider the bipolar transistor shown in Fig. P5-2 with no bias.

   (*a*) Calculate the Fermi energy levels in all three regions, using E_i as a reference.

   (*b*) Plot the energy-band diagram of (*a*) and sketch the charge density $\rho(x)$ vs. x.

   (*c*) Now assuming there is a bias, $V_B = +0.7\,\text{V}$ and $V_C = +1.5\,\text{V}$, plot the energy-band diagram.

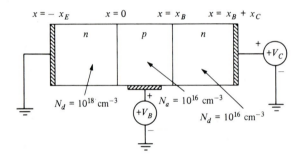

FIGURE P5-2

5-3. (*a*) Derive an expression for the carrier concentration in the base in an *npn* transistor if x_B and L_n are of the same order of magnitude.

   (*b*) Simplify the result in (*a*) for $x_B \ll L_n$ to obtain Eq. (5-20).

5-4. Consider a silicon *npn* transistor with these parameters: $x_B = 2\,\mu\text{m}$, $N_a = 5 \times 10^{16}\,\text{cm}^{-3}$ in a uniformly doped base, $\tau_n = 1\,\mu\text{s}$, and $A = 0.01\,\text{cm}^2$. If the collector is reverse-biased and $I_{nE} = 1\,\text{mA}$, calculate the excess-electron density at the base side of the emitter junction, the emitter-junction voltage, and the base transport factor.

5-5. In the transistor of Prob. 5-4, assume that the emitter is doped with $10^{18}\,\text{cm}^{-3}$, $x_E = 0.5\,\mu\text{m}$, $\tau_{pE} = 10\,\text{ns}$, and $\tau_o = 0.1\,\mu\text{s}$ in the emitter space-charge region. Calculate the emitter efficiency and h_{FE} at $I_{nE} = 1\,\text{mA}$.

5-6. In the transistor of the first example in Sec. 5-3, the base doping was changed to $5 \times 10^{18}\,\text{cm}^{-3}$. Calculate the emitter efficiency and current gain.

5-7. An *npn* transistor has the following specifications: emitter width $= 2\,\mu\text{m}$, base width $= 1\,\mu\text{m}$, emitter resistivity $= 4 \times 10^{-4}\,\Omega\cdot\text{cm}$, base sheet resistivity $= 0.06\,\Omega\cdot\text{cm}$, collector resistivity $= 0.3\,\Omega\cdot\text{cm}$, hole lifetime in emitter $= 1\,\text{ns}$, electron lifetime in base $= 100\,\text{ns}$. It is assumed that the emitter recombination current is constant and equal to 1 μA. Step junctions and uniform doping are also assumed. Calculate h_{FE} for $I_E = 10 \, \mu A$, $100 \, \mu A$, 1 mA, and 10 mA. Plot on semilog axes.

5-8. Verify that the exponential factor of the nonidealed base current in Fig. 5-5 is 1.6.

5-9. Derive Eq. (5-34), starting from the continuity equations in the base and emitter.

5-10. (a) Ignoring the space-charge recombination currents, show that the exact expression for the common-emitter output characteristics of a transistor is

$$-V_{CE} = \phi_T \ln \frac{I_{CO} + \alpha_N I_B - I_C(1 - \alpha_N)}{I_{EO} + I_B + I_C(1 - \alpha_I)} + \phi_T \ln \frac{\alpha_I}{\alpha_N}$$

Note: First solve explicitly for the junction voltages in terms of the currents.

(b) If $I_B \gg I_{EO}$ and $I_B \gg I_{CO}/\alpha_N$, show that the foregoing equation reduces to

$$V_{CE} = \phi_T \ln \frac{1/\alpha_I + I_C/I_B h_{FEI}}{1 - I_C/I_B h_{FEN}}$$

5-11. The circuit in Fig. P5-11 shows an ideal op-amp connected up as a logarithmic amplifier. Using the Ebers-Moll model for the *pnp* transistor, find an expression for V_A in terms of V_s, R, ϕ_T, and relevant Ebers-Moll parameters. Are there any restrictions on V_s? Under what conditions would it be a good log amp?

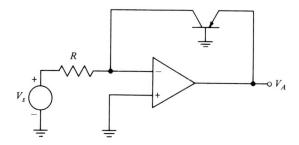

FIGURE P5-11

5-12. In a silicon *pnp* transistor, the doping concentrations for the emitter, base, and collector are 10^{19}, 5×10^{17}, and $10^{15} \, \text{cm}^{-3}$. The emitter and base widths are 1 and 0.7 μm, respectively, and the diffusion length is 10 μm for both electrons and holes. The collector width is 50 μm and the cross-sectional area is $10^{-4} \, \text{cm}^2$. Calculate (a) normal alpha, (b) inverse alpha, (c) I_{EO}, (d) I_{CO}, and (e) $V_{CE}(\text{sat})$ at $I_C = 1 \, \text{mA}$ and $I_B = 100 \, \mu A$.

5-13. For the transistor in Prob. 5-12, determine the hybrid-pi parameters at $I_C = 1 \, \text{mA}$ and $V_C = 5 \, \text{V}$.

5-14. Show that the emitter voltage-current characteristic of a transistor in the active region is given by

$$-I_E \approx \frac{I_{EO}}{1 - \alpha_N \alpha_I} e^{V_E/\phi_T} + \frac{q A n_i x_{dE}}{2\tau_o} e^{V_E/2\phi_T}$$

5-15. Consider a silicon *npn* transistor with the following parameters: $N_B = 1.2 \times 10^{17} \, \text{cm}^{-3}$, $x_B = 0.9 \, \mu m$; $N_C = 10^{15} \, \text{cm}^{-3}$; $N_E = 10^{19} \, \text{cm}^{-3}$, $x_E = 0.5 \, \mu m$. In addition, the minority-carrier diffusion length is much larger than the base and emitter widths.

Estimate the hybrid-pi parameters for the transistor at $I_C = 2$ mA, $V_C = 10$ V, and $h_{FE} = 50$. Let $A = 10$ square mils and assume abrupt junctions.

5-16. The effect of dc base spreading resistance on the collector current can be expressed by $I_C = I_o \exp[(V_E - I_B r_x)/\phi_T]$. Estimate r_x by using this equation and the data shown in Fig. 5-5.

5-17. Calculate the base resistance of the stripe-transistor structure shown in Fig. P5-17. Assume a uniformly doped base with $N_a = 10^{17}$ cm^{-3}.

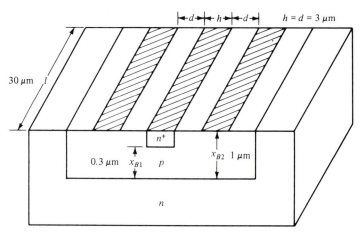

FIGURE P5-17

5-18. Using the transistor described in Prob. 5-15, calculate the cutoff frequency ω_α at 300 K. *Note:* Use $\tau_B = x_B^2/4D_n$ for a double-diffused transistor. Let $r_{sc} = 0$ and $v_{th} = 8 \times 10^6$ cm/s.

5-19. If the actual CB current gain is given by $\alpha = \alpha_o e^{jm\omega/\omega_\alpha}/(1 + j\omega/\omega_\alpha)$, show that $\omega_T \approx \omega_\alpha/(1 + 2\alpha_o m)$, where ω_T is the frequency at which the common-emitter current gain is unity. Assume $m\omega_T < \omega\alpha$.

5-20. Derive an expression of the base transit time for the transistor with uniformly doped base. Assume $x_B/L_n \ll 1$.

5-21. The silicon transistor shown in the circuit (Fig. P5-21) has $h_{FE} = 50$ and $\tau_s = 20$ ns Estimate the storage time. Assume $V_{BE}(\text{sat}) = 0.8$ V and it remains the same during storage period.

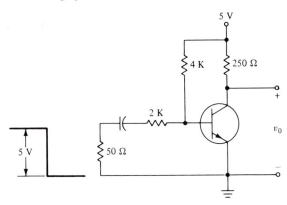

FIGURE P5-21

5-22. Find BV_{CBO} and BV_{CEO} for the transistor of Prob. 5-7. *Note*: Use curves in Fig. 3-18 to obtain the avalanche voltage and compare it with the punch-through voltage.

5-23. Derive an expression of the anode current I_A as a function of the gate current by using the two-transistor analog in an SCR.

5-24. A small-area SCR can be turned off by a negative gate current. The turn-off gain is defined as I_A/I_G, where I_A is the anode ON current and I_G is the minimum gate current that turns off the device.

(*a*) Derive an expression for the turn-off gain.

(*b*) Specify the alphas that give rise to a high turn-off gain.

CHAPTER
6

BIPOLAR TECHNOLOGY AND MINIATURIZATION

Modern electronic systems are built on the foundation of semiconductor devices and integrated circuits. More than any other electronic component, the integrated circuit (IC) is responsible for such low-cost products as the hand-held calculator, digital wristwatch, and personal computer. An IC consists of both active and passive elements formed on a single-crystal silicon substrate. Each circuit element is electrically isolated from the others, and interconnections are provided by a metallization pattern.

In this chapter, we begin with a description of the isolation technology for bipolar integrated circuits. Two approaches, using a reverse-biased *pn* junction or a dielectric film to isolate individual components, are presented. Structural designs for overcoming or utilizing the limitations imposed by the isolation technology are then given. These structures include the multiple-emitter transistor, lateral *pnp* transistor, and integrated injection logic. The last device eliminates the isolation of components in the same logic gate.

An important challenge of bipolar transistor design is how to miniaturize a device's dimensions without sacrificing its performance. Making use of new technology, self-aligned structures have been fabricated in which efficient use of chip area is realized. Polycrystalline silicon is found to be useful as a material for contacts, masking, and interconnections. An unexpected phenomenon has been observed when an arsenic-doped polysilicon is used as the emitter contact in that the current gain is increased by a factor of 3 or more. These recent developments will be discussed in the latter part of this chapter.

6-1 ISOLATION TECHNOLOGY FOR BIPOLAR INTEGRATED DEVICES

In an IC, devices on the same substrate are usually isolated from one another so that there is no current conduction between them. Such an isolation may be provided by (1) a reverse-biased *pn* junction or (2) isolating dielectric materials. There are a number of methods reported in the literature, all of which use either the junction or dielectric technique or a combination of both. In this section, we present two approaches that have been the standards of the bipolar technology.

Junction Isolation

Basically, junction isolation uses an impurity diffusion to produce *n*-type islands surrounded by *p*-type materials. The standard industrial technology is the *epitaxial-diffused process* (EDP). Fabrication using EDP begins with a *p*-type silicon substrate upon which an *n*-type epitaxial layer, typically 3 to 15 μm thick, is grown (Fig. 6-1a). By means of the oxidation and photoresist steps described in Sec. 3-1, areas of *n*-type islands are defined, and acceptor impurity atoms are diffused into regions between the islands. The acceptor concentration in the newly diffused regions, known as the *isolation regions*, must be higher than the donor concentration in the epitaxial layer. In addition, acceptors must diffuse through the epitaxial layer so that the isolation regions join with the *p*-type substrate. After the isolation diffusion, an *n* island is electrically isolated, as shown in Fig. 6-1b, and is ready to have components built in it. For example, let us consider a transistor fabricated in the island shown. The *n*-type island is used as the collector region, and two diffusion steps, a *p* diffusion followed by an n^+ diffusion, are made to form, respectively, the base and emitter regions. During the emitter diffusion, additional donor impurities are diffused directly into the *n* collector to facilitate ohmic contacts (Sec. 7-6) to the collector. The final structure after the deposition of metal contacts is shown in Fig. 6-1c.

A major disadvantage of the integrated transistor shown in Fig. 6-1c is its high collector series resistance since the collector current is horizontally routed through the lightly doped collector. This difficulty is overcome by making an n^+ diffusion before growing the epitaxial layer, as illustrated in Fig. 6-1d. The n^+ layer, known as the *buried layer*, provides a low-resistance path from the collector contact to the active portion of the transistor (Fig. 6-1e). Presently, the buried layer is used in all bipolar ICs fabricated by the EDP process. The thickness of the epitaxial layer is usually less than 10 μm.

Since the integrated transistor is formed by base and emitter diffusions, the impurity profile is not uniform. A typical impurity distribution is shown in Fig. 6-2 for a transistor with a buried layer. Note that the dashed curves represent the

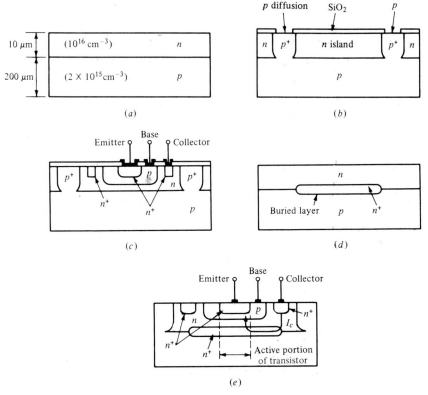

FIGURE 6-1
Fabrication of IC using epitaxial-diffused process: (*a*) substrate with *n* epitaxial layer, (*b*) *p*-type isolation diffusion, (*c*) completed transistor, (*d*) buried n^+ layer, and (*e*) integrated transistor with buried layer.

individual base and emitter diffusion profiles, and the solid curve is obtained after taking the absolute value of $N_d - N_a$. More will be said about the effect of nonuniform profiles on the transistor's characteristics in the next section. The dotted curve in the collector indicates the result of the out-diffusion of arsenic atoms in the buried layer. This out-diffusion puts a constraint on the epitaxial-layer thickness. During the isolation diffusion, acceptor atoms diffuse in both the vertical and lateral directions, the lateral diffusion being about 75 to 80 percent of the vertical penetration. For this reason, the width of the isolation region is approximately twice the epitaxial-layer thickness in order to have the acceptors penetrate through the epitaxial layer. Since the isolation regions do not play an active role electronically, they should be minimized.

Oxide Isolation

Junction isolation takes up a large area for the inactive isolation regions, and it introduces large parasitic capacitances. The former is a handicap for high packing

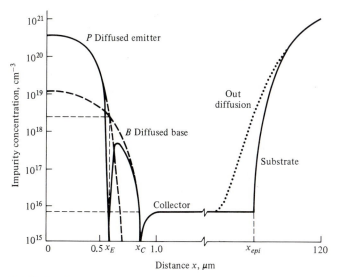

FIGURE 6-2
Impurity profile of an *npn* transistor.

density, and the latter degrades the speed performance. These disadvantages can be minimized with devices isolated by *local oxidation of silicon* (LOCOS), also known as *recessed oxide* (ROX) or *field oxide* (FOX). The LOCOS process makes use of silicon nitride as a mask against thermal oxidation. Figure 6-3 illustrates how the isolation is formed. We begin with the wafer with a buried layer, as shown in Fig. 6-1a. After growing a thin oxide (40 nm) on the silicon, a patterned film of silicon nitride (80 nm) is deposited as seen in Fig. 6-3a. The pattern is produced by the photolithography explained in Chap. 3. The thin oxide is needed to release stress at the Si_3N_4–Si interface to prevent defect formation during subsequent high-temperature cycling. This is followed by an etching step using the nitride as the mask (Fig. 6-3b). This step is not necessary, but it will achieve a smoother surface of the final structure. An implantation of boron ions is made to produce a p^+ layer which will be underneath the field oxide. This is used as the *channel stopper*. It is found that the thermally grown oxide tends to have positive charge resided within so that it induces negative charge in the silicon beneath the surface. As a result, an n layer may be formed there to connect the n^+ buried layers, destroying the electrical isolation. The channel stop prevents this from happening. Now, the wafer is placed in a high-temperature furnace for thermal oxidation, typically in a steam environment at 1000°C for 2 to 3 hours. Since about 0.45 monolayer of silicon is consumed to produce a monolayer of silicon dioxide, the oxide grows downward and expands upward to fill up the etched space, as depicted in Fig. 6-3c. The nitride may now be removed, and boron ions are implanted to form the base. Further processing steps can be added to fabricate the bipolar transistor, as shown in Fig. 6-3d and e. This is a straightforward implementation of the oxide isolation. More complicated procedures are usually employed to take

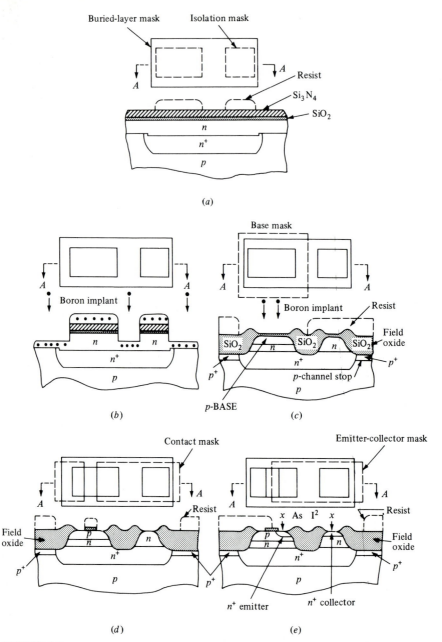

FIGURE 6-3
Top and cross-section views of bipolar transistor fabrication. (*a*) Buried-layer mask and isolation masks and cross section with isolation resist mask on nitride-pad oxide. (*b*) Cross section after nitride-pad oxide-silicon etch and channel-stop implant. (*c*) Base mask and cross section after isolation-oxide growth and boron base implant. (*d*) Contact mask and cross section after base-emitter-, and collector-contact opening. (*e*) Emitter-collector mask and cross section after arsenic emitter-collector ion implant. (*After Parrilio in Sze [1].*)

full advantage of the technique, which will be discussed in self-aligned structures later.

6-2 GRADED IMPURITY PROFILE AND CURRENT GAIN

In the last chapter, our analyses were based on the assumption that the impurity distribution is homogeneous in the base and emitter regions. In a practical integrated transistor, the impurity profile is nonuniform, as depicted in Fig. 6-2. The graded impurity distribution in the base, according to Sec. 2-10, introduces a built-in field

$$\mathscr{E} = \frac{\phi_T}{N_a(x)} \frac{dN_a(x)}{dx} \tag{6-1}$$

This equation is obtained by using Eq. (2-55) with N_d replaced by N_a, the impurity distribution in the base, and removing the negative sign.

This electric field, which keeps the holes in their place, is in such a direction that it aids the transport of injected electrons in most of the base region. The electrons are now moved by both diffusion and drift across the base layer so that the transport factor is increased. Let us now derive an expression of the transport factor for the graded base transistor.

We obtain the following equation by substituting Eq. (6-1) into Eq. (2-17) and making use of Einstein's relationship:[1]

$$\frac{dn}{dx} + \frac{n}{N_a} \frac{dN_a}{dx} = \frac{-I_n}{qAD_n} \tag{6-2}$$

Multiplying both sides of Eq. (6-2) by N_a and integrating yields

$$N_a n(x) = \int \frac{I_n N_a}{qAD_n} dx + C \tag{6-3}$$

where C is an integration constant. In most practical planar transistors, the base recombination is negligible, so that I_n is constant in Eq. (6-3). We can now substitute the boundary condition $n_p(x_B) = 0$ [Eq. (5-19)] into Eq. (6-3), leading to

$$n(x) = \frac{I_n}{qAD_n N_a} \int_x^{x_B} N_a dx \tag{6-4}$$

The current I_n is obtained by making use of Eq. (5-18) at $x = 0$ with n_{po} replaced by $n_i^2/N_a(0)$:

$$I_n = \frac{qAD_n n_i^2}{\int_0^{x_B} N_a dx} e^{V_{E}/\phi_T} = \frac{qAD_n n_i^2}{G_B} e^{V_{E}/\phi_T} \tag{6-5}$$

---

[1] All currents in this section are considered positive to eliminate the confusion of polarity.

The integral in the denominator of Eq. (6-5) represents the number of impurity atoms in the base and is known as the *Gummel number* G_B.† Notice that Eq. (6-5) reduces to Eq. (5-21) for a base with uniform doping. In addition a larger electron current is realized with a smaller Gummel number, which corresponds to a narrower base width.

The base transport factor can be estimated by first taking the total base recombination current as

$$I_B = \frac{qA}{\tau_n} \int_0^{x_B} n(x)\,dx$$

Thus, from the definition of the base transport factor, we obtain

$$\beta_T = \frac{I_n - I_B}{I_n}$$

$$= 1 - \frac{qA}{I_n \tau_n} \int_0^{x_B} n(x)\,dx \tag{6-6}$$

Substituting Eq. (6-4) into (6-6) and setting $D_n \tau_n = L_n^2$, we have

$$\beta_T = 1 - \frac{1}{L_n^2} \int_0^{x_B} \left(\frac{1}{N_a} \int_x^{x_B} N_a dx \right) dx \tag{6-7}$$

This is the general expression of the base transport factor for arbitrary base impurity distribution. For a uniform base distribution, Eq. (6-7) reduces to Eq. (5-27).

Let us now consider the carrier injection from the base to the nonuniformly doped emitter. Although the gradient of the impurity profile sets up a retarding field, the general formulation of the current flow is the same as that of the base minority-carrier current. Therefore, we can write the hole current as

$$I_p = \frac{qAD_p n_i^2}{G_E} e^{V_{E}/\phi_T} \tag{6-8}$$

and

$$G_E = \int_0^{x_E} N_d dx \tag{6-9}$$

where G_E represents the total number of active impurity atoms in the emitter and is known as the emitter Gummel number. The emitter injection efficiency may be obtained by

$$\gamma = \frac{I_n}{I_n + I_p} = \left(1 + \frac{D_p G_B}{D_n G_E} \right)^{-1} \tag{6-10}$$

---

†An alternative definition is the effective Gummel number given by $G_B D_n$. This definition takes into account the dependence of D_n on the doping concentration and is practical in device design.

In most advanced bipolar transistors, the base and space-charge recombination currents are negligible. Thus, the current gain is specified by the emitter injection efficiency which is determined by the ratio of the Gummel numbers. The base Gummel number can be obtained from the semilog plot of the I-V characteristics of the collector current, and the G_E can be calculated from the base-current plot. Assuming the base transport factor is unity and the space-charge recombination current is negligible, we obtain the common-emitter current gain as

$$h_{FE} = \frac{\gamma}{1 - \gamma} = \frac{I_n}{I_p} = \frac{D_n G_E}{D_p G_B} \tag{6-11}$$

This equation is useful for first-order gain calculation.

Example. An integrated transistor has an impurity profile of Fig. 6-2 and the I-V plots of Fig. 5-5. Let $D_n = 35 \text{ cm}^2/\text{s}$, $D_p = 12 \text{ cm}^2/\text{s}$ and an emitter area of 0.1 cm^2. Estimate the Gummel numbers in the base and the emitter and calculate the emitter injection efficiency.

Solution. In Fig. 5-5, the $\ln I_C$ vs. V_E plot is a straight line which can be extrapolated to $V_E = 0$, yielding

$$I_n(0) = I_C(0) = 10^{-14} \text{ A}$$

Since the base Gummel number is given by Eq. (6-5), we obtain

$$G_B = \frac{qAD_n n_i^2}{I_n(0)} = 1.4 \times 10^{19} \text{ cm}^{-2}$$

The emitter Gummel number can be estimated by using the base current where the steepest slope is in parallel with the collector current in Fig. 5-5. The base current extrapolated to $V_E = 0$ is

$$I_B(0) = 10^{-16} \text{ A} = I_p(0)$$

Using Eq. (6-8), we have

$$G_E = \frac{qAD_p n_i^2}{I_p(0)} = 4.7 \times 10^{20} \text{ cm}^{-2}$$

The emitter injection efficiency is, using Eq. (6-10),

$$\gamma = 0.99$$

6-3   BASE-WIDTH STRETCHING

Previously, we have assumed that the carrier concentration is zero at the base side of the collector depletion layer. This is a reasonable approximation when the collector current is small. At high currents, the injected electrons traversing the collector depletion may become comparable to the space-charge density there. Thus, the collector junction and the base width are modified.

Let us consider a typical integrated transistor in which the base doping

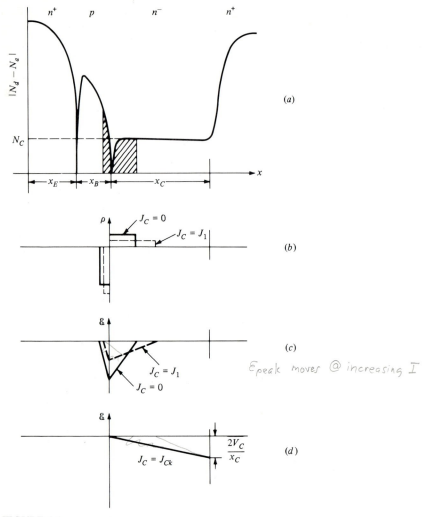

FIGURE 6-4
Base-width-widening effect in an $n^+pn^-n^+$ transistor. (a) Impurity profile. (b) Space-charge region of the collector junction at $J_C = 0$ and $J_C = J_1$. (This diagram shows the increase of negative charge due to spreading of base region's electrons in the base side. The same electron spreading in the depletion region of n^- causes charge compensation and depletion-layer widening). (c) Electric fields of (b). (d) Electric field at J_{Ck}.

is significantly higher than that of the collector (Fig. 6-4). The depletion layer and the electric field distribution are shown for a given collector voltage and zero collector current in Fig. 6-4b and c. The electric current crossing the collector junction is given by

$$J_C = qv(x)n(x) \tag{6-12}$$

where $v(x)$ is the carrier velocity and $n(x)$ is the electron density. Inside the depletion layer, the electric field is high so that the velocity may be assumed as having reached the saturation velocity v_s. Therefore

$$n(x) = \frac{J_C}{qv_s} \qquad (6\text{-}13)$$

This equation states that the electron concentration increases with the collector current. As the current is increased, the electron density will first approach the density of the lightly doped epitaxial collector. The net space charge there, $N_C - n(x)$, will decrease, leading to a lower electric field gradient

$$\frac{d\mathscr{E}}{dx} = \frac{q}{K_s \varepsilon_o}\left(N_C - \frac{J_C}{qv_s}\right) \qquad (6\text{-}14)$$

Since the collector voltage is a constant, the integration of the electric field will remain the same with or without current. In other words, the two areas under the electric field curves are identical. Therefore, the new electric field distribution is changed to the dashed line depicted in Fig. 6-4c. With a lower $\mathscr{E}(0)$, the depletion-layer edge on the base side moves to the right, effectively increasing the base width. However, this stretching of the base is quite small. The collector current is estimated to be $J_C = J_1 = qv_s N_C$, where N_C is the collector doping density.

As the current is further increased, the electric field $\mathscr{E}(0)$ continues to drop. Eventually, it becomes zero (Fig. 6-4d), which corresponds to the collapse of the original collector space-charge layer. That is to say, the original base-to-epitaxial collector is forward-biased. The base spreads over the collector epitaxial layer, and its width becomes very large. By solving Eq. (6-14) with the boundary conditions of $\mathscr{E}(0) = 0$ and $\mathscr{E}(x_C) = -2V_C/x_C$ (see Prob. 6-8a), we obtain the collector current as [2]

$$J_{Ck} = qv_s\left(N_C + \frac{2K_s \varepsilon_o V_C}{qx_C}\right) \qquad (6\text{-}15)$$

This is the critical current density that gives rise to the large base stretching and is known as the *Kirk effect*. As the base width is increased, the base Gummel number increases. Therefore, the current gain decreases at high collector currents. At the same time, the increased base width reduces the cutoff frequency and the transistor becomes slower. For these reasons, high-current operation should be avoided. If a transistor has to be operated at a high current, the impurity profile should be designed to take the base-width widening into consideration.

6-4   THE INTEGRATED BIPOLAR TRANSISTOR

In this section, an integrated bipolar transistor is described to highlight the effects of parasitic transistors, series resistances, and junction capacitances. Most designs are attempts to minimize these parasitic components.

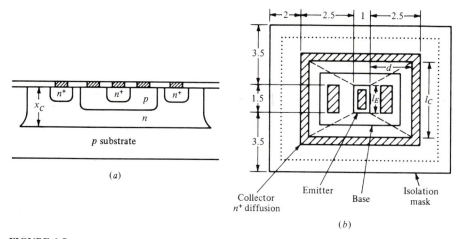

FIGURE 6-5
Layout of the integrated transistor. All dimensions are in mils, and all clearances are 0.5 mil. Dashed lines denote boundaries of trapezoids for r_{cs} calculation.

An integrated transistor fabricated by the EDP is redrawn in Fig. 6-5, which includes the top view of the transistor to show its standard layout. The EDP without a buried layer needs five masks for the following functions: isolation, base diffusion, emitter and n^+ diffusion, contact-window opening, and metallization pattern etching. The dotted line in Fig. 6-5b indicates the lateral diffusion of the isolation step so that the actual isolation island is smaller than the isolation oxide mask. Lateral diffusions of other steps are not significant and are neglected here. Typically, the width of aluminum stripes is 3 to 5 μm, and the clearance between lines is about the same. In general, the theory developed in Chap. 5 is applicable to the integrated transistor, whose impurity profile is essentially the same as the double-diffused transistor. The differences are mostly parasitic effects introduced by the isolation.

When the substrate is included, the integrated transistor has a four-layer structure. The npn section is the desired transistor, and the pnp section is a parasitic element. It was shown in Sec. 5-10 that the $pnpn$ structure has a built-in positive feedback action such that it latches onto an ON position (all junctions become forward-biased) if the emitter and substrate-collector junctions are allowed to be forward-biased simultaneously. Therefore, a reverse bias must be applied between the collector and substrate. This is accomplished by connecting the substrate to the most negative dc voltage in the electronic circuit. The second parasitic element is the isolation-junction capacitance partly due to the sidewall junction and partly to the bottom junction. In a practical integrated transistor, the bottom junction is a step junction, and the capacitance is determined by the substrate doping. The sidewall junction can be approximated by a linearly graded junction.

The requirement for making all contacts on the surface of the integrated transistor forces the collector current to traverse the lightly doped collector region laterally. As a result, the collector series resistance r_{CS} in Fig. 5-1c can be quite large. A large r_{CS} decreases the cutoff frequency and introduces an ohmic drop which increases the saturation voltage. Since both effects degrade the transistor performance, it is necessary to estimate the value of r_{CS}. One of the techniques used to estimate the collector resistance is known as the *trapezoid method*. The collector region is divided into four trapezoids, shown by the dashed lines in Fig. 6-5b. The resistance of each trapezoid is approximated by

$$r = \frac{\rho_C d}{A_{\text{eff}}} \tag{6-16}$$

where ρ_C = collector resistivity
$\quad d$ = distance between emitter-base junction and collector contact
$\quad A_{\text{eff}}$ = cross-sectional area at middle of trapezoid

Thus,

$$A_{\text{eff}} = \frac{l_E + l_C}{2} x_C \tag{6-17}$$

where l_E = top length of trapezoid
$\quad l_C$ = bottom length of trapezoid
$\quad x_C$ = thickness of collector layer

Now the collector resistance is given by the resistance of the trapezoids in parallel. The collector resistance can be reduced significantly if a buried layer is incorporated. The lateral collector current now flows through the heavily doped buried layer, which has negligible resistance. Thus, r_{CS} is given by the sum of series resistances in the vertical direction under the emitter and collector contacts:

$$r_{CS} = \rho_C \left(\frac{x_C}{A_E} + \frac{x_C}{A_{CC}} \right) \tag{6-18}$$

where A_E and A_{CC} are the emitter and collector-contact areas, respectively. The buried layer is used in practically all industrial bipolar processes in which the effect of r_{CS} becomes negligible.

The parasitic capacitance can be reduced by using the oxide isolation method which eliminates the capacitances of the sidewall. An additional benefit of oxide isolation is the reduction of the current gain of the parasitic *pnp* transistor, making the *pnpn* switching action extremely unlikely.

Multiple-Transistor Structures

In the previous paragraphs, we considered the integrated transistor, which occupies an isolation island all by itself. Using such transistors, we can build integrated circuits by making use of a metallization and etching step to provide interconnections. However, it is possible to combine two or more transistors into one isolation island in circuits in which collectors are connected electrically. For

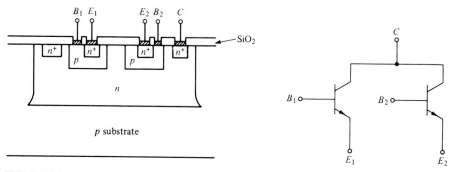

FIGURE 6-6
Integrated common-collector transistors and equivalent circuit.

example, two transistors with a common collector are shown in Fig. 6-6 along with the equivalent circuit. Note that the base and emitter diffusions are made in two regions in the same isolation island. When both the collector and base regions are shared by a number of transistors, the *multiple-emitter transistor* is obtained, as depicted in Fig. 6-7.

By combining the collector region or the collector and base regions, we achieve a reduction of the total isolation area, resulting in smaller isolation capacitance, simpler diffusion masks, and fewer interconnections. This design technique lowers the cost and improves the reliability of integrated circuits and should be employed whenever possible. At present, most commercial digital bipolar ICs are of the type known as *transistor-transistor logic* (TTL), which utilizes the multiple-emitter transistor as its basic building block. A simple TTL circuit is shown in Fig. 6-8 along with its IC implementation by means of the EDP. The integrated circuit shown is the top view, excluding the interconnection pattern. Note that the base diffusion is used to define the resistor stripe since the base-sheet resistance is relatively high (typically 200 $\Omega/\square$).[1] If there is more than one resistor, all resistors may be formed in the same isolation island and isolation between the

---

[1] Defined by ρ_B/x_{jB} where the symbol $\square$ represents a square.

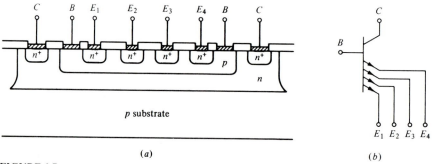

(a)

(b)

FIGURE 6-7
Multiple-emitter transistor: (a) cross section and (b) equivalent circuit.

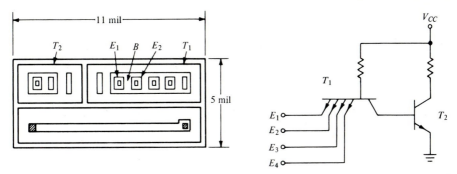

FIGURE 6-8
Transistor-transistor logic IC implementation and its equivalent circuit.

resistors is realized by reverse biasing the base-collector junction of this isolation island. The emitters are usually considered to be independent although interaction does take place when they are physically close. For example, if E_1 is forward-biased and E_2 is reverse-biased, E_1-B-E_2 forms an *npn* transistor with E_2 as the collector. It results in current flow in E_2, which destroys the logic function of E_2. For this reason, the gain of the E_1-B-E_2 transistor should be made very small.

6-5 THE LATERAL *pnp* TRANSISTOR

In an integrated circuit, the basic structure is the vertical *npn* transistor described in the previous sections. Occasionally, it is desirable to have both *npn* and *pnp* transistors fabricated on the same chip. This complementary circuit can be realized by various techniques. The most common method utilizes the lateral *pnp* structure shown in Fig. 6-9. The *p*-type emitter and collector regions are formed simultaneously during the base diffusion of the *npn* transistor, and the epitaxial *n* layer provides the base region. Since the current flow is along the lateral direction, it is known as the *lateral transistor*. The advantage of the lateral structure is that it does not require any processing steps beyond those necessary for the fabrication

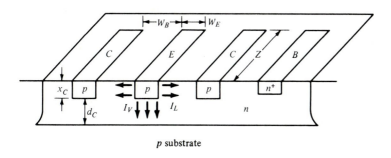

FIGURE 6-9
The lateral *pnp* transistor.

of the *npn* transistor. However, the performance of the *pnp* device is significantly inferior to that of the vertical *npn* transistor.

Let us examine the idealized lateral *pnp* transistor illustrated in Fig. 6-9. With the emitter junction forward-biased and the collector junction reverse-biased, the injected current is divided into vertical and lateral components I_V and I_L. By assuming that $W_B \ll L_p \ll d_C$, where L_p is the hole diffusion length in the *n*-type base, and making use of

$$I_p = -qAD_p \left. \frac{dp}{dx} \right|_{x=0} \tag{6-19}$$

we derive (Prob. 6-14)

$$I_L = 2qZx_C D_p \frac{P_o}{W_B} \tag{6-20}$$

$$I_V = qZW_E D_p \frac{P_o}{L_p} \tag{6-21}$$

where P_o is the injected-hole density at the edge of the depletion layer of the emitter junction. Since the vertically injected carriers recombine in the base, they are not collected by the collector and the current I_V constitutes the base current. Therefore, the common-emitter current gain is given by

$$h_{FE} \equiv \frac{I_C}{I_B} \approx \frac{I_L}{I_V} = \frac{2L_p x_C}{W_E W_B} \tag{6-22}$$

In a practical device, the current gain is typically less than 20. In addition to reducing the current gain, the vertical injection current also degrades the frequency response of the transistor. An equivalent circuit of the lateral structure is shown in Fig. 6-10, where a shunting diode represents the effect of vertical hole injection. This shunting diode acts as a large capacitance and reduces the cutoff frequency of the composite structure. The gain-frequency characteristic shows a 3-dB/octave slope.

The vertical carrier injection can be reduced significantly if the lateral transistor is fabricated on a region with an n^+ buried layer, as shown in Fig. 6-11. Since the doping level of n^+ is higher than that of the *p* emitter, hole injection into the n^+ layer becomes very small. However, there will be electron injection from the n^+ layer into the *p* emitter, and the injection efficiency is reduced. The overall gain of this modified lateral transistor is improved to about 50, and the frequency response follows a 6-dB/octave decrease, as in the vertical double-diffused *npn* transistor.

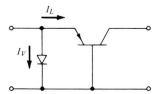

FIGURE 6-10
Equivalent circuit of the lateral transistor.

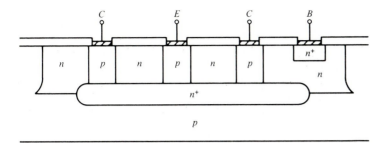

FIGURE 6-11
An improved lateral-transistor structure.

6-6 INTEGRATED INJECTION LOGIC

Integrated injection logic (I^2L), also known as *merged-transistor logic*, is one approach to bipolar large-scale-integration chip design. The I^2L gate is laid out in such a fashion that the individual transistor-isolation step is eliminated and the large-area passive resistors are replaced by small-area lateral transistors in the form of current sources. The device's self-isolation and elimination of resistors are the keys to high functional density and low power consumption.

To understand the operation of an I^2L, let us first examine the logic circuit shown in Fig. 6-12a. This is basically a direct-coupled transistor logic (DCTL) with the resistors replaced by constant current sources. The current sources may be provided by lateral *pnp* transistors, as shown in Fig. 6-12b. Implementation of the logic gate in integrated form is illustrated in Fig. 6-13. Fabrication of this device begins with an n^+ substrate on which an *n*-type epitaxial layer is grown. Subsequently, two diffusions are made to obtain the structure shown. The *n*-on-n^+ layer serves as the emitter and common ground plane. The *p*-diffused region serves as both the base of the vertical *npn* transistor and the emitter and collector of the lateral *pnp* transistor. Thus, the lateral *pnp* transistor is integrated into the vertical *npn* transistor and does not exist as a separate component. The n^+-diffused regions are the multiple collectors. In terms of the standard double-diffused transistor, the vertical *npn* transistor here is operated upside down. Special care is necessary to obtain high current gain for the upside-down transistor.

The previous paragraph describes the I^2L circuit in its nonisolated form. It is possible to isolate the I^2L gate using junction or dielectric isolation. The isolated integrated injection logic allows all other standard bipolar and MOS circuits to be combined directly with the I^2L gates on the same chip.

6-7 BIPOLAR TRANSISTOR SCALING

To increase the functional density of a silicon chip, the size of a transistor has to be minimized. The dimensions and doping concentrations are usually changed proportionally for the down-sized device, and the technique is known as *scaling*. To make sure that the small transistor performs as well as its larger counterpart,

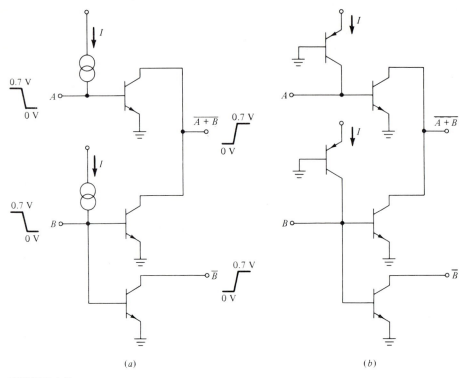

FIGURE 6-12
A direct-coupled transistor logic gate with (a) base-current sources and (b) *pnp* transistor base drive.

a few basic guidelines, known as *scaling rules*, are used. In general, there are three parameters under the designer's control: the lateral and vertical dimensions, the doping concentrations, and the bias voltages. A smaller bias voltage is desirable since it reduces the power consumption and the internal electric field that could cause junction breakdowns. But the bias voltages of a bipolar circuit are already near their lower limit at room temperature, and decreasing the voltages will not be practical. For this reason, the voltages are usually kept the same as devices are scaled down. This is known as *constant-voltge scaling*, and all the voltage-dependent elements are treated as constant.

The decrease of lateral dimensions is advantageous since it reduces the parasitic resistances and capacitances so that the switching speed is improved. However, this is limited by the resolution of lithography, and the small dimensions would impact on the fabrication yield and device reliability. It is also important to keep a small contact resistance, which requires a large contact area.

When the vertical dimension is reduced, the base width becomes narrower and the current density is increased. The base and collector doping concentrations must be increased to avoid punch-through and base stretching. Using Eq. (3-32),

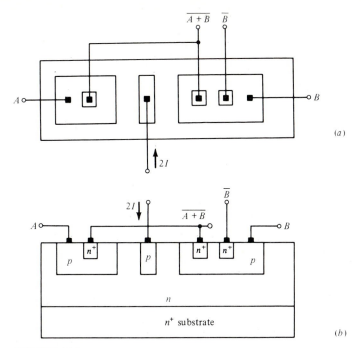

FIGURE 6-13
Implementation of the circuit shown in Fig. 6-12b in integrated form: the I^2L gate. (a) Top view and (b) cross-sectional view.

the punch-through voltage is

$$V_{PT} + \psi_o = \frac{q x_B^2 N_a}{2 K_s \varepsilon_o} \qquad (6\text{-}23)$$

If the punch-through voltage of the scaled transistor is the same, the base doping has to vary as

$$N_a' = \lambda^2 N_a \qquad (6\text{-}24)$$

since

$$x_B' = \frac{x_B}{\lambda} \qquad (6\text{-}25)$$

where the primed symbols represent the scaled device and λ is the scaling factor. To prevent base stretching, the collector doping N_C should be kept proportional to the current density. Since the critical doping density for zero field gradient is, using Eq. (6-14),

$$N_C = \frac{J}{q v_s} \qquad (6\text{-}26)$$

TABLE 6-1
Scaling rules

Parameters	Scaling factor
X_j, W, L	λ^{-1}
N	λ^2
J	λ^2
V	Constant
P	λ^2
Delay	λ^{-1}

the collector impurity concentration should be chosen accordingly. In most cases, the current density is limited by the power dissipation and the total current is specified by the logic fanout. A set of rules is listed in Table 6-1, and details are given in Ref. 3.

From simple calculation with Eq. (5-21), the current density should be scaled by λ^{-1} rather than λ^2 as shown in Table 6-1. However, the current must be increased by λ^2 if the time delay is to scale with λ^{-1} [3]. This is accomplished by slightly increasing the bias voltage across the emitter junction through circuit arrangement. Because of the exponential dependence of the current on the voltage, a slight increase in V_E will push the current up to the desired level.

6-8   SELF-ALIGNED STRUCTURES

For applications in very large scale integrated circuits, the packing density of bipolar transistors has to increase. The first thing to do is to make use of better lithography to achieve smaller feature size, a topic that has been discussed in bipolar scaling. In addition, the inactive areas of the chip, i.e., the spacing between devices and contacts, must be minimized. On the one hand, these inactive regions are necessary to allow for misalignment of masks and to provide electrical isolation. But on the other hand, the same regions give rise to parasitic capacitances and resistances that degrade the transistor performance as well as waste valuable real estate. By using the technique known as *self-alignment*, these unused regions can be reduced significantly. Self-alignment usually means that a contact is automatically aligned with the region for electrical connection, or the contact is used as an implantation mask to protect the area under its cover. In other words, a structure developed by one processing step is used for a subsequent step without using an additional mask, avoiding the problem of misalignment between the two steps.

There are a number of variations on the fabrication technology; two of these will be described here. The basic processing steps of a self-aligned bipolar transistor are shown in Fig. 6-14. After local oxidation, a boron-doped p^+ polysilicon is deposited by chemical vapor deposition (CVD), which is then covered by a CVD

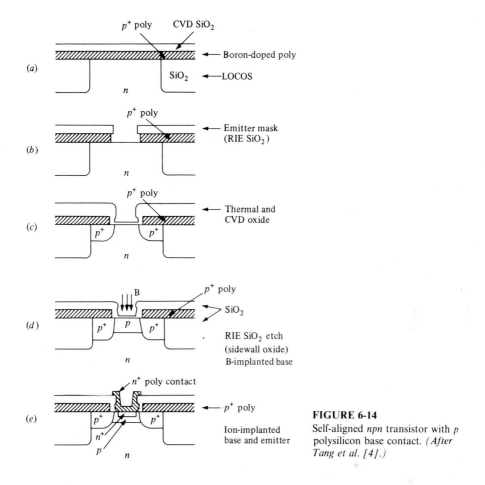

FIGURE 6-14
Self-aligned *npn* transistor with *p* polysilicon base contact. *(After Tang et al. [4].)*

oxide (Fig. 6-14a). Using the emitter mask, a photoresist step is taken to open the oxide and then to undercut the polysilicon by reactive ion etching (RIE) (Fig. 6-14b). A high-temperature thermal-oxidation step is employed to produce the structure with vertical sidewall oxide, as shown in Fig. 6-14c. At the same time, the extrinsic base (p^+) regions are formed by boron diffusion with the p^+ polysilicon as the diffusion source. The base region under the emitter, known as the *intrinsic base*, is produced by boron implantation (Fig. 6-14d). The self-alignment step essentially eliminates the extrinsic base resistance. With the oxide cover as shown in Fig. 6-14c, it avoids the large spacing between the base and emitter contacts which would have to be necessary for all metal contacts. The emitter is formed by first depositing an n^+ polysilicon as the emitter contact, which is also employed as the diffusion source to produce the n^+ emitter (Fig. 6-14e). Notice that the emitter is very shallow and separated from the p^+ base under the oxide vertical sidewall. This is a critical step, and the drive-in diffusion must be under strict control to avoid direct contact between the n^+ and p^+ regions. When these two are joined, they form a p^+n^+ tunneling junction, introducing large leakage current and capacitance as well as low breakdown voltage for the emitter junction. The polysilicon emitter contact has two other advantages. As a diffusion source, it eliminates the defect generation and the anomalous diffusion effects associated with high concentration of phosphorus diffusion or implantation-induced dislocation. As an emitter contact, it increases the current gain h_{FE} by a factor of 3 or more. The physics of the polysilicon emitter will be discussed in the next section.

An alternative method of self-alignment is to make use of the concentration-dependent oxidation to selectively oxidize the polysilicon emitter which is heavily doped with arsenic. Experimentally, the oxidation rate of arsenic-doped polysilicon is found to be 3 to 4 times higher than that of silicon [5]. This effect is utilized to produce the self-aligned structure shown in Fig. 6-15. After the arsenic implantation in Fig. 6-15a, the polysilicon is etched selectively to form the emitter contact. The structure is then oxidized, resulting in a thick oxide on the polysilicon and a thin oxide over the silicon, as shown in Fig. 6-15b. Now, a high-energy boron implantation is made to produce the p^+ layer. After a high-temperature diffusion step, the emitter and the extrinsic base are developed as shown in Fig. 6-15c.

Although self-aligned structures are preferred over the traditional methods, they increase the process steps and, more often, the smaller tolerances make them difficult to be used in production. These techniques are attractive, but one must not be overly aggressive in using them.

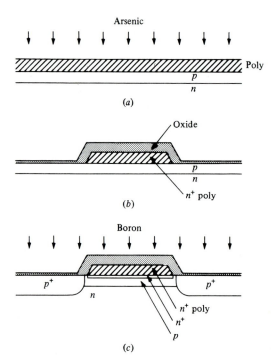

FIGURE 6-15
Schematics outlining processing steps for fabricating self-aligned transistors with polysilicon-contacted emitters.   *(After Cuthbertson and Ashburn [5].)*

6-9 POLYSILICON EMITTER CONTACT

To achieve a high current gain, the base of a transistor is usually made very thin and the emitter is heavily doped. Typically, the base Gummel number is 10^{12} cm^{-2}, and the emitter Gummel number is about 10^{16} cm^{-2}. The theoretical limit of the current gain should therefore be 10^4. Such a high current gain, however, has never been realized because the heavy emitter doping tends to reduce the band gap and increase the recombination rate, both of which degrade the emitter injection efficiency. As a result, continued increases of the emitter Gummel number become ineffective in boosting the current gain. Besides, the high emitter doping has the undesirable consequence of reducing the breakdown voltage and increasing the junction capacitance.

In the last few years, researchers have been exploring a different way to arrive at a high current gain. It came as a surprise when an arsenic-doped polysilicon layer was inserted between the doped emitter and the metal contact and the current gain increased by more than a factor of 3. In this section, the influence of the emitter contact to the transistor current gain is first presented to show the unusual performance of the polysilicon emitter contact. Subsequently, the physics of the polysilicon-silicon interface is discussed based on the latest experimental data. Understanding of the polysilicon emitter, as it turns out, requires collaborative data from material, processing, and electrical measurements. It is a good example of how modern device research is related to material science and processing to uncover the basic mechanism of an unusual transistor.

Three different methods of contact formation for the emitter are shown in Fig. 6-16. The silicon transistors are fabricated with a base width of 100 nm and an emitter depth of 200 nm using boron and arsenic implantation. The aluminum contact was deposited by vacuum evaporation. For the silicide contact, 50 nm of Pd was deposited which consumed 25 nm of silicon, forming the Pd$_2$Si layer. The n^+ polysilicon was produced by low-pressure chemical vapor deposition (LPCVD) at 650°C. The current gain as a function of the collector current is shown in Fig. 6-17. With the aluminum contact, the minority-carrier concentration at the Al–Si interface is essentially zero because of the large recombination velocity there. Assuming negligible recombination inside the emitter, the injected hole current is proportional to $p_E(0)/x_E$, where $p_E(0)$ represents the hole concentration at the emitter junction. In the case of the silicide contact, the single-crystal emitter is narrower since part of the silicon has been converted to silicide. The effective emitter depth is 175 nm instead of 200 nm. The emitter hole current, which is inversely proportional to the emitter depth, increases, giving rise to a decrease in the current gain in agreement with the experimental data. The enhancement of current gain with the polysilicon emitter is more complicated and is explained in the following paragraphs.

A key to the understanding of the polysilicon emitter comes from the impurity profile of experimental devices measured by a technique known as secondary ion mass spectroscopy (SIMS). This method can measure the impurity concentration down to 10^{17} cm^{-3}, and a typical profile of the polysilicon emitter region is shown in

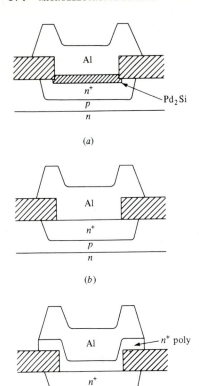

(a)

(b)

(c)

FIGURE 6-16
Schematics illustrating the three types of emitter contacts used in this study: (a) $Pd_2Si + Al$, (b) Al, (c) n^+ polysilicon $+ Al$. (*After Ning and Isaac [6].*)

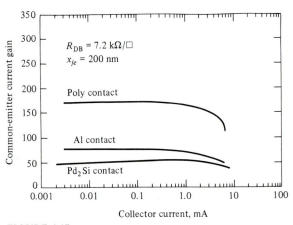

FIGURE 6-17
Current gain as a function of collector current for an emitter 200 nm deep contacted by either Pd_2Si, Al, or 100 nm polysilicon. (*After Ning and Isaac [6].*)

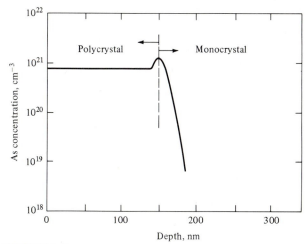

FIGURE 6-18
Experimental profile of arsenic concentration in a polysilicon-silicon emitter contact.

Fig. 6-18. At the polysilicon-silicon interface, arsenic atoms segregate to produce a sharp peak giving rise to a potential barrier for the holes (Fig. 6-19). The barrier is defined as $E_f - E_v = q\phi_B$ and is given by Ref. 7 as

$$e^{-q\phi_B/kT} = \frac{n_i^2}{N_v N_{d,\,max}} \tag{6-27}$$

where $N_{d,\,max}$ is the peak doping concentration. Under forward bias, the injected holes from the emitter-base junction diffuse toward the metal contact. When the holes arrive at the polysilicon-silicon interface, they must overcome the potential

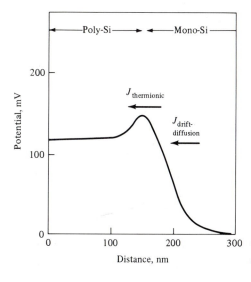

FIGURE 6-19
Typical potential distribution in the polysilicon emitter. (*After Ng and Yang [7].*)

barrier to reach the polysilicon side. Figure 6-20 illustrates the effect of the potential barrier on hole distribution. Without the polysilicon emitter, the metal contact drains all the holes rapidly so that the hole concentration is zero at the contact interface (solid line). With the polysilicon emitter, only those holes with energy greater than the potential barrier can reach the contact. Thus, the hole concentration at the interface is not zero, and it increases with the potential-barrier height. The blocking action of the barrier decreases the slope of the hole concentration, as indicated in Fig. 6-20. Consequently, the hole current is reduced and the current gain is increased.

Since only those holes with high-enough thermal energy can overcome the barrier, the current crossing the interface is called the *thermionic-emission current*. The physical principle of thermionic emission will be discussed in the next chapter. We shall make use of the result and write the hole thermionic-emission current as

$$I_{th} = A A^{**} T^2 e^{-\phi_B/\phi_T} (e^{V_E/\phi_T} - 1) \tag{6-28}$$

where A^{**} is the Richardson's constant. The diffusion current inside the crystalline

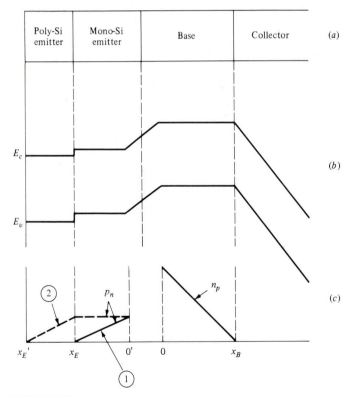

FIGURE 6-20
The polysilicon-emitter bipolar transistor: (a) structure, (b) energy-band diagram under normal biases, (c) minority-carrier distribution ① without polysilicon contact layer and ② with polysilicon contact layer.

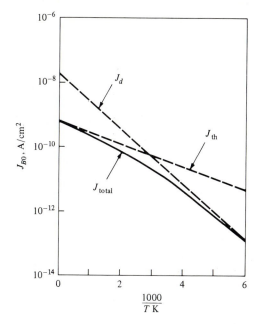

FIGURE 6-21
Nonlinear temperature dependence of base reverse-saturation current. *(After Ng and Yang [7].)*

emitter is given by Eq. (6-8). Since the two currents are in series, the actual minority-carrier current is

$$\frac{1}{I} = \frac{1}{I_{th}} + \frac{1}{I_p} \tag{6-29}$$

The smaller current of the two will therefore limit the current flow. For example, if the thermionic-emission current is very large compared with the diffusion current, the limiting process is carrier diffusion and the polysilicon emitter has negligible effect on current gain. On the other hand, if the thermionic-emission current is the smaller one, it limits the hole current and thus produces a **higher current** gain. Experimental results seem to confirm this finding. In particular, the temperature dependence of the current gain shown in Fig. 6-21 follows the theory proposed here.

REFERENCES

1. Sze, S. M. (ed.): "VLSI Technology," McGraw-Hill, New York, 1983.
2. Ghandhi, S. K.: "Semiconductor Power Devices," Wiley, New York, 1977.
3. Solomon, P. M., and D. D. Tang: Bipolar Circuit Scaling, *ISSCC Digest*, 86 (1979).
4. Tang, D. D., et al.: Subnanosecond Self-Aligned I²L/MTL Circuits, *IEEE J. Solid-State Circuits*, **SC-15**:444 (1980)
5. Cuthbertson, A., and P. Ashburn: Self-Aligned Transistors with Polysilicon Emitter for Bipolar VLSI, *IEEE Trans. Electron. Devices*, **ED-32**:242 (1985).
6. Ning, T. H., and R. D. Isaac: Effect of Emitter Contact on Current Gain of Silicon Bipolar Devices, *IEEE Trans. Electron. Devices*, **ED-27**:2051 (1980).
7. Ng, C. C., and E. S. Yang: A Thermionic-Diffusion Model of the Polysilicon Emitter, *Proc. IEDM*, 36 (1986).

PROBLEMS

6-1. An n-type silicon substrate of $0.5\,\Omega\cdot$cm is first subjected to a boron diffusion with constant surface concentration of $5 \times 10^{18}\,\mathrm{cm}^{-3}$ and a junction depth of $1.5\,\mu$m, and then phosphorus impurities are added to a surface concentration of $10^{21}\,\mathrm{cm}^{-3}$ with a new junction depth of $1.0\,\mu$m

    (a) Plot impurity concentrations (log scale) vs. distance from surface (linear scale) for the completed structure. Assume that the phosphorus diffusion does not affect the impurity profile of boron.

    (b) Indicate emitter, base, and collector on this drawing.

    (c) If the phosphorus diffusion is conducted at 1100°C, how long should be allowed?

    (d) Estimate the new position of the base-collector junction by assuming a constant boron surface concentration during the emitter diffusion.

6-2. Show that Eq. (6-7) reduces to Eq. (5-27) for a uniformly doped base.

6-3. Derive an expression of the base transport factor for a base with exponential distribution; that is, $N_a = N_o e^{-ax/x_B}$.

6-4. Derive the electron distribution in the base for the transistor of Prob. 6-3 under the normal active mode of operation. Plot the electron concentration for $a = 1$ and 10. Explain the physical meaning of the difference between these two cases.

6-5. (a) Show that the punch-through voltage of a planar epitaxial double-diffused transistor is given by

$$BV = \frac{qG_B}{2K_s\varepsilon_0}\left(x_B + \frac{G_B}{N_{dC}}\right)$$

    (b) Find the punch-through voltage of the transistor in Prob. 5-2, with $x_B = 0.5\,\mu$m and $x_C = 20\,\mu$m

6-6. For the transistor described in Prob. 6-1, calculate (a) the base Gummel number, (b) the emitter Gummel number, and (c) the emitter injection efficiency.

6-7. Consider an npn transistor having an emitter Gummel number of $2 \times 10^{16}\,\mathrm{cm}^{-2}$ and $x_E = 1\,\mu$m, and the base doping is controlled by boron implantation with a base width of $0.5\,\mu$m. The epitaxial collector is doped with $10^{16}\,\mathrm{cm}^{-3}$. Calculate the emitter injection efficiency and the punch-through voltage (use formula in Prob. 6-5a) and plot them as a function of the base Gummel number.

6-8. (a) Use the relationship between the electric field and potential to show that $\mathscr{E}(x_c) = -2V_C/x_C$ is the correct boundary condition to derive Eq. (6-15).

    (b) Estimate the doping level in the 2-μm n-epitaxial collector needed to prevent base stretching at $I_C = 2\,$mA. Assume that the emitter area is $2 \times 4\,\mu$m^2 and the collector voltage is 5 V.

6-9. Using the transistor of Fig. 6-5 with a buried layer, draw a monolithic layout of the circuit shown in Fig. P6-9. Draw a complete set of (individual) masks. Darken areas which are not to pass light. (Tracing from your layout suggested.) Use rectangular coordinate paper with 20 lines to the inch and a scale of $\frac{1}{20}$ in per 0.1 mil. You may use any necessary chip size, but about 20 square mils is reasonable. In layout use green for isolation-diffusion-mask boundaries, red for base, blue for emitter, and black for contact windows and metal.

6-10. For the integrated transistor of Fig. 6-5 with an impurity profile shown in Fig. 6-2 (but changing it to a p substrate of $N_a = 10^{15}\,\mathrm{cm}^{-3}$ and an n-type collector of $N_d = 1.2 \times 10^{16}\,\mathrm{cm}^{-3}$), estimate the collector-base and collector-substrate capa-

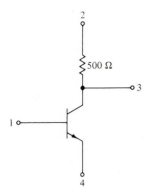

FIGURE P6-9.

citances at 5 V. Also find the collector series resistance and breakdown voltages of the collector, emitter, and substrate junctions. Let $h_{FE} = 100$, $N_B = 10^{17}\,\text{cm}^{-3}$, and the collector depth be 1 μm.

6-11. Calculate the cutoff frequency of the transistor of Prob. 6-10 at $I_E = 1\,\text{mA}$ and $V_C = -1\,\text{V}$.

6-12. (a) In integrated circuits, a transistor may be connected as a diode. Sketch five possible diode configurations along with the minority-carrier distribution in the base for $V \gg \phi_T$.

(b) Use the doping profile and carrier distribution to compare the diode performance in speed, conductance, and breakdown voltage if the impurity profile of the double-diffused transistor is assumed.

6-13. Derive Eqs. (6-20) and (6-21).

6-14. (a) The following parameters are given for a lateral transistor: the epitaxial layer is 20 μm, the junction depth of the p diffusion is 3 μm, $W_B = 3\,\mu$m, $W_E = 1.5$ mils, and $L_p = 15\,\mu$m. Calculate h_{FE}.

(b) Repeat (a) for $W_B = 10\,\mu$m.

6-15. (a) Given $N_d = 10^{18}\,\text{cm}^{-3}$ at $x = 0'$ (see Fig. 6-20c) and $10^{20}\,\text{cm}^{-3}$ at $x = x_E$, calculate the built-in potential $q\psi_o$ between $x = 0'$ and $x = x_E$ using the Boltzmann statistics.

(b) Repeat part (a) if $N_d = 10^{19}\,\text{cm}^{-3}$ at $x = 0'$.

(c) Calculate the ratio of the thermionic-emission current at $x = x_E$ between (a) and (b).

CHAPTER

7

METAL-
SEMICONDUCTOR
JUNCTIONS

Historically, the first practical semiconductor device was the metal-semiconductor diode. In the early development of electronics, rectifiers were made by pressing a metallic whisker to a semiconductor crystal to form a contact. Since the characteristics of point-contact diodes were not reproducible, in most cases they were replaced by the *pn* junction diode in the 1950s. Recently, modern semiconductor and vacuum technology has been employed to fabricate reproducible metal-semiconductor contacts so that it is now possible to obtain both rectifying and nonrectifying metal-semiconductor (M-S) junctions. The nonrectifying junction has a low ohmic drop regardless of the polarity of the externally applied voltage and is called the *ohmic contact*. All semiconductor devices need ohmic contacts to make connections to other devices or circuit elements. The rectifying junction is commonly known as the *Schottky-barrier* diode. In this chapter, the energy-band diagram and the rectifying properties of the Schottky barrier are presented. In addition, the current-voltage characteristics are given, and the application of the M-S contact to some device structures is discussed.

7-1 THE IDEAL SCHOTTKY BARRIER

The ideal energy-band diagram for a metal and an *n*-type semiconductor before making contact is shown in Fig. 7-1*a*, with $\phi_m > \phi_s$, where $q\phi_m$ and $q\phi_s$ represent the *work functions* of the metal and the semiconductor, respectively. As shown in

180

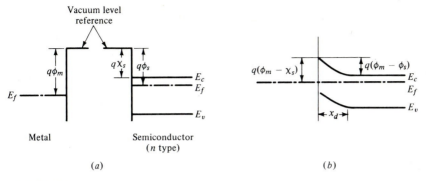

FIGURE 7-1

Energy-band diagram of a metal-semiconductor contact with $\phi_m > \phi_s$: (a) before contact and (b) after contact and at thermal equilibrium.

the diagram, the work function is defined as the work required to bring an electron from the Fermi level of the material to the vacuum level. The symbol χ_s, called the *electron affinity*, specifies the energy required to release an electron from the bottom of the conduction band to the vacuum level. Since we have assumed that ϕ_m is greater than ϕ_s, the Fermi level in the semiconductor is higher than that of the metal. When the metal and the semiconductor are joined, electrons from the semiconductor, having higher energy, cross over to the metal until the Fermi level of the M-S system is aligned. This is the condition of thermal equilibrium depicted in Fig. 7-1b, which corresponds to a lowering of the semiconductor's Fermi level by $q(\phi_m - \phi_s)$. The migration of the electrons from the semiconductor to the metal leaves behind ionized donors as fixed positive charges creating a depletion layer similar to that of the one-sided step junction. The depletion of electrons in the vicinity of the interface produces an upward band bending near the surface so that a built-in potential ψ_o is established. The built-in voltage thus accounts for the difference between the original Fermi levels so that we have

$$\psi_o = \phi_m - \phi_s \tag{7-1}$$

This potential ψ_o is supported by a space-charge layer with a width x_d, as depicted in Fig. 7-1b. The barrier for electrons to flow from the metal to the semiconductor is given by

$$q\phi_b = q(\phi_m - \chi_s) \tag{7-2}$$

where ϕ_b is called the barrier height of the M-S contact. The barrier height given by Eq. (7-2) is a basic assumption of the physical model proposed by Schottky in 1938, and the rectifying M-S contact is known as the *Schottky barrier*.

If we apply a negative voltage V to the semiconductor with respect to the metal, the semiconductor-to-metal potential is reduced to $\psi_o - V$ while ϕ_b remains unchanged (Fig. 7-2b). The reduction of the barrier potential on the semiconductor side makes it easier for electrons in the semiconductor to move to the metal. This is the forward-bias condition, which enables a large current to flow. On the other

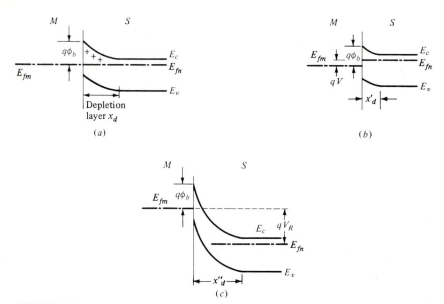

FIGURE 7-2
Energy-band diagram of a Schottky barrier: (a) without bias, (b) with forward bias, and (c) with reverse bias, where $x'_d < x_d < x''_d$.

hand, if a positive voltage is applied to the semiconductor, the potential barrier is raised to prevent current conduction, and the reverse bias is achieved (Fig. 7-2c). With uniform doping in the semiconductor, the space-charge-layer width of the Schottky barrier is identical to that of a one-sided step *pn* junction:

$$x_d = \left[\frac{2K_s \varepsilon_o (\psi_o + V_R)}{q N_d} \right]^{1/2} \tag{7-3}$$

where N_d is the doping level in the semiconductor and V_R is the reverse-bias voltage. Therefore, the capacitance of the junction is given by

$$C = \frac{K_s \varepsilon_o A}{x_d} = \left[\frac{q K_s \varepsilon_o N_d}{2(\psi_o + V_R)} \right]^{1/2} A \tag{7-4}$$

which can be rewritten as

$$\frac{1}{C^2} = \frac{2}{q K_s \varepsilon_o N_d A^2} (\psi_o + V_R) \tag{7-5}$$

As for a *pn* junction, we can plot $1/C^2$ vs. V_R to obtain a straight-line relationship (Fig. 7-3). The built-in potential and the semiconductor doping can be calculated from the slope and intercept of such a capacitance-voltage plot. Once ψ_o is determined, the barrier height ϕ_b can be calculated using

$$\phi_b = \psi_o + V_n \tag{7-6}$$

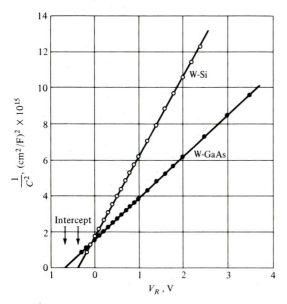

FIGURE 7-3

$1/C^2$ vs. applied voltage for tungsten-silicon and tungsten-gallium arsenide diodes. *(From Sze [1].)*

where V_n is the potential difference between the Fermi level and E_c; its value can be deduced from the impurity concentration by using Eq. (1-14) and setting $n = N_d$.

Example. Calculate the donor concentration, built-in potential, and barrier height of the silicon Schottky diode from Fig. 7-3.

Solution. Using Eq. (7-5), we write

$$N_d = \frac{2}{qK_s\varepsilon_o A^2}\frac{d(\psi_o + V_R)}{d(1/C^2)} = \frac{2}{qK_s\varepsilon_o A^2}\frac{\Delta V_R}{\Delta(1/C^2)}$$

In Fig. 7-3, the capacitance is for a unit area so that $A = 1$. We find $1/C^2 = 6 \times 10^{15}$ at $V_R = 1$ V and $1/C^2 = 10.6 \times 10^{15}$ at $V_R = 2$ V. Thus,

$$\frac{\Delta V_R}{\Delta(1/C^2)} = \frac{1}{4.6 \times 10^{15}} = 2.17 \times 10^{-16} \text{ V} \cdot \text{F}^2/\text{cm}^4$$

and

$$N_d = \frac{(2)(2.17 \times 10^{-16})}{(1.6 \times 10^{-19})(11.8)(8.84 \times 10^{-14})} = 2.6 \times 10^{15} \text{ cm}^{-3}$$

Using Eq. (1-14) and the table on the back end paper, we find

$$V_n = \phi_T \ln\frac{N_c}{N_d} = 0.026 \ln\frac{2.8 \times 10^{19}}{2.6 \times 10^{15}} = 0.24 \text{ V}$$

Since $\psi_o = 0.4$ V from Fig. 7-3, we obtain

$$\phi_b = \psi_o + V_n = 0.4 + 0.24 = 0.64 \text{ V}$$

The ideal model presented describes the Schottky-barrier characteristics well except that the barrier height ϕ_b obtained from the $C\text{-}V$ measurement is not necessarily the difference of the work functions. The difficulty is usually attributed to charge states at the interface known as *Bardeen states* or *metal-induced gap states* (MIGS), a subject under intense study and not yet fully understood [2].

7-2 CURRENT-VOLTAGE CHARACTERISTIC

The current flow in the M-S barrier is governed by the carrier transport from the edge of the space-charge layer to the metal. In Fig. 7-2b, the forward bias reduces the electric field and the potential barrier in the depletion layer. As a result, electrons diffuse through the depletion layer with a velocity v_D and are emitted from the semiconductor to the metal with a velocity v_E. If electrons are emitted rapidly into the metal, then $v_E \gg v_D$ and the current is controlled by the diffusion of carriers across the space-charge layer. On the other hand, if $v_E \ll v_D$, the current is governed by the emission process near the M-S interface. It turns out that at room temperature in most practical Schottky-barrier diodes, diffusion is much faster than emission so the current transport mechanism is limited by the emission process [3, 4]. The diffusion effect of carriers across the space-charge layer will be neglected in our discussion.

When the electrons are emitted from the semiconductor to the metal, they have an energy greater than that of the metal electrons by approximately $q\phi_b$. Before they collide in the metal to give up this extra energy, these electrons are considered to be *hot* because their equivalent temperature is higher than that of the electrons in the metal. For this reason the Schottky-barrier diode is sometimes called the *hot-carrier diode*. These hot carriers reach equilibrium with the metal electrons in a very short time, typically less than 0.1 ns.

To understand the emission process, let us consider a tungsten filament inside a high-vacuum chamber. The potential barrier at the metal-vacuum interface is the metal work function $q\phi_m$, as shown in Fig. 7-4a. The Fermi-Dirac distribution function $f(E)$ and the density-of-state function $N(E)$ are also shown in Fig. 7-4b and c. The distribution of electrons in energy is given by the product $f(E)N(E)$. If this product yields electrons in energy equal to or greater than the vacuum level $E_f + q\phi_m$, electrons are emitted. If not, there is no electron emission. At room temperature, the Fermi-Dirac function is sketched as the dashed line so that the product $f(E)N(E)$ (also dashed) is clearly below the vacuum level. Physically, electrons in the metal cannot go beyond the metal surface because of the confinement of the surface-potential barrier. Consequently, no electrons are emitted. If a current is passed through the filament to heat it up sufficiently, the Fermi-Dirac function will be seen as the solid curve. The $f(E)N(E)$ product will give rise to a "tail" reaching beyond the vacuum level, as indicated by the shaded area in Fig. 7-4d. Those electrons in the tail have a thermal energy high enough to escape into

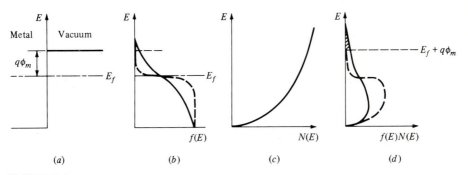

FIGURE 7-4
Graphical method of determining the electron density with energy greater than $E_f + q\phi_m$. The dashed lines represent room temperature conditions, and the solid curves in (b) and (d) are for high temperatures.

the vacuum. This phenomenon of electrons escaping from a hot surface is called *thermionic emission*. As shown in Fig. 7-4, the number of electrons occupying the energy level between E and $E + dE$ is

$$dn = N(E)f(E) \, dE \qquad (7\text{-}7)$$

where $N(E)$ is given by Eq. (1-8) with E_c set to zero and $f(E)$ is given by Eq. (1-12). Since only electrons with an energy greater than $E_f + q\phi_m$ and having a velocity component normal to the surface can escape the solid,[1] the thermionic current is

$$I = A \int q v_x \, dn$$

$$= \int_{E_f + q\phi_m}^{\infty} \frac{A4\pi q(2m)^{3/2}}{h^3} v_x E^{1/2} e^{-(E - E_f)/kT} \, dE \qquad (7\text{-}8)$$

where v_x is the component of velocity normal to the surface of the metal. To integrate Eq. (7-8), we must use the relation $p_x = mv_x$ and Eq. (1-5) to convert the energy into the momentum variable. In addition, the momentum components must be changed to rectangular coordinates. After integration, we obtain the Richardson-Dushman equation for the thermionic current (see Appendix D):

$$I_o = AA^*T^2 \exp\!\left(\frac{-q\phi_m}{kT}\right) \qquad (7\text{-}9)$$

where $A^* = 4\pi q m k^2/h^3 = 120 \text{ A/K}^2 \cdot \text{cm}^2 = $ Richardson constant
$A = $ junction area
$m = $ free-electron mass

---

[1] If an electron has an energy of $E_f + q\phi_m$ but is traveling, for instance, in a direction of $45°$ from normal to the surface, it cannot escape into the vacuum.

In analogy with the metal-vacuum system, electrons in the semiconductor of the M-S junction shown in Fig. 7-2a may overcome the potential barrier $q\psi_o$ to reach the metal and produce a current flow. The equation governing this current component can be derived and shown to be identical to Eq. (7-9) except that ϕ_m is replaced by ϕ_b and m is replaced by m_e, the effective electron mass. Under the thermal-equilibrium condition, however, equal numbers of electrons in the metal and semiconductor surmount the potential barrier to reach the other side, so that the net current flow across the M-S junction is zero.

When a forward-bias voltage is applied, the potential barrier from the semiconductor to metal is reduced to $\psi_o - V$. The reduced potential barrier favors the electron transport from the semiconductor to the metal. By substituting $\phi_m = \phi_b - V$ into Eq. (7-9), we obtain the forward current as

$$I_F = AA^{**}T^2 \exp\left[\frac{-q(\phi_b - V)}{kT}\right] = I_o e^{V/\phi_T} \tag{7-10}$$

where

$$I_o = AA^{**}T^2 \exp\left(\frac{-\phi_b}{\phi_T}\right) \tag{7-11}$$

In addition $\phi_T = kT/q$, and A^{**} is the effective Richardson constant to indicate the use of the effective mass of electrons.

When a reverse-bias voltage V_R is applied, the potential barrier for electrons from the semiconductor to the metal is increased to $\psi_o + V_R$ but the potential barrier from metal to semiconductor remains unchanged, as shown in Fig. 7-2c. Under this condition, the current is given by electron transport from the metal to the semiconductor which constitutes the reverse current. The magnitude of the reverse current is obtained by setting a zero voltage in Eq. (7-10), yielding I_o, the reverse saturation current. Thus, the current-voltage relationship of the M-S junction for both forward and reverse bias is given by

$$I = I_o(e^{V/n\phi_T} - 1) \tag{7-12}$$

where n is known as the ideality factor, which is introduced by nonideal effects. For an ideal Schottky diode, we have $n = 1$. Experimental I-V characteristics of two Schottky diodes are shown in Fig. 7-5. By extrapolating the forward I-V curve to $V = 0$, we can find the parameter I_o which, along with Eq. (7-11), can be used to obtain the barrier height. The ideality factor is calculated by using the slope of the semilog plot, giving $n = 1.02$ for the silicon diode and $n = 1.04$ for the GaAs diode.

Example. The following parameters of a Schottky diode are given: $\phi_m = 4.7$ V, $\chi_s = 4.0$ V, $N_c = 10^{19}$ cm^{-3}, $N_d = 10^{16}$ cm^{-3}, and $K_s = 12$. Assume that the density of interface states is negligible. At 300 K calculate (a) the barrier height, built-in potential, and depletion-layer width at zero bias and (b) the thermionic-emisson current at a forward bias of 0.2 V.

Solution

(a)
$$\phi_b = \phi_m - \chi_s = 4.7 - 4.0 = 0.7 \text{ V}$$

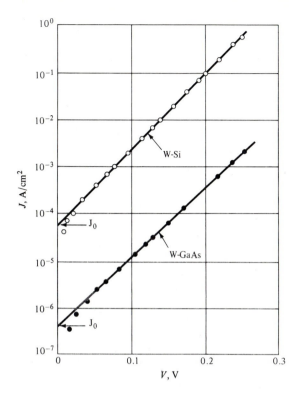

FIGURE 7-5
Forward current density vs. voltage of W–Si and W–GaAs Schottky diodes. *(After Sze [1].)*

Use Eq. (1-14) and rewrite it as

$$n = N_d = N_c e^{-(E_c - E_f)/kT} = N_c e^{-V_n/\phi_T}$$

Therefore

$$V_n = \phi_T \ln \frac{N_c}{N_d} = 0.17 \text{ V}$$

Thus

$$\psi_o = 0.7 - 0.17 = 0.53 \text{ V}$$

and

$$x_d = \left(\frac{2K_s \varepsilon_o \psi_o}{qN_d}\right)^{1/2} = 2.6 \times 10^{-5} \text{ cm}$$

(b)

$$J = A^{**}T^2 e^{-\phi_b/\phi_T}(e^{V/\phi_T} - 1)$$

$$= 1.2 \times 10^2 (300)^2 e^{-0.7/0.026}(e^{0.2/0.026} - 1)$$

$$= 4 \times 10^{-2} \text{ A/cm}^2$$

where A^{**} is taken as $120 \text{ A/cm}^2 \cdot \text{K}^2$.

The thermionic-emission current describes the majority-carrier (electron) current across the Schottky barrier. In addition to the majority-carrier current, a minority-carrier current exists as a result of hole injection from the metal to the semiconductor. The hole injection is the same as in a *pn* junction; the current is given by Eq. (4-18) and is rewritten as

$$I_p = I_{po}(e^{V/\phi_T} - 1) \tag{7-13}$$

with

$$I_{po} = \frac{qAD_pN_vN_c}{N_dL_p} \exp\left(\frac{-E_g}{kT}\right) \tag{7-14}$$

where Eqs. (1-19) and (1-21) have been employed. We may note that the minority-carrier current expressed in Eq. (7-13) has the same form as the majority-carrier current in Eq. (7-12). Equation (7-14) is written in a form permitting the magnitudes of I_{po} and I_o to be compared conveniently. In a covalent semiconductor such as Si, the junction barrier potential ϕ_b is always smaller than the band-gap energy E_g. As a result, the thermionic-emission current is usually much larger than the minority-carrier current. As a numerical example consider an aluminum-on-nSi diode with $N_d = 10^{16}$ cm^{-3}, $q\phi_b = 0.69$ eV, and $L_p = 10$ μm At room temperature we find

$$I_o = 3 \times 10^{-5} A \quad A \quad \text{and} \quad I_{po} = 3 \times 10^{-11} A \quad A \tag{7-15}$$

From this comparison, it is seen that the majority-carrier current is about 6 orders of magnitude greater than the minority-carrier current. Thus, the minority-carrier current can be neglected in most cases for Schottky barriers.

The reverse current, according to Eq. (7-11), should be a constant and equal to I_o. However, the top of the barrier at $x = 0$ in Fig. 7-2a will be slightly rounded off because of the image-force lowering (see next section). In addition, carrier generation is also taking place in the space-charge region. As a result, the reverse current will be seen as slightly voltage-dependent. For a very high reverse bias, avalanche breakdown will occur just as it does in a one-sided step pn junction. As in a pn junction, the avalanche-breakdown voltage increases with semiconductor resistivity.

7-3   IMAGE-FORCE LOWERING OF THE BARRIER HEIGHT

In this section, we consider the effects of the image charge and the application of an electric field to the metal surface at a metal-vacuum interface. The result is then used for an M-S junction.

An electron with a charge $-q$ located at a distance x from the metal surface establishes electric field lines, as shown in Fig. 7-6a. All the field lines must be normal to the surface since the metal is a good conductor. These field lines act as if an image charge $+q$ were situated at $-x$ (Fig. 7-6a). This means that the force on the electron at x is the same if the metal surface were replaced by $+q$ at $-x$. By Coulomb's force of attraction, the force on the electron is

$$F = -q\mathscr{E} = -\frac{q^2}{4\pi\varepsilon_o(2x)^2} \tag{7-16}$$

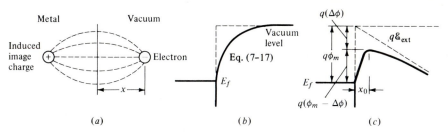

FIGURE 7-6
Image-force lowering of a metal-vacuum barrier.

Since $\mathscr{E} = -d\psi/dx$ and the potential energy is zero at $x = \infty$, we obtain from Eq. (7-16)

$$-\psi(x) = \frac{1}{q}\int_x^\infty F\,dx = -\frac{q}{16\pi\varepsilon_o x} \tag{7-17}$$

If there is no external electric field applied to the emitting surface, the only contribution to ψ is the image force expressed in Eq. (7-17), and its effect is illustrated in Fig. 7-6b. When an external field $\mathscr{E}_{\text{ext}}$ is applied, the potential becomes

$$-\psi(x) = -\frac{q}{16\pi\varepsilon_o x} - \mathscr{E}_{\text{ext}}x \tag{7-18}$$

This potential energy for an electron, $-q\psi(x)$, is plotted in Fig. 7-6c. It can be shown that the maximum potential barrier is located at (Prob. 7-7):

$$x_o = \left(\frac{q}{16\pi\varepsilon_o\mathscr{E}_{\text{ext}}}\right)^{1/2} \tag{7-19}$$

and

$$\Delta\phi = \left(\frac{q\mathscr{E}_{\text{ext}}}{4\pi\varepsilon_o}\right)^{1/2} \tag{7-20}$$

where $q\,\Delta\phi$ is the amount by which the work function $q\phi_m$ is lowered when an external field is applied to the surface. The reduction of the surface work function gives rise to a higher thermionic-emission current. In other words, electron emission from the metal can be enhanced by increasing either the temperature of the metal or the external electric field normal to the surface, i.e., one by increasing the energy of the metal electrons and the other by lowering the potential barrier at the surface. For example, if $\mathscr{E}_{\text{ext}} = 104$ V/cm, we obtain $\Delta\phi = 0.039$ eV and $x_o = 19$ nm.

At an M-S junction, the effect of barrier-height lowering is comparable with that of the metal-vacuum surface when the dielectric constant K_s is included in the numerator of Eq. (7-20). Therefore, the saturation current I_o is modified to be

$$I_o = AA^{**}T^2\,\exp\left[\frac{-q(\phi_b - \Delta\phi)}{kT}\right] \tag{7-21}$$

It has been found experimentally that this equation is more accurate in describing the current-voltage characteristics of a Schottky diode, particularly in the reverse-bias direction.

7-4   METAL-INSULATOR-SEMICONDUCTOR SCHOTTKY DIODE

In practice, when a metal contact is evaporated onto a chemically prepared silicon surface, an interfacial oxide layer is likely to exist between the metal and semiconductor. The oxide layer is very thin, typically on the order of 0.5 to 1.5 nm. The energy-band diagram of such a metal-insulator-semiconductor (MIS) structure is shown in Fig. 7-7. The energy-band gap of the thick oxide is 9 eV, as seen in Fig. 1-10*b*, but its value for a thin oxide less than 3 nm is not known. For this reason, the diagram in Fig. 7-7 should be considered as a schematic representation. Under thermal equilibrium, there is a potential drop across the oxide layer so that the barrier height is modified. The current conduction in the MIS Schottky diode results from carrier tunneling through the oxide layer and is described by [5]

$$I = AA^{**}T^2 e^{-\chi^{(1/2)}\delta} e^{-q\phi_b/kT} e^{qV/nkT} \tag{7-22}$$

where χ is the mean barrier height, in electronvolts, from the conduction-band edge and δ is the oxide thickness, in nanometers. The product $\chi^{1/2}\delta$ is normalized so that it is dimensionless. The factor n in the last exponent results from the partial drop of the applied voltage across the oxide layer so that the voltage across the semiconductor is reduced. Note that Eq. (7-22) reduces to Eq. (7-10) for $\delta = 0$ and $n = 1$. In general, the thin oxide layer reduces the majority-carrier current but not the minority-carrier current if the same voltage is applied. This leads to an increase of the ratio of the minority-carrier current to majority-carrier current. It turns out that the increase in minority-carrier injection ratio is beneficial to device applications such as solar cells.

7-5   COMPARISON BETWEEN A SCHOTTKY-BARRIER AND A *pn* JUNCTION DIODE

As described in Sec. 7-2, the current in a Schottky barrier is carried by majority carriers, whereas in a *pn* junction it is by minority carriers. When a *pn* junction is switched from a forward bias to a reverse bias abruptly, the minority carriers cannot be removed instantaneously, and the switching speed is limited by this minority-carrier storage effect. In a Schottky barrier, the storage time is negligible because there is no minority-carrier storage. Therefore, the frequency response is limited by the *RC* time constant rather than the charge storage. For this reason, Schottky-barrier diodes are ideal for high-frequency and fast-switching applications.

The saturation current in a Schottky barrier is much higher than that of a *pn* junction diode for diodes of same area since the majority-carrier current is much higher than the minority-carrier current [Eq. (7-15)]. Consequently, the

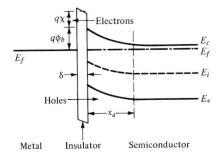

FIGURE 7-7
Energy-band diagram of MIS structure.

forward-voltage drop will be much less in a Schottky barrier than in a *pn* junction for the same current. Figure 7-8 shows the current-voltage characteristics of a Pd–*n*Si Schottky barrier and that of a *pn* diode. The typical cut-in voltage or turn-on voltage (the knee of the *I-V* curve) of a Schottky-barrier diode is 0.3 V and that of the Si *pn* junction is 0.6 V. The low cut-in voltage makes Schottky diodes attractive for clamping and clipping applications. Under reverse bias, however, the Schottky diode has a higher reverse current which does not saturate. Furthermore, additional leakage current and soft breakdown usually exist in the Schottky diode, and special care must be taken in device fabrication. The nonideal reverse characteristics can be eliminated by using the guard-ring or overlapped-metal structure to be discussed in a later section.

The temperature dependence of Schottky barriers and *pn* junctions under forward bias is different. Experimental results are plotted in Fig. 7-9, where a difference of 0.4 mV/°C in the temperature coefficients is observed. This difference should be taken into account in the design of circuits utilizing both types of diodes.

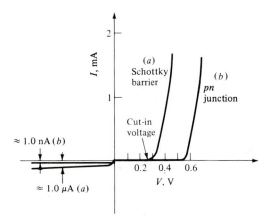

FIGURE 7-8
Current-voltage characteristics of a *pn* junction and a Schottky-barrier diode.

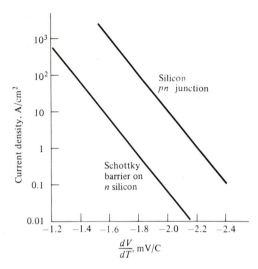

FIGURE 7-9

Temperature coefficient of forward voltage as a function of current density. *(After Yu [6].)*

7-6 OHMIC CONTACTS

An ohmic contact is defined as a junction that will not add a significant parasitic impedance to the structure on which it is used and will not sufficiently change the equilibrium-carrier densities within the semiconductor to affect the device characteristics. Ideally, the current passing through the contact should be a linear function of the applied voltage regardless of the voltage's polarity; i.e., the *I-V* characteristic should be a straight line passing through the origin. Unfortunately, the potential barrier at an M-S interface blocks current in one direction, giving rise to a nonlinear *I-V* relation. Let us define a *specific contact resistance* R_c by the slope of the *J-V* curve, *J* being the current density:

$$\frac{1}{R_c} \equiv \frac{\partial J}{\partial V}\bigg|_{V=0} \qquad (\Omega \cdot \text{cm}^2)^{-1} \qquad (7\text{-}23)$$

A good omhic contact should have a small specific contact resistance, typically $10^{-6}\,\Omega \cdot \text{cm}^2$ or less for micrometer-size devices. Using the foregoing definition and differentiating Eq. (7-10), we have

$$R_c = \frac{\phi_T}{A^{**}T^2}\, e^{\phi_b/\phi_T} \qquad (7\text{-}24)$$

Note that R_c is exponentially related to ϕ_b, so a small ϕ_b is desirable for a low contact resistance. A good ohmic contact is achieved with a platinum-on-*p*Si contact since the barrier height is 0.25 eV. On the other hand, the smallest barrier height for an M–*n*Si junction is greater than 0.5 eV, so R_c is about $0.1\,\Omega \cdot \text{cm}^2$ for lightly doped silicon, too high for any practical purpose. For this reason, most practical ohmic contacts make use of a heavily doped n^+ or p^+ surface layer, as explained below.

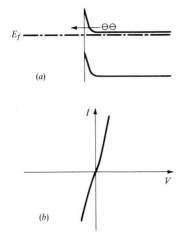

FIGURE 7-10
Energy-band diagram and current-voltage curve of a
metal-on-n^+ semiconductor contact.

When the semiconductor is heavily doped, with an impurity density of 10^{19} cm^{-3} or higher, the depletion layer of the junction becomes very thin so that carriers can tunnel through instead of going over the potential barrier. The energy-band diagram of such a junction is shown in Fig. 7-10a, where electrons in either side of the barrier may tunnel across to the other side. Thus, an essentially symmetrical I-V curve for both forward and reverse bias is realized, as shown in Fig. 7-10b. The barrier is nonrectifying, and a low contact resistance is achieved. If we make use of the tunneling probability expressed in Eq. (4-43) and assume the tunneling barrier to be ϕ_b, then the specific contact resistance is given by (Prob. 7-16)

$$R_c \sim \exp\left(\frac{4\pi\sqrt{K_s\varepsilon_o m_e}}{h} \frac{\phi_b}{\sqrt{N_d}}\right) \qquad (7\text{-}25)$$

Equation (7-25) and experimental data are plotted in Fig. 7-11. While the Schottky-barrier height is not adjustable, the doping level can be controlled. An impurity density of 10^{20} cm^{-3} reduces the contact resistance to less than 10^{-6} $\Omega \cdot$ cm^2 with little influence by the barrier height. Obviously, the best combination is a low barrier height and a high doping concentration for a good ohmic contact.

7-7 SILICIDE CONTACTS AND INTERCONNECTS [8]

In most integrated circuits, aluminum is the metal used for contacts and interconnects. Aluminum is also the material used for the metal gate of earlier families of MOS devices as well as the Schottky barrier for bipolar circuits. Unfortunately, the aluminum-silicon system has a low eutectic temperature (577°C), and the interface atoms tend to interdiffuse at an even lower temperature (400°C). This atomic migration produces a large leakage current in shallow junction devices where contacts are only 0.1 to 0.2 μm from the junctions. How this leakage occurs

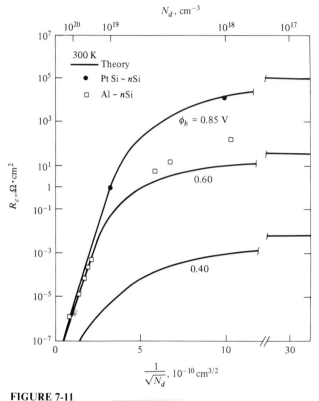

FIGURE 7-11
Theoretical and experimental values of specific contact resistance. *(After Chang, Fang, and Sze [7].)*

will be discussed in Sec. 7-9 on contact reliability. This situation is compounded by the new deposition and etching techniques, e.g., electron-beam evaporation, plasma and reactive ion etching, which create defects or traps that cannot be removed by annealing below 550°C. For these reasons, it is necessary to have an alternative contact material with low resistivity and high-temperature stability. Metal *silicides* are found to be most suitable for the application on hand.

Silicide is a metal-silicon compound with a specific ratio of composition and other chemical properties. About half the elements in the periodic table react with silicon to form one or more phases of silicides [9]. Of these compounds, the most important ones are those of the refractory metal family (Mo, Ta, Ti, W) and the near-noble metal family (Co, Ni, Pd, Pt). Besides its low resistivity and high stability, a good silicide should be easy to form and etch, strongly adhesive, and capable of forming an oxide. It should have a smooth surface and not react with aluminum. The most stable silicides are silicon-rich and of the form MSi_2, which is known as disilicide. Some important properties of widely used silicides are shown in Table 7-1.

The use of silicides as Schottky barriers was first reported by Lepselter and his coworkers. Typically, a pure-metal film is evaporated onto the silicon substrate

TABLE 7-1
Properties of selected silicides

Metal	Silicide	Binary eutectic, °C	Reaction temp., °C[†]	Resistivity,[‡] $\mu\Omega \cdot cm$	Barrier height to nSi, eV
Co	$CoSi_2$	1195	350	18–20	0.65
Mo	$MoSi_2$	1410	500	40	0.64
Ni[§]	$NiSi_2$	966	400	12–15	0.66
Pd	Pd_2Si	720	200	30–35	0.72
Pt	PtSi	830	300	28–35	0.84
Ta	$TaSi_2$	1385	550	35–45	0.59
Ti	$TiSi_2$	1330	450	13–16	0.58
W	WSi_2	1440	650	30–40	0.67

[†] For 30 to 60 min depending on interface cleanliness and metal deposition method. Industrial applications usually require higher temperature to assure complete reaction.
[‡] The lowest resistivity reported. These may require longer annealing time at a higher reaction temperature.
[§] Two other nickel silicides, Ni_2Si and NiSi, are obtained at reaction temperatures of 300 and 350°C, respectively.

first and then the reaction is accomplished in a heated oven with an inert gas. For an n-type silicon, the Schottky-barrier height ranges from 0.5 eV for $TiSi_2$ to 0.93 eV for $IrSi_3$. When the silicide is formed, the silicon is partly consumed so that the silicide-silicon interface moves into the bulk of the silicon. As a result, the defects and contaminations of the original surface are left behind, and they do not influence the interface properties (Fig. 7-12). Consequently, the interface is very clean and nearly perfect, so the Schottky-barrier height is easily reproducible. The Schottky-barrier height as a function of the metal electronegativity, a chemical property which specifies the ability of the metal to attract an electron, is plotted in Fig. 7-13. The systematic change implies that the chemical nature of the metal and the interface bonding configuration are important in determining the barrier height. The best linear approximation is represented by the dashed line, whose bending indicates that the interface states may play a role in pinning the Fermi level [2].

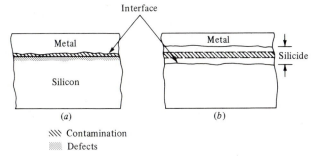

FIGURE 7-12
Reaction at transition metal–silicon interfaces (a) as deposited and (b) after reaction. (*After Ho [9].*)

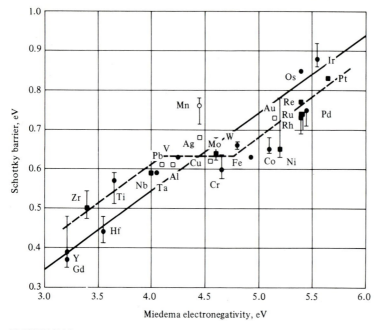

FIGURE 7-13
Barrier height of transition metal silicides and nonreacting metal Schottky barriers on *n*-type silicon. Solid circles and squares: silicides; open squares: nonreacting metals; solid circles: data from the literature. *(After Schmid [10].)*

Recently, nickel and cobalt silicides have been grown epitaxially on silicon [11]. These are the most perfect metal-semiconductor interfaces, and they may open a new area of device applications, e.g., metal-base transistors and other structures. At present, the most widely used silicide Schottky barrier is the PtSi–Si system which is employed in bipolar integrated circuits and infrared photodetectors.

7-8 FABRICATION OF SCHOTTKY BARRIERS

Although a Schottky diode is the simplest device in structure, its characteristics depend a great deal on the chemical reactivity of the metal and surface cleanliness of the semiconductor. From the processing point of view, we may separate the devices into chemically cleaned and vacuum-cleaned samples. The chemically cleaned devices go through a standard degreasing and oxide removal procedure before being loaded into the vacuum chamber for metal evaporation. Details of cleaning processes are given in the literature (see e.g., Ref. 2 of Chap. 2). Since a silicon wafer oxidizes quickly upon exposure to the atmosphere, a very thin surface oxide (~ 1 nm) is almost unavoidable.

The sample's surface may be cleaned inside the vacuum chamber by sputtering, which makes use of an ion beam to physically remove the surface oxide and other contaminants. However, an annealing step is necessary to eliminate ion beam-induced damages. An alternate method is first to chemically clean the wafer and then to heat it inside the vacuum chamber momentarily to a high temperature, typically 1000 °C for Si and 550 °C for GaAs, to evaporate the native oxide. In the case of GaAs, the crystal has crystal planes that are weakly bonded and can be cleaved to produce an extremely clean surface in an ultrahigh vacuum system. Deposition of metal can then proceed within the superclean environment. Researchers have gone to these extremely carefully prepared samples in order to learn more about the basic physics of the interface. Clean and controlled M-S contacts are useful for new device structures such as the metal-base transistor discussed in Sec. 7-10.

From the chemical point of view, we may divide Schottky barriers into reactive and nonreactive contacts, neglecting the interface oxide temporarily. All metals that form silicides with silicon are reactive, but their reaction temperatures vary greatly. For example, Pd–Si contacts form the silicide Pd_2Si at less than 250 °C, and the chemical reaction of the interface takes place at room temperature for the first few monolayers. On the other hand, tungsten silicide is produced at a temperature greater than 800 °C. As pointed out in the last section, a major advantage of silicides is that once the compound is formed, it becomes very stable. Furthermore, the silicide formation moves the interface deep inside the bulk of the silicon so that the interface is clean. As a result, silicide-silicon Schottky barriers are very reproducible and reliable.

An example of the nonreactive contact is the Au–Si system. Although there is no chemical reaction at the interface, Au tends to diffuse into silicon to form deep traps, adversely affecting device performance. In the Al–Si interface, interdiffusion is the most serious problem. The contact is not stable even at low temperatures. The as-deposited Al–pSi is a Schottky barrier with a large barrier height and can be used as a solar cell if the native oxide is kept. On the other hand, Al–nSi is an ohmic contact as deposited. By heating the Al–nSi to near eutectic, an alloy junction is formed with an equivalent barrier height as high as 1 eV. However, the poor thermal stability prevents its use in direct contact.

In the case of GaAs, Pd and Pt are relatively reactive and Au is less reactive. The most stable Schottky barriers are made with sputtered WSi_2 which can go through a high-temperature annealing cycle, an important step in self-aligned field-effect-transistor fabrication. The barrier height of the most frequently used metal-semiconductor contacts are given in Tables 7-2 and 7-3.

Three commonly used Schottky-barrier structures are shown in Fig. 7-14. In Fig. 7-14a, the n-type epitaxial film on an n^+ Si substrate is cleansed and thermally oxidized. Subsequently, windows are opened by using the standard photoresist technique, and the metal is deposited by either evaporation or sputtering in a vacuum system. The metal geometry is defined by another photoresist step. Unfortunately, this simple structure does not give ideal Schottky-barrier characteristics because of the sharp edge and positive fixed charge Q_{ss} that exists at the Si–SiO_2 interface. These conditions establish a high electric field in the depletion region in the semiconductor near the periphery, leading to excess current at the

TABLE 7-2
Schottky-barrier height of ideal metal contacts to GaAs as measured by I-V **and** C-V **methods**

Metal	ϕ_{Bn}^{IV} (eV)	ϕ_{Bn}^{CV} (eV)	ϕ_{Bp}^{IV} (eV)	$(\phi_{Bn}^{IV} + \phi_{Bp}^{IV})$ (eV)	ϕ_m (eV)
Cu	0.96	0.96	0.45	1.41	4.65
Pd	0.91	0.93	0.50	1.41	5.12
Ag	0.90	0.89	0.50	1.40	4.26
Au	0.89	0.87	0.50	1.39	5.1
Al	0.85	0.84	0.61	1.46	4.28
Ti	0.83	0.83	0.56	1.39	4.33
Mn	0.81	0.89	0.56	1.37	4.1
Pb	0.80	0.91	0.52	1.32	4.25
Bi	0.77	0.79	0.61	1.38	4.22
Ni	0.77	0.91	0.62	1.39	5.15
Cr	0.77	0.81	0.56	1.33	4.5
Co	0.76	0.86	0.61	1.37	5.0
Fe	0.72	0.75	0.60	1.32	4.5
Mg	0.62	0.66	0.55	1.17	3.66

Source: After Waldrop [12].

TABLE 7-3
Barrier height of metal-silicon contacts

	nSi	pSi
Al	0.62	0.7
Au	0.74	0.34
Cr	0.61	0.50
Mo	0.68	0.42
Ni	0.57	0.51
Pd	0.75	0.38
Pt	0.87	0.25
Ta	0.55	
Ti	0.50	0.6
W	0.67	0.45

corners, a soft reverse characteristic, and low breakdown voltage as well as poor noise properties. The periphery effect can be eliminated by allowing the metal to overlap the oxide, as shown in Fig. 7-14b. The depletion regions under the metal-oxide-semiconductor (MOS) capacitances are now rounded off, and the sharp edge which causes the soft breakdown is eliminated. The overlapped regions should be small; otherwise the added capacitance would degrade the high-frequency response of the diode. The guard-ring structure shown in Fig. 7-14c uses an

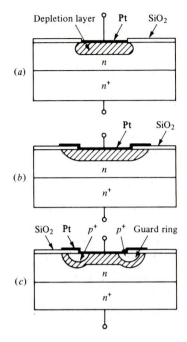

(a)

(b)

(c)

FIGURE 7-14
Practical Schottky-diode structures: (a) simple contact,
(b) with metal overlap, and (c) with guard-ring diode.

additional p^+ diffusion ring to reduce the edge effect in order to obtain an ideal
I-V characteristic. Usually, the overlapped-metal structure is preferred in integrated
circuits because of its simplicity.

7-9 CONTACT RELIABILITY

Probably the most likely failure mode in integrated circuits is the ohmic contact.
The basic issue is interdiffusion of Al and Si at contact interfaces at temperatures
below 400°C. Silicon also has a high solubility in aluminum at relatively low
temperatures. As current is passed through the contact, silicon is dissolved into
aluminum so that a void is created below the interface where the Al atoms tend
to fill in. This is known as *spiking*, a process in which aluminum forms sharp spikes,
as illustrated in Fig. 7-15. For modern devices with shallow junction depths, the

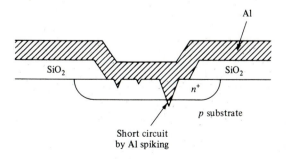

FIGURE 7-15
Aluminum spikes caused by silicon dis-
solution in Al and subsequent Al
atomic migration.

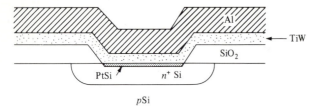

FIGURE 7-16
A typical IC contact with TiW as diffusion barrier to prevent reaction between Al and PtSi.

spikes could penetrate the junction and form a large leakage path or even a short circuit, destroying the device function of the structure.

If a silicide such as PtSi is inserted between the Al and Si, the interdiffusion of Al and Si can be significantly reduced. The PtSi is known as a *diffusion barrier* and is useful up to about 400°C, though there is a slight increase in the specific contact resistance. For high-temperature annealing, the most popular diffusion barriers are TiW (with 30% Ti) as shown in Fig. 7-16, and Mo. These barriers essentially stop the interdiffusion of Al and Si, and the interface thermal stability is excellent for annealing temperatures greater than 550°C. In addition, the specific contact resistance on n^+ or p^+ may be as low as $10^{-7} \, \Omega \cdot cm^2$. For these reasons, the contact structure of a typical advanced chip tends to be multilayered and very complicated.

7-10 APPLICATIONS OF SCHOTTKY-BARRIER DIODES

As a majority-carrier device, the Schottky diode does not have the minority-carrier storage effect, and it can be turned off in less than 1 ns. The simplicity in fabricating a Schottky barrier makes it possible to produce devices with a very small area for high-frequency operation, possibly up to 100 GHz (1 GHz = 10^9 Hz). In this section, we describe some practical applications; the Schottky solar cell and field-effect transistor will be discussed in later chapters.

Schottky-Barrier Detector or Mixer

In small-signal operation, the Schottky diode can be represented by the equivalent circuit shown in Fig. 7-17. In this figure, C_d is the junction capacitance, and r_s is the ohmic series resistance. The diode-junction resistance is defined as

$$r_d \equiv \frac{dV}{dI} \tag{7-26}$$

An efficient detector or mixer requires that the radio-frequency power be absorbed by the diode resistance r_d and that the power dissipation in r_s be small. Usually

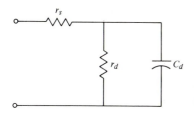

FIGURE 7-17
Equivalent circuit of a Schottky diode.

we have $r_s \ll r_d$, so that the effect of r_s at low frequency is negligible. However, increasing the operating frequency reduces the junction impedance with respect to r_s. Eventually, a frequency is reached at which point the power dissipation in r_s is equal to that of the junction, namely,

$$r_s = \frac{r_d}{1 + \omega_c^2 C_d^2 r_d^2}$$

(7-27)

where ω_c is called the *cutoff frequency*. Since $r_d \gg r_s$, we have

$$\omega_c^2 \approx \frac{1}{C_d^2 r_d r_s}$$

(7-28)

For high-frequency operation, C_d, r_d, and r_s should all be small. A small r_s is realizable if the semiconductor has a high impurity concentration and a high mobility. By using GaAs materials, an operating frequency near 100 GHz appears to be possible.

Schottky-Barrier Clamped Transistor

Because of its fast switching response, a Schottky barrier can be connected in parallel with the collector-base junction of an *npn* transistor, as shown in Fig. 7-18*a*, to reduce the storage time of the transistor. When the transistor is saturated, the collector junction is forward-biased to approximately 0.5 V. If the forward-voltage drop in the Schottky diode (typically 0.3 V) is lower than the base-collector ON voltage of the transistor, most of the excess base current flows through the diode, which has no minority-carrier storage effect. Thus the storage time of the composite device is reduced markedly compared with that of the transistor alone. The measured storage time can be less than 1 ns. The Schottky-barrier clamped transistor is realized in integrated-circuit form by the structure shown in Fig. 7-18*b*. The platinum forms an excellent Schottky barrier over the lightly doped collector *n* region and at the same time forms a good ohmic contact over the heavily doped base *p̃* region; these two contacts can be made by one single metallization step, and no extra processing is needed.

Metal-Base Transistor

In scaling down the dimensions of a bipolar transistor for high-speed operation, the base width is made very thin and the doping is increased to avoid punch-through

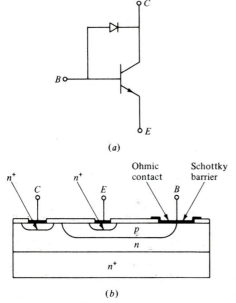

(a)

(b)

FIGURE 7-18
Schottky clamped transistor: (a) circuit representation and (b) integrated structure. All contacts are metal silicide junctions.

and to have a low base resistance. The higher base doping reduces the emitter injection efficiency, which is not desirable. The conflicting requirements can be accommodated if the base is a thin metal film rather than a semiconductor. This is known as the *metal-base transistor*. A practical structure is shown in Fig. 7-19a where the cobalt silicide constitutes the base [13]. Because the cobalt silicide is crystalline with a lattice constant almost the same as silicon, the material can be grown epitaxially, resulting in nearly ideal interfaces and few defects. Under normal

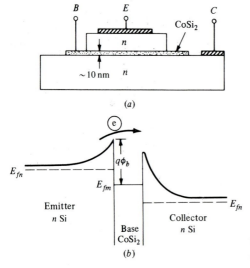

(a)

(b)

FIGURE 7-19
The metal-base transistor: (a) physical structure and b) energy-band diagram under normal biases.

operation, the emitter is forward-biased and the collector is reverse-biased. The energy-band diagram is shown in Fig. 7-19b. Because of the image force barrier lowering, the base-collector barrier is slightly rounded off and lowered at the peak. Electrons from the emitter are thermionically injected across the emitter Schottky barrier. If the base is thin enough, i.e., less than the electron's mean free path in the metal, some of the injected hot electrons will be able to reach the collector to produce the collector current. This is known as *ballistic transport* that could give rise to very fast switching time. At the present, high-quality transistors have not been fabricated although preliminary data appear to be very encouraging.

7-11 HETEROJUNCTIONS†

A junction formed by two semiconductor materials, e.g., silicon and germanium, is called a *heterojunction*. In contrast, the *pn* junction described in Chap. 3 is a *homojunction*, in which both sides of the junction are made of the same material. Since the energy-band gaps of the two materials are different, the energy-band diagram of a heterojunction exhibits a discontinuity at the junction interface and the theory is more complicated than that of the homojunction. In this section, we shall not delve into the various complexities of the problem; instead we shall introduce some concepts that are useful for device applications. We shall find that the construction of the heterojunction energy-band diagram is similar to that of a Schottky barrier, but the current transport may follow either the *pn* homojunction or the M-S barrier.

In the following discussion, germanium and gallium arsenide (Ge–GaAs) are chosen as the semiconductor pair for two reasons: (1) The material properties of Ge and GaAs are well-understood, and the fabrication technology is well under control, and (2) the lattice constants (table on inside back cover) of these materials are matched to within 1 percent. The latter criterion is of primary importance because the lattice mismatch introduces a large number of interface states and degrades the heterojunction characteristics. Since each semiconductor may be either *n* type or *p* type, there are four possible heterojunction combinatons: *n*Ge–*p*GaAs, *n*Ge–*n*GaAs, *p*Ge–*n*GaAs, and *p*Ge–*p*GaAs. We shall limit our discussion to the *n*Ge–*p*GaAs and *n*Ge–*n*GaAs junctions.

The energy-band diagrams for isolated *n*Ge and *p*GaAs are shown in Fig. 7-20, where the vacuum level is used as the reference. The two semiconductors are distinctly different in band-gap energy, dielectric constant, work function, and electron affinity. The subscripts 1 and 2 refer to Ge and GaAs, respectively. The difference in energy of the conduction-band edge is represented by ΔE_c and that in the valence-band edge by ΔE_v; they are obtained from the diagram as

$$\Delta E_c = q(\chi_1 - \chi_2) \tag{7-29}$$

$$\Delta E_v = (E_{g2} - E_{g1}) - \Delta E_c \tag{7-30}$$

---

† This section may be skipped without loss of continuity.

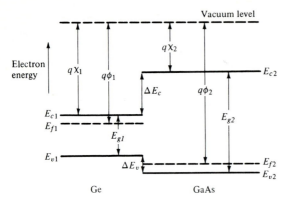

FIGURE 7-20
Energy-band diagram for two isolated
semiconductors. (*After Anderson [14].*)

As the two semiconductors are brought into contact, constancy of the Fermi level must be satisfied at equilibrium. This is accomplished when electrons in the nGe are transferred to the pGaAs and holes are transferred in the opposite direction until the Fermi level is aligned. The energy-band diagram is shown in Fig. 7-21. As in homojunction, the redistribution of charges creates a depletion layer on each side of the junction. Within the depletion layer, the energy band bends down on the p side and up on the n side, indicating the depletion of free carriers. The bending of the energy-band edges also indicates the existence of a built-in voltage in both sides of the junction. The total built-in voltage ψ_o equals the sum of the partial built-in voltages:

$$\psi_o = \psi_{o1} + \psi_{o2} \tag{7-31}$$

where ψ_{o1} and ψ_{o2} are the proportion of the built-in voltage in sides 1 and 2, respectively.

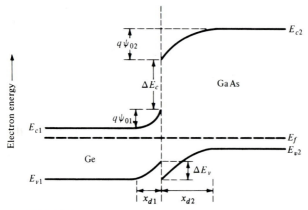

FIGURE 7-21
Energy-band diagram for np heterojunction at equilibrium. (*After Anderson [14].*)

The depletion widths can be obtained by solving Poisson's equation on both sides of the heterojunction. One of the boundary conditions is that the electric displacement be continuous, that is,

$$K_1 \mathscr{E}_1 = K_2 \mathscr{E}_2 \qquad (7\text{-}32)$$

where K and $\mathscr{E}$ are the dielectric constant and electric field, respectively. In addition, the condition of charge neutrality leads to

$$\frac{x_{d2}}{x_{d1}} = \frac{N_{d1}}{N_{a2}} \qquad (7\text{-}33)$$

as in a *pn* homojunction. The derivation of x_{d1} and x_{d2} is left as an exercise for the reader. We wish to find the portions of an applied voltage across the two sides of the depletion region. The electric fields in Ge and GaAs are

$$\mathscr{E}_1 = \frac{\psi_{o1} - V_1}{x_{d1}} \qquad (7\text{-}34)$$

$$\mathscr{E}_2 = \frac{\psi_{o2} - V_2}{x_{d2}} \qquad (7\text{-}35)$$

where V_1 and V_2 are the portions of the applied voltage appearing on the Ge side and on the GaAs side, respectively. Solving Eqs. (7-32) to (7-35), we obtain

$$\frac{\psi_{o1} - V_1}{\psi_{o2} - V_2} = \frac{K_2 N_{a2}}{K_1 N_{d1}} \qquad (7\text{-}36)$$

When the doping levels on the two sides are significantly different, Eq. (7-36) indicates that the external bias voltage is across the lightly doped side, just as in the one-sided step *pn* homojunction.

The relative magnitudes of the current components in a heterojunction are determined by the potential barriers involved. For the heterojunction shown in Fig. 7-21, the hole current from GaAs to Ge is expected to dominate because of the low potential barrier ψ_{o2} for hole injection and the high potential barrier $(\psi_{o1} + \psi_{o2} + \Delta E_c)$ for electron injection. The current density is therefore given by

$$J = \frac{qD_p p_{no}}{L_p} (e^{qV/kT} - 1) \qquad (7\text{-}37)$$

where D_p, L_p, and p_{no} are, respectively, the diffusion constant, diffusion length, and equilibrium density for holes in Ge. In contrast with the homojunction, the dominant current component is not necessarily the minority-carrier current in the lightly doped side. In fact, the discontinuity in the band diagram favors the injection of majority carriers from the larger band-gap material regardless of the doping levels.

The model presented in the foregoing paragraph is a first-order approximation and has been found to be valid in Ge–GaAs and GaAs–AlGaAs *pn*

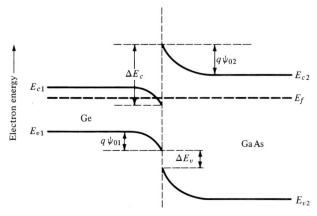

FIGURE 7-22
Energy-band diagram for *nn* heterojunction at equilibrium. *(After Anderson [14].)*

heterojunctions where the lattice mismatch is small. In other heterojunctions, such as the Ge–Si pair, a large lattice mismatch (4 percent in Ge–Si) leads to a high interface state density. As a result, recombination- and tunneling-current components have to be included.

The energy-band diagram for an nGe–nGaAs heterojunction is shown in Fig. 7-22. The barrier here acts like the metal-semiconductor contact, and the current is due to thermionic emission of electrons from GaAs to Ge.

REFERENCES

1. Sze, S. M.: "Physics of Semiconductor Devices," 2d ed., Wiley, New York, 1981.
2. Yang, E. S., X. Wu, H. L. Evans, and P. S. Ho: Silicide-Silicon Interface States, *Mater. Res. Soc. Symp. Proc.*, **54** (1986).
3. Crowell, C. R., and M. Beguwala: Recombination Velocity Effects on Current Diffusion and Imref in Schottky Barriers, *Solid-State Electron.*, **14**:1149 (1971).
4. Rhoderick, E. H.: Comments on the Conduction Mechanism in Schottky Diodes, *J. Phys. D: Appl. Phys.*, **5**:1920 (1972).
5. Card, H. C., and E. H. Rhoderick: Studies of Tunnel MOS Diodes, I: Interface Effects in Silicon Schottky Diodes, *J. Phys. D: Appl. Phys.*, **4**:1589 (1971).
6. Yu, A. Y. C.: The Metal-Semiconductor Contact: An Old Device with a New Future, *IEEE Spectrum*, **7**:83 (March 1970).
7. Chang, C. Y., Y. K. Fang, and S. M. Sze: Specific Contact Resistance of Metal-Semiconductor Barriers, *Solid-State Electron.*, **14**:541 (1971).
8. Muraska, S. P.: "Silicides for VLSI Applications," Academic, New York, 1983.
9. Ho, P. S.: Chemical Bonding and Schottky Barrier Formation at Transition Metal-Silicon Interfaces, *J. Vacuum Sci. Technol.*, **A1**:745 (1983).
10. Schmid, P. E.: *Helv. Phys. Acta*, **58**:371 (1985).
11. Tung, R.: Schottky Barrier Formation at Single Crystal Metal-Semiconductor Interfaces, *Phys. Rev. Lett.*, **52**:461 (1984).
12. Waldrop, J. R.: Electrical Properties of Ideal Metal Contacts to GaAs: Schottky-Barrier Height, *J. Vacuum Sci. Technol.*, **B2**:445 (1984).
13. Hensel, J. C.: Operation of the Si/CoSi$_2$/Si Heterostructure Transistor, *Appl. Phys. Lett.*, **49**:522 (1986).
14. Anderson, R. L.: Experiments on Ge–As Heterojunctions, *Solid-State Electron.*, **5**:341 (1962).

ADDITIONAL READINGS

Milnes, A. G. and D. L. Feucht: "Heterojunction and Metal-Semiconductor Junctions," Academic, New York, 1972.

Rhoderick, E. H.: "Metal-Semiconductor Contacts," Clarendon Press, Oxford, 1978.

Rideout, V. L.: A Review of the Theory, Technology, and Applications of Metal-Semiconductor Rectifiers, *Thin Solid Films*, **48**:261 (1978).

Rubloff, G. W.: Microscopic Properties and Behavior of Silicide Interface, *Surface. Sci.*, **132**:268 (1983).

Sze, S. M.: "Physics of Semiconductor Devices," 2d ed., Wiley, New York, 1981, chapter 5 and references.

PROBLEMS

7-1. A silicon Schottky-barrier diode has a contact area of 0.01 cm^2, and the donor concentration in the semiconductor is 10^{16} cm^{-3}. Let $\psi_o = 0.7 \text{ V}$ and $V_R = 10.3 \text{ V}$. Calculate (a) the thickness of the depletion layer, (b) the barrier capacitance, and (c) the field strength at the surface.

7-2. (a) Obtain the donor concentration, built-in potential, and barrier height of the GaAs Schottky diode from its capacitance-voltage plot shown in Fig. 7-3.

(b) Calculate the barrier height from Fig. 7-5 and compare your result with that of part (a).

7-3. Construct the energy-band diagram from a metal-on-p-type Schottky barrier with negligible surface states for (a) $\phi_m > \phi_s$ and (b) $\phi_m < \phi_s$. Indicate whether it is a rectifying or nonrectifying junction and specify the built-in potential and barrier height.

7-4. The following parameters of a Schottky diode are given: $\phi_m = 5.0 \text{ V}$, $\chi_s = 4.05 \text{ V}$, $N_c = 10^{19} \text{ cm}^{-3}$, $N_d = 10^{15} \text{ cm}^{-3}$, and $K = 11.8$. Assume that the density of interface states is negligible. At 300 K, calculate (a) the barrier height, the built-in potential, and the depletion-layer width at zero bias and (b) the thermionic-emission-current density at a forward bias of 0.3 V.

7-5. In a metal-silicon contact, the barrier height is $q\phi_b = 0.8 \text{ eV}$, and the effective Richardson constant $A^{**} = 10^2 \text{ A/cm}^2 \cdot \text{K}^2$; $E_g = 1.1 \text{ eV}$, $N_d = 10^{16} \text{ cm}^{-3}$, and $N_c = 10^{19} \text{ cm}^{-3}$.

(a) Calculate the bulk potential V_n and the built-in potential of the semiconductor at zero bias at 300 K.

(b) Assuming $D_p = 15 \text{ cm}^2/\text{s}$ and $L_p = 10 \mu\text{m}$, calculate the ratio of injected majority-carrier current to the minority-carrier current.

7-6. Calculate the ratio of the majority-carrier current to the minority-carrier current of a gold-on-nGaAs Schottky barrier at room temperature. The donor concentration is 10^{15} cm^{-3}, $L_p = 1 \mu\text{m}$, and $A^{**} = 0.068A^*$.

7-7. Derive Eqs. (7-19) and (7-20).

7-8. (a) Sketch the energy-band diagram for an ideal Schottky barrier with $q\phi_m = 4.8 \text{ eV}$ and $q\chi_s = 4.05 \text{ eV}$. Calculate the barrier height and the built-in potential for $N_d = 10^{15} \text{ cm}^{-3}$.

(b) Repeat (a) for $N_a = 10^{15} \text{ cm}^{-3}$.

(c) A 1-nm layer of $N_a = 10^{17} \text{ cm}^{-3}$ is inserted between the metal and silicon in part (a). Sketch the new energy-band diagram.

7-9. Calculate $\Delta\phi$ and x_o for a metal-insulator barrier for $\mathscr{E}_{ext} = 10^4 \text{ V/cm}$ and a dielectric constant of (a) $K = 4$ and (b) $K = 12$. Compare your results with the example given in Sec. 7-3.

7-10. (a) Derive an expression of dV/dT as a function of the current density in a Schottky diode. Assume that the minority-carrier current is negligible.

(b) Estimate the temperature coefficient if typically $V = 0.25$ V and $\phi_b = 0.7$ V at 300 K.

7-11. Frequently, the active area of a Schottky diode is not known, which leads to uncertainty in the I-V method to determine the barrier height. Using Eq. (7-10), show that the barrier height can be calculated accurately by using temperature data at a given forward-bias voltage without knowing the active area of the diode.

7-12. Calculate the cutoff frequency of a Schottky detector with a capacitance of 10 pF, a series resistance of 10 Ω, and a diode resistance of 100 Ω.

7-13. Construct the energy-band diagram of (a) a pGe-nGaAs heterojunction and (b) a pGe-pGaAs heterojunction.

7-14. (a) Sketch the charge density and electric field distributions in the np heterojunction shown in Fig. 7-21.

(b) Derive an expression for the junction capacitance.

7-15. In an nGaAs–pAlGaAs heterojunction, the band-gap offsets are $\Delta E_c = 0.3$ eV and $\Delta E_v = 0.15$ eV. The doping concentrations are $N_d = 10^{18}$ cm^{-3} and $N_a = 10^{16}$ cm^{-3}. The minority-carrier diffusion lengths are 1 μm and the widths of the p side and n side are 10 μm each.

(a) Calculate the AlGaAs band gap and the injection ratio.

(b) Explain the basic difference of this pn junction from a homojunction.

7-16. Derive Eq. (7-25) using the expression

$$T_t \sim \exp\left[-\frac{4\pi}{h} \int_0^{x_d} \sqrt{2m_e q x_B(x)} \; dx \right]$$

where $q x_B(x)$ is the energy-barrier profile of the tunneling barrier.

CHAPTER
8

JFET
AND
MESFET

The *junction field-effect transistor* (JFET) is a three-terminal semiconductor device in which the *lateral* current flow is controlled by an externally applied *vertical* electric field. It can be used as a switch or an amplifier. In its early development, the JFET was known as the *unipolar* transistor because the current is transported by carriers of one polarity, namely, the majority carriers. This is in contrast with the bipolar junction transistor discussed previously in which both majority- and minority-carrier currents are important. This chapter begins with a qualitative description of the JFET followed by the derivation of the current-voltage characteristics. The small-signal parameters and equivalent circuits are given, and the cutoff frequency is derived. The *metal-semiconductor FET* (MESFET) is introduced with emphasis on GaAs applications. The concept of the enhancement-mode FETs and fabrication techniques are then presented. The question of current conduction beyond pinchoff and channel-length modulation is addressed in a simplified form, and the permeable-base transistor is described in the last section.

8-1 PHYSICAL DESCRIPTION OF THE JFET

A typical JFET fabricated by the standard planar epitaxial process is shown in Fig. 8-1. Because of the need of observing the cross-sectional view, we have divided the JFET into two halves and shown half of it. The device can be better understood if a section of Fig. 8-1, represented schematically in Fig. 8-2a, is used. The active region of the device consists of a lightly doped *n*-type *channel* sandwiched between

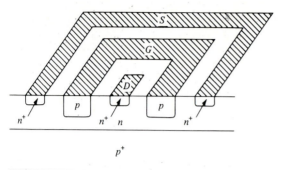

FIGURE 8-1
A junction field-effect transistor. The n^+ regions are added to provide good ohmic contacts.

two heavily doped p^+-gate regions. The lower p^+ layer is the substrate, and the upper p^+ region is formed by boron diffusion into the epitaxially grown n-type channel. The p^+ regions are connected either internally or externally to form the *gate* terminal. Ohmic contacts attached to the two ends of the channel are known as the *drain* and *source* terminals through which the channel current flows. Alternatively, the JFET may be fabricated by the double-diffused technique with diffused channel and upper gate as illustrated in Fig. 8-2b. The structures shown are *n-channel* JFETs since the channel is doped with donor impurities and the channel current consists of electrons. If the channel is doped with acceptor atoms and the gate regions are n^+ type, the channel current consists of holes and the device is a *p-channel* JFET. Since electrons have higher mobility than holes, an n-channel device provides higher conductivity and higher speed and is preferred in most applications. In the following sections, our discussion will be limited to the n-channel JFET.

Under normal operating conditions, a reverse bias is applied across the *pn* gate junctions so that free carriers are depleted from the channel and space-charge regions extending into the channel are produced. Consequently, the cross-sectional

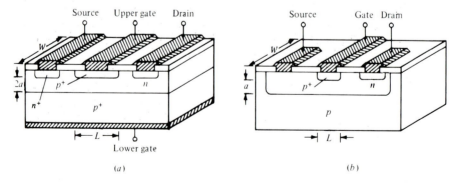

(a)   (b)

FIGURE 8-2
An n-channel JFET fabricated by (a) epitaxial-diffused process and (b) double-diffused process.

area of the channel is reduced, and the channel resistance is increased. Thus the current flow between the source and the drain is modulated by the gate voltage.

Let us connect the source and gate terminals to the ground potential and set the drain voltage at V_D (Fig. 8-3). Under these conditions, the voltage across the gate junctions at $x = 0$ is zero, but the full value of V_D is across the junctions at $x = L$. As a result, the space-charge regions extend farther into the channel at the drain end, as depicted in Fig. 8-3a. When we increase V_D, the bottleneck in the channel becomes smaller and the channel resistance is increased. If the drain voltage is further increased, a condition will eventually be reached, as shown in Fig. 8-3b, in which the space-charge regions join and all free carriers are completely depleted in the joining region. This condition is called *pinchoff*. Further increase of the drain voltage beyond pinchoff would not increase the drain current significantly. Therefore, the current is *saturated*, and the channel resistance becomes very large. The current-voltage characteristic of the drain with respect to ground is shown in Fig. 8-3c, where I_{DSS} specifies the *saturation drain current* and V_P is called the *pinchoff voltage*.

Without a gate bias voltage, the transistor has a conducting channel between the source and the drain terminals. This is the ON state, and the transistor is called a *normally on* JFET. To reach the OFF state, a gate voltage must be applied to *deplete* all carriers in the channel. For this reason, the device is also known as the *depletion-mode* JFET, or D-JFET.

8-2 THEORY OF THE JFET

With applied voltages at both the gate and the drain, the JFET shown in Fig. 8-3a is reproduced in Fig. 8-4 in an enlarged scale. The space-charge layer profile indicates that the electric field and carrier distributions are of a two-dimensional nature. This is a difficult analytical problem to solve, and it must be simplified in order to obtain some physical insight into the current-conduction mechanism. Fortunately in most practical JFETs, the channel length L is 2 or more times greater than that of the channel height $2a$. In these *long* devices, the change of the channel height along the channel is small in comparison with the width of the channel. Therefore, the electric field in the space-charge layer may be considered as in the

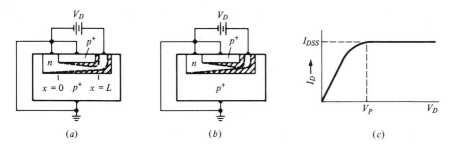

FIGURE 8-3
The JFET with $V_G = 0$ and (a) $V_D < V_P$, (b) $V_D = V_P$, and (c) idealized drain characteristic.

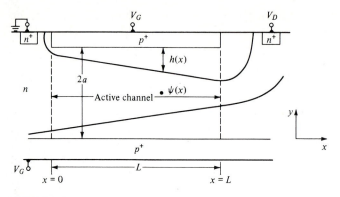

FIGURE 8-4
Enlarged schematic diagram of Fig. 8-3a, showing gradual change of the space-charge regions in the active channel. The n^+ regions are added to provide good ohmic contacts.

y direction, i.e., normal to the gate junctions. At the same time, the electric field in the neutral channel may be assumed to be in the x direction only. The separation of the electric field in the space-charge layer and the channel allows us to solve the one-dimensional Poisson's equation in each region. This technique was developed by Shockley and is known as the *gradual-channel approximation* [1]. Using this approach, we can obtain the important features of the JFET without resorting to the solution of two-dimensional partial differential equations.

By assuming gate junctions as one-sided step junctions, the depletion-layer width in the JFET is obtained by using Eq. (3-32):

$$V_{C-G} = \frac{qN_d h^2}{2K_s \varepsilon_o} \tag{8-1}$$

where V_{C-G} is the channel-to-gate voltage and the depletion-layer width x_d has been replaced by the symbol h, representing the channel height. Since the built-in potential is ψ_o and the externally applied voltage is $\psi(x) - V_G$, the channel-to-gate voltage is given by

$$V_{C-G} = \psi_o + \psi(x) - V_G \tag{8-2}$$

The space-charge-layer width at the pinchoff point is exactly equal to the channel height. Thus, the pinchoff voltage is obtained by setting $h = a$ and $\psi - V_G = V_P$ in Eqs. (8-1) and (8-2):

$$V_P + \psi_o = \frac{qa^2 N_d}{2K_s \varepsilon_o} = V_{PO} \tag{8-3}$$

where V_P is the externally applied voltage to reach the pinchoff condition and is the pinchoff voltage. V_{PO} is the sum of the pinchoff voltage and the built-in potential, and it may be referred to as the *internal* pinchoff voltage. The depletion-layer width as a function of x may now be written as

$$\frac{h(x)}{a} = \left[\frac{\psi(x) + \psi_o - V_G}{V_{PO}}\right]^{1/2} \tag{8-4}$$

where Eqs. (8-1) to (8-3) have been used.

The drain current can be obtained by making use of Eq. (2-69). Since the electron distribution in the neutral channel is assumed to be uniform, the gradient of electrons is zero and the diffusion current component can be neglected. Thus, the drain current consists of the electron-drift component only, and Eq. (2-69) becomes[1]

$$I_D = -q\mu_n nA\mathscr{E} = 2q\mu_n N_d(a-h)W\frac{d\psi}{dx} \tag{8-5}$$

where I_D denotes the drain current and $2(a-h)W$ is the cross-sectional area. Substituting Eq. (8-4) into (8-5) and integrating, we have

$$\int_0^L \frac{I_D\,dx}{2q\mu_n N_d Wa} = \int_0^{V_D}\left[1 - \sqrt{\frac{1}{V_{PO}}(\psi + \psi_o - V_G)}\right]d\psi \tag{8-6}$$

The limits of integration are defined by the length of the active region from $x = 0$ to $x = L$ with the corresponding voltage from 0 to V_D. Performing the integration yields

$$I_D = G_o\left\{V_D - \frac{2}{3}\sqrt{\frac{1}{V_{PO}}}\left[(V_D + \psi_o - V_G)^{3/2} - (\psi_o - V_G)^{3/2}\right]\right\} \tag{8-7}$$

where

$$G_o = \frac{2qaW\mu_n N_d}{L} \tag{8-8}$$

G_o is the channel conductance without any depletion layers. Equation (8-7) describes the drain current as a function of both the drain and gate voltages before the pinchoff condition is reached. The I-V characteristics of the drain are plotted in Fig. 8-5. Curves in Fig. 8-5a are based on the theoretical model expressed by Eq. (8-7), and it is assumed that the drain current is constant after pinchoff. We have also plotted data of an experimental device in Fig. 8-5b for comparison. The discrepancy can be accounted for when the effect of series resistance is included, and it will be explained in a later section.

Example. For an n-channel silicon JFET with $K_s = 12$, $N_d = 5 \times 10^{15}\,\text{cm}^{-3}$, $N_a = 10^{19}\,\text{cm}^{-3}$, $a = 1\,\mu\text{m}$, $L = 30\,\mu\text{m}$, $W = 0.1\,\text{cm}$, and $\mu_n = 1350\,\text{cm}^2/\text{V}\cdot\text{s}$, find (a) the pinchoff voltages V_{PO} and V_P and (b) the drain current at $V_D = V_P$ with both gate and source grounded.

---

[1] The drain current is defined as positive in the direction against the x axis in Fig. 8-4, thus yielding the negative sign in Eq. (8-5).

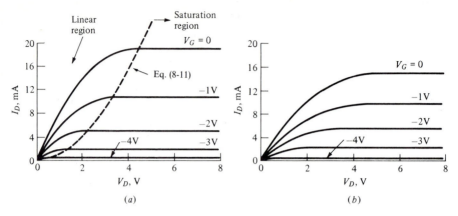

FIGURE 8-5
Current-voltage characteristics of a silicon n-channel JFET with $a = 1.5\,\mu\text{m}$, $W/L = 170$, and $N_d = 2.5 \times 10^{15}\,\text{cm}^{-3}$: (a) theoretical plot of Eq. (8-7) with $R_s = 0$, and (b) experimental results. *(After Grove [2].)*

Solution

(a) $V_{PO} = \dfrac{qa^2 N_d}{2K_s \varepsilon_0} = \dfrac{(1.6 \times 10^{-19})(10^{-8})(5 \times 10^{15})}{(2)(12)(8.85 \times 10^{-14})} = 3.77\ \text{V}$ from Eq. (8-3)

$\psi_o = \phi_T \ln \dfrac{N_a N_d}{n_i^2} = 0.026 \ln \dfrac{(5 \times 10^{15})(10^{19})}{2.25 \times 10^{20}} = 0.86\ \text{V}$ from Eq. (3-15)

$V_P = V_{PO} - \psi_o = 3.77 - 0.86 = 2.91\ \text{V}$

(b) $G_o = \dfrac{2qaW\mu_n N_d}{L} = 7.2 \times 10^{-3}$ S   from Eq. (8-8)

$I_D = G_o \left\{ V_P - \dfrac{2}{3\sqrt{V_{PO}}} [(V_P + \psi_o)^{3/2} - (\psi_o)^{3/2}] \right\} = 4.8\ \text{mA}$ from Eq. (8-7)

8-3 STATIC CHARACTERISTICS

The current-voltage characteristics displayed in Fig. 8-5 can be divided into linear and saturation regions with the pinchoff condition as the boundary. In this section we examine the I-V characteristics in these two regions and discuss the significance of the gate leakage current, breakdown voltage, and series resistances.

Linear Region

Examining the current-voltage characteristics shown in Fig. 8-5, we find that the drain current is proportional to the drain voltage at small drain voltages.

In addition, the slope of the I-V curves near the origin is a function of the gate voltage. This region of operation is known as the *linear region*. Mathematically, the current-voltage relationship of the linear region is obtained by first letting $V_D \ll \psi_o - V_G$ in Eq. (8-7). Using the binomial series expansion, we can write the second term in Eq. (8-7) as

$$(\psi_o - V_G)^{3/2}\left(1 + \frac{V_D}{\psi_o - V_G}\right)^{3/2} \approx (\psi_o - V_G)^{3/2}\left(1 + \frac{3}{2}\frac{V_D}{\psi_o - V_G}\right) \tag{8-9}$$

Substituting Eq. (8-9) into (8-7) and simplifying the expression, we obtain

$$I_D = G_o\left(1 - \sqrt{\frac{\psi_o - V_G}{V_{PO}}}\right)V_D \tag{8-10}$$

This equation shows indeed the linear dependence of the drain current on the drain voltage. The effect of the gate voltage on the slope of the I-V curves is also apparent in Eq. (8-10).

Saturation Region[1]

At the point of pinchoff, the magnitudes of the biasing drain and gate voltages satisfy the condition

$$V_D - V_G = V_P \tag{8-11}$$

Thus, the drain voltage needed to reach the pinchoff condition is different for each gate voltage. Equation (8-11) is plotted in Fig. 8-5a and is known as the *pinchoff curve*. The current-voltage characteristics beyond pinchoff are called the *saturation region* because the drain current is saturated. The magnitude of the drain current in saturation I_{DS} is derived by substituting Eq. (8-11) into Eq. (8-7):

$$I_{DS} = G_o\left(\frac{2}{3}\sqrt{\frac{\psi_o - V_G}{V_{PO}}} - 1\right)(\psi_o - V_G) + \frac{G_o V_{PO}}{3} \tag{8-12}$$

Equation (8-12) expresses the saturation drain current as a function of the gate voltage. It is called the *transfer characteristic* and is plotted in Fig. 8-6, where we have also plotted the parabola

$$I_{DS} = I_{DSS}\left(1 - \frac{V_G}{V_P}\right)^2 \tag{8-13}$$

where I_{DSS} denotes the drain saturation current at zero gate voltage; i.e., the gate is short-circuited to the source. Note that the simple square law expressed in Eq. (8-13) approximates Eq. (8-12) very well. It has been found experimentally that even with arbitrary nonuniform impurity distribution in the y direction, the transfer

---

[1] The reader is cautioned here that the meaning of *saturation* in FETs is completely different from that of a bipolar transistor (Chap. 5).

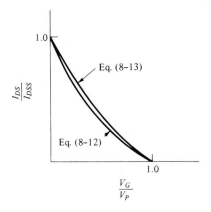

FIGURE 8-6
Transfer characteristic of a JFET.

characteristic of any JFET falls within the boundaries set by the two curves shown in Fig. 8-6. In amplifying applications, the JFET is usually operated in the saturation region, and the transfer characteristic is used to find the output drain current for a given input gate-voltage signal.

Gate Leakage Current

The gate leakage current is the sum of the reverse saturation, generation, and surface leakage currents. In a planar JFET, the surface leakage component is usually small. The saturation current and the generation current are given by Eqs. (4-25) and (4-41), respectively. In a typical device, the magnitude of the gate leakage current is between 10^{-9} and 10^{-12} A, yielding an input impedance greater than 10^8 Ω. However, the surface leakage current due to poor fabrication control can significantly degrade the input impedance of a JFET.

Breakdown Voltage

When the drain voltage is increased, the gate leakage current remains small until avalanche breakdown at the gate junction takes place. The breakdown occurs at the drain end of the channel because it has the highest reverse-bias voltage. This breakdown voltage is given by

$$V_B = V_{DO} + V_G \tag{8-14}$$

where V_{DO} is the drain breakdown voltage for zero gate voltage. Similar to that of the *pn* junction diode, the drain current of a JFET exhibits an abrupt increase at breakdown. Notice that the gate bias voltage is negative so that the breakdown voltage is reduced as the magnitude of V_G is increased. In other words, maximum V_B occurs at zero gate voltage, as shown in Fig. 8-7, and it decreases with increasing negative gate voltage.

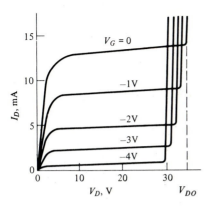

FIGURE 8-7
Breakdown at high V_D in a JFET.

Series Resistances

The sections between the ohmic contacts and the active region of the channel introduce ohmic series resistances R_S and R_D. These resistances, shown in Fig. 8-8, are constant since they are not affected by the gate or drain voltages. In the linear region of operation, both R_S and R_D contribute to ohmic drops. In the saturation region the effect of R_D is negligible, but R_S has a strong influence because of the negative feedback it produces (see next section). The source resistance reduces the drain current (Fig. 8-5) as well as the gain of the amplifier. For this reason, it should be kept as small as possible.

8-4 SMALL-SIGNAL PARAMETERS AND EQUIVALENT CIRCUITS

In the linear region, the drain conductance is obtained by differentiating Eq. (8-10). The result is

$$g_{dl} \equiv \left. \frac{\partial I_D}{\partial V_D} \right|_{V_G} = G_o\left(1 - \sqrt{\frac{\psi_o - V_G}{V_{PO}}}\right) \qquad \text{for} \quad V_D \ll V_{PO} \qquad (8\text{-}15)$$

where the drain conductance is seen as a function of the applied gate voltage. This feature makes the JFET suitable for applications as a voltage-controlled variable

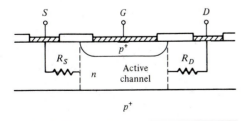

FIGURE 8-8
Series resistances in a JFET.

resistance. The transconductance of the linear region is given by

$$g_{ml} \equiv \left.\frac{\partial I_D}{\partial V_G}\right|_{V_D} = \frac{G_o}{2} \frac{V_D}{\sqrt{V_{PO}(\psi_o - V_G)}} \tag{8-16}$$

derived by using Eq. (8-10).

In the saturation region, the transconductance is derived by differentiating Eq. (8-12). The result is

$$g_m \equiv \frac{\partial I_{DS}}{\partial V_G} = G_o\left(1 - \sqrt{\frac{\psi_o - V_G}{V_{PO}}}\right) \tag{8-17}$$

Note that Eqs. (8-15) and (8-17) are identical, thus, the linear output conductance is equal to the saturation transconductance. Experimental data for the transconductance of a typical JFET are plotted in Fig. 8-9. It is seen that the experimental data agree with Eq. (8-17) when g_m is small. At large values of g_m, Eq. (8-17) does not give an accurate description of the JFET because of the series resistance R_S. Using circuit analysis, we find that the negative-feedback effect of R_S leads to an effective transconductance g_m' (Prob. 8-6):

$$g_m' = \frac{g_m}{1 + g_m R_S} \tag{8-18}$$

Thus, the effective transconductance is equal to g_m for small values of $g_m R_S$, but it is reduced when $g_m R_S$ is comparable to unity.

The pn junction between the gate and the channel has a junction capacitance

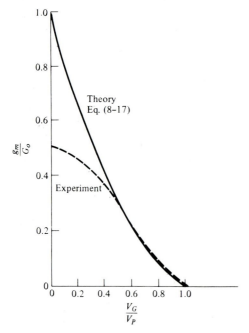

FIGURE 8-9
Theoretical and experimental curves of the transconductance.

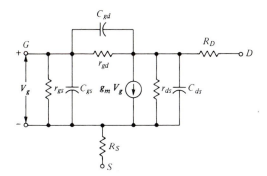

FIGURE 8-10
A small-signal equivalent circuit.

under reverse bias. Let $\bar{h}$ be the average depletion-layer width; the total gate capacitance is given by

$$C_G = 2WL \frac{K_s \varepsilon_o}{\bar{h}} \tag{8-19}$$

The factor of 2 accounts for the two gate junctions, each having an area of WL. With $V_G = 0$ and at pinchoff, the average depletion-layer height is $a/2$. Therefore, the gate capacitance at pinchoff is

$$C_G = 4WL \frac{K_s \varepsilon_o}{a} \tag{8-20}$$

Despite the distributed nature of the gate capacitance, for simplicity it is usually represented by two lumped capacitances, namely, the gate-to-drain capacitance C_{gd} and the gate-to-source capacitance G_{gs}. These two capacitances are voltage-dependent. In addition, a small capacitance C_{ds} is introduced by the device package between the drain and the source.

A small-signal equivalent circuit for the JFET incorporating the parameters described in the preceding paragraphs is shown in Fig. 8-10. The resistances r_{gd} and r_{gs} represent the effect of the gate leakage current. They are usually very large and can be neglected for most purposes. The resistance r_{ds} is the finite drain resistance introduced by the channel-length modulation (discussed later). Its typical value is between 100 kΩ and 2 MΩ. A simplified equivalent circuit is given in Fig. 8-11. This simple representation is sufficient for most applications. In addition, the capacitances may be ignored for low-frequency operation.

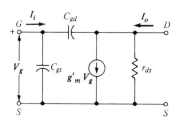

FIGURE 8-11
A simplified small-signal equivalent circuit.

8-5   CUTOFF FREQUENCY

The maximum operating frequency f_{co} is defined as the condition when the JFET is no longer amplifying the input signal. Let us use the equivalent circuit shown in Fig. 8-11 with output short-circuited and replacing g'_m by g_m. The unity-gain condition is reached when the current through the input capacitance is equal to the output drain current. The input current is given by

$$I_i = 2\pi f_{co}(C_{gs} + C_{gd})V_g = 2\pi f_{co}C_G V_g \qquad (8\text{-}21)$$

and the output current is

$$I_o = g_m V_g \qquad (8\text{-}22)$$

Equating Eqs. (8-21) and (8-22), we obtain the cutoff frequency as

$$f_{co} = \frac{g_m}{2\pi C_G} \leqslant \frac{qa^2\mu_n N_d}{4\pi K_s \varepsilon_o L^2} \qquad (8\text{-}23)$$

The last expression in Eq. (8-23) is obtained by using Eqs. (8-20) and (8-17) with $g_m \leqslant G_o$. If Eq. (8-19) is used with $\bar{h} = a$ for minimum gate capacitance, the derived cutoff frequency will double the value given by Eq. (8-23). We notice that the term $qa^2 N_d/K_s \varepsilon_o$ in Eq. (8-23) is equal to $2V_{PO}$ [Eq. (8-3)] and therefore is determined by the pinchoff voltage. Usually, this term cannot be adjusted for maximum frequency consideration. The other variables in Eq. (8-23) are the mobility and channel length. To achieve the best high-frequency performance, we should have large mobility and short channel length.

8-6   THE MESFET

When the *pn* junction gate in the JFET is replaced by a Schottky barrier, a metal-semiconductor field-effect transistor, better known by its acronym MESFET, is realized [3]. The fabrication procedure and the final MESFET structure are shown in Fig. 8-12. It should be pointed out that in silicon technology, the JFET or MESFET is seldom used in integrated circuits because a better device is available in the form of the MOSFET, a subject to be discussed in great detail in following chapters. In GaAs, there is no practical way to form a usable oxide, so the GaAs MOSFET has not yet been realized. Because of the higher electron mobility and saturation velocity, GaAs devices are expected to be faster than silicon devices, and most works in MESFET concentrate on compound semiconductors. GaAs JFETs can be fabricated, but the device dimension is larger because of lateral diffusion during gate formation. This drawback is removed when a Schottky barrier is used as the gate junction. To fabricate a MESFET, a typical sequence is shown in Fig. 8-12. A Si_3N_4 layer of about 100 nm in thickness is first deposited on a (100) GaAs semi-insulating (S.I.) substrate. The nitride functions as a surface encapsulation during subsequent ion implantation and annealing steps. This is followed by the deposition of a 250-m-thick phosphosilicate glass (PSG) which is selectively removed by photolithography to produce a desired pattern. Using

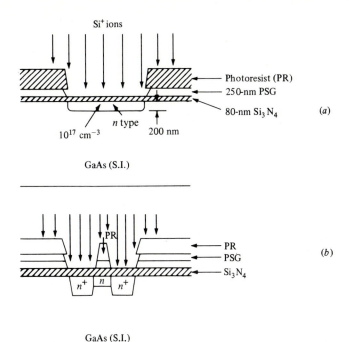

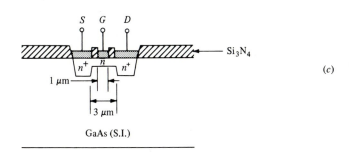

FIGURE 8-12
The fabrication sequence of a GaAs MESFET: (*a*) channel implantation, (*b*) source and drain n^+ implantation, and (*c*) after metalization.

the PSG and photoresist as the mask, silicon ions are implanted to form the channel, as depicted in Fig. 8-12*a*. The process is then repeated to produce the n^+ regions which will be needed for the source and drain ohmic contacts (Fig. 8-12*b*). A high-temperature annealing cycle, typically at 850 to 900°C, is necessary to recover the crystalline quality and to activate the dopants electrically. Then, the nitride layer is removed and metal films are deposited to form the Schottky barrier and the ohmic contacts, as shown in Fig. 8-12*c*. Typically, metalization of a

Au–Ge–Ni layer is used to make an alloyed contact, and a Ti–Pt–Au layer serves as the Schottky gate barrier. The Ti–Pt–Au is also employed as interconnects between devices. The metal line width of the gate is usually less than 1 μm. The MESFET shown is operated in the depletion mode and is known as the D-MESFET. Other gate-metalization schemes include Pt, W, and WSi_2. The channel doping is about 10^{17} cm^{-3}, and the channel height and length are typically 0.2 and 1 μm, respectively. The characteristics of the GaAs MESFET is similar to the silicon JFET except that the transconductance is higher and the input gate capacitance is lower. Since it is built on a semi-insulating substrate, the MESFET is electrically isolated and the parasitic capacitance is small. For these reasons, the cutoff frequency of the MESFET approaches 30 GHz and the delay per gate is near 10 ps, that is, about a factor of 2 to 5 better than silicon devices. The other advantages of GaAs MESFETs include lower noise and higher tolerance of radiation. However, the GaAs technology is not yet mature, and the higher fabrication costs will make it less competitive than silicon. In the next decade, GaAs devices should play a more important role in high-speed logic circuits and systems.

8-7   ENHANCEMENT JFET AND MESFET

In the depletion-mode operation, the FET (JFET or MESFET) is normally on at zero gate bias. The drain current is turned off by applying a negative gate voltage sufficiently large enough to deplete the channel. If the FET has a narrow conducting channel which is lightly doped, it is possible that the depletion layer established by the built-in potential is wide enough to pinch off the channel without a gate voltage. In other words, the channel is shut off at zero gate bias, and it is necessary to apply a positive gate bias to induce a channel. This is known as the *normally off* or *enhancement* FET. The term "enhancement" is used because the channel's carrier concentration is enhanced with the positive gate voltage.

The enhancement JFET, or E-JFET, is a practical approach to designing GaAs integrated circuits, and an experimental structure is shown in Fig. 8-13.

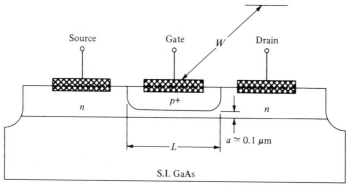

FIGURE 8-13
A typical GaAs enhancement JFET. *(After Zuleeg [4].)*

The source and drain n^+ regions are obtained by silicon or selenium ion implantation. The p^+ gate is produced either by Zn diffusion or by beryllium or magnesium ion implantation. Zinc diffusion is relatively easy to perform, but the lateral diffusion poses a limit on the minimum channel length. For channel length less than a micrometer, ion implantation is necessary.

When the conducting channel is small compared with the depletion-layer width, the drain current in saturation may be approximated by

$$I_{DS} = \frac{K_s \varepsilon_o \mu_n W}{2aL} (V_G - V_T)^2 \tag{8-24}$$

where V_T is the *threshold voltage* specified by the applied gate voltage that is sufficient to open the channel and allow drain current to flow. The derivation of Eq. (8-24) is the same as that of an MOS transistor, and it will be given in the next chapter. An experimental plot of the drain current vs. the gate voltage is shown in Fig. 8-14, where the dashed line represents the ideal transfer characteristic given by Eq. (8-24). Notice that the intercept on the horizontal axis defines the threshold voltage. In this graph, Eq. (8-24) is a good approximation for a gate voltage between 0.4 and 1.0 V, but it does not describe the data well for V_G smaller than 0.4 V. The output characteristics of the E-JFET are shown in Fig. 8-15. It is seen that the drain current is negligible for a gate voltage less than 0.4 V.

The threshold voltage is an important parameter, and its value can be controlled by the channel impurity density and the channel height a. Since the

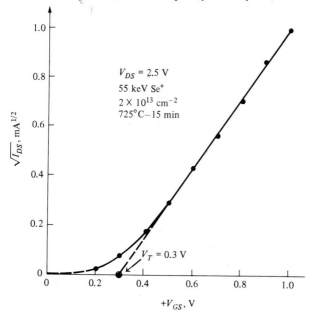

FIGURE 8-14

Experimental $\sqrt{I_{DS}}$ vs. V_{GS} characteristic for an ion-implanted shallow-junction E-JFET. (*After Zuleeg* [4].)

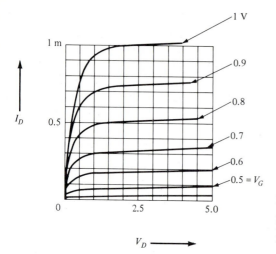

FIGURE 8-15

I-V characteristics of a GaAs E-JFET. Device dimensions are $L = 4\,\mu m$ and $W = 350\,\mu m$. *(After Zuleeg[4].)*

E-JFET is a normally off device, the metallurgical channel height a has to be smaller than the depletion layer established by the built-in potential. We may define the pinchoff voltage V_{PO} as

$$V_{PO} = \frac{qa^2 N_d}{2K_s \varepsilon_o} \tag{8-25}$$

The threshold voltage is then given by

$$V_T = \psi_o - V_{PO} \tag{8-26}$$

where ψ_o is the built-in potential. At a gate bias voltage of V_T, the channel becomes conducting between the source and the drain. Therefore, the threshold voltage is a critical parameter in E-FET devices.

An enhancement-mode MESFET, or E-MESFET, can be fabricated if the built-in potential of the Schottky barrier is sufficient to cut off channel conduction. An E-MESFET with WSi_2 as its metal gate is shown in Fig. 8-16. The tungsten silicide is employed as the mask for the source and drain n^+ implantation. Using the T-bar structure for the gate, the source and drain are self-aligned to reduce the source resistance and, at the same time, to prevent n^+ regions encroaching the areas under the gate. Tungsten silicide is a very stable compound, and it maintains its integrity during high-temperature post-implantation annealing.

There are quite a few advantages of using E-MESFET in integrated circuits. These include a single power supply, direct-coupled logic with level shifting, low power consumption, and high gain bandwidth product. However, since the enhancement device operates in the forward-bias region of the gate junction, the logic swing is small. Too large a bias would produce the forward gate current, which is undesirable. Typically, the threshold voltage is less than 0.2 V and the logic swing is less than 0.5 V. These values impose an extremely difficult criterion for the control of the Schottky-barrier height. To build a large number of devices

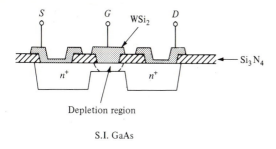

S.I. GaAs

FIGURE 8-16
An enhancement MESFET with T-bar
WSi_2 for self-alignment.

on the same chip with small tolerances in characteristics, the Schottky-barrier-height variation must be within 25 meV. This condition cannot be satisfied with the present technology of E-MESFET. The E-JFET and D-MESFET appear to be more mature and should be available in the near future.

8-8 BEHAVIOR OF THE JFET BEYOND PINCHOFF

The pinchoff condition specifies that the two space-charge regions meet at the center of the channel, as shown by the solid line in Fig. 8-17. When the drain voltage is increased further, more free carriers are depleted from the channel. As a result, the length of the depleted region is increased, and the length of the neutral channel is decreased. This is called *channel-length modulation*, and the change of the depletion-layer profile is drawn as the dashed line in Fig. 8-17. In the center of the channel, the applied drain voltage is now shared by the depletion and neutral region, with the neutral-channel region supporting the potential V_P and the depleted-channel region supporting the potential $V_D - V_P$. Since the reduced neutral-channel length supports the same V_P, the drain current will increase slightly for any drain voltage beyond pinchoff. For this reason, the drain current beyond pinchoff is not saturated, and the drain resistance is finite.

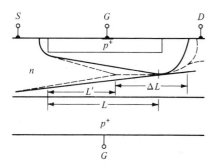

FIGURE 8-17
Channel-length modulation beyond pinchoff.

With this physical picture in mind, let us derive the drain resistance in the saturation region. The drain current beyond pinchoff is obtained by modifying Eq. (8-12) to

$$I'_{DS} = G'_o\left(\frac{2}{3}\sqrt{\frac{\psi_o - V_G}{V_{PO}}} - 1\right)(\psi_o - V_G) + \frac{G'_o V_{PO}}{3} \tag{8-27}$$

where

$$G'_o = \frac{2qaW\mu_n N_d}{L'} \tag{8-28}$$

The new channel length L' supports the pinchoff voltage V_P. According to Eq. (3-32), the drain voltage beyond pinchoff increases the length of the depleted channel by

$$\Delta L = \left[\frac{2K_s\varepsilon_o(V_D - V_P)}{qN_d}\right]^{1/2} \tag{8-29}$$

Assuming that the depleted channel extends equally toward the source and the drain, we obtain L' as

$$L' = L - \tfrac{1}{2}\Delta L = \frac{1}{2}\left[\frac{2K_s\varepsilon_o(V_D - V_P)}{qN_d}\right]^{1/2} \tag{8-30}$$

The small-signal drain resistance at pinchoff is approximately given by the slope of the current-voltage characteristics of the drain. Therefore, we have

$$r_{ds} = \frac{\Delta V}{\Delta I} = \frac{V_D - V_P}{I'_{DS} - I_{DS}} \tag{8-31}$$

Since the current is not linearly related to the drain voltage, the drain resistance r_{ds} must be calculated for each drain voltage to obtain the variation of the drain resistance. The procedure of calculating r_{ds} is demonstrated in the following example.

Example. Consider the JFET in the example of Sec. 8-2. Find the drain resistance at $V_D = V_P + 2$ V and $V_G = 0$.

Solution. Using Eq. (8-12), we may rewrite Eq. (8-27) as

$$I'_{DS} = I_{DS}\frac{L}{L'} = I_{DS}\frac{L}{L - \Delta L/2}$$

and

$$\Delta L = \left[\frac{(2)(12)(8.85 \times 10^{-14})\,\Delta V}{(1.6 \times 10^{-19})(5 \times 10^{15})}\right]^{1/2}$$

Let us take $V'_D = V_P + 1$ V and $V''_D = V_P + 3$ V as the two points on the V-I curve for our calculation. We find

$$\Delta L' = 0.52\ \mu\text{m} \qquad I'_{DS} = I_{DS}\frac{30}{29.74} \qquad \text{at } V'_D$$

$$\Delta L'' = 0.9 \, \mu m \qquad I'_{DS} = I_{DS} \frac{30}{29.55} \qquad \text{at } V''_D$$

Equation (8-31) is rewritten as

$$r_{ds} = \frac{V''_D - V'_D}{I''_{DS} - I'_{DS}} = \frac{2}{(30/29.55 - 30/29.74)(4.8 \times 10^{-3})} = 64 \, k\Omega$$

where $I_{DS} = 4.8 \, mA$ is used.

The model presented here is valid for long JFETs, i.e., with a channel length-to-width ratio greater than 4, and is applicable to most commercial general-purpose JFETs. In a *short* device ($L/a < 2$), however, the saturation mechanism is more involved and is beyond the scope of this text [5].

8-9 THE PERMEABLE-BASE TRANSISTOR [6]

The *permeable-base transistor* is a vertical MESFET where the metal gate (called the *base* by the inventors) is a grating embedded inside a single-crystal semi-conductor. A cross-sectional schematic diagram is shown in Fig. 8-18. The gate consists of tungsten stripes surrounded by an *n*-type single-crystal GaAs. Since the W–*n*GaAs contact forms a Schottky barrier with a barrier height of 0.8 eV, a depletion layer is created around the gate's metal stripes. If the gate stripes have small spacings between them, the depletion layers of neighboring stripes are joined together at zero bias, Under this condition, the permeable-base transistor is operated in the enhancement mode of the MESFET. A forward bias on the metal stripes will reduce the depletion-layer width and open the channel for current conduction between the emitter and the collector.

As a vertical MESFET, the channel length is very short because it is specified by the thickness of the metal stripes. In other words, the channel length can easily be made to less than 100 nm. Using Eq. (8-23) to estimate the speed, the cutoff frequency can be as high as 200 GHz.

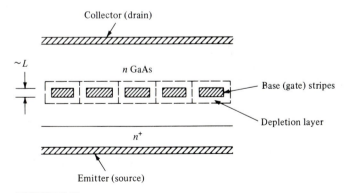

FIGURE 8-18
Cross section of the permeable-base transistor. Metal stripes are connected together.

REFERENCES

1. Shockley, W.: A Unipolar Field-Effect Transistor, *Proc. IRE*, **40**:1365 (1952).
2. Grove, A. S.: "Physics and Technology of Semiconductor Devices," chap. 8, Wiley, New York, 1967.
3. Eden, R. C., et al.: Integrated Circuits: The Case for GaAs, *IEEE Spectrum,* 30 (1983).
4. Zuleeg, R., et al.: Femtojoule High-Speed Planar GaAs E-JFET Logic, *IEEE Trans. Electron Devices,* **ED-25**:628 (1978).
5. Yang, E. S.: Current Saturation Mechanisms in JFETs, *Adv. Electron. Electron Phys.*, **31**:247 (1972).
6. Bozler, C. O., and G. D. Alley: The Permeable Base Transistor and Its Application to Logic Circuits, *Proc. IEEE*, **70**:46 (1982).

PROBLEMS

8-1. A silicon n-channel JFET has the structure of Fig. 8-2a and the following parameters: $N_a = 10^{18}$ cm^{-3}, $N_d = 10^{15}$ cm^{-3}, $a = 2\,\mu m$, $L = 20\,\mu m$, and $W = 0.2$ cm. Calculate (a) the built-in potential ψ_o, (b) the pinchoff voltages V_{PO} and V_P, (c) the conductance G_o, and (d) the actual channel conductance with zero bias at the gate and drain terminals.

8-2. Derive the current-voltage relationship of the JFET having the n channel with a cross section of $2a$ by $2a$ surrounded by the p^+ region. The length of the device is L.

8-3. Derive the current-voltage relationship of the junction field-effect tetrode in which the two gates are separated. The applied voltages at the two gates are V_{G1} and V_{G2}, respectively. Assume one-sided step junctions.

8-4. Calculate and plot the transfer characteristics of the JFET in Prob. 8-1 at 25, 150, and $-50°C$. Use the electron-mobility data given in Chap. 2. Use 0.5-V increments for the gate voltage.

8-5. (a) Calculate and plot the small-signal saturation transconductance of the JFET of Prob. 8-1 at 25°C.

(b) Repeat (a) if $R_s = 50\,\Omega$.

8-6. Derive Eq. (8-18) by using the simplified small-signal equivalent circuit shown in Fig. 8-10. Neglect the capacitances and r_{gd}. Assume $r_{ds} \gg R_S + R_D$.

8-7. Derive the cutoff frequency by estimating the transit time.

8-8. Determine the relationship between the pinchoff voltage and the channel-implantation dosage Q cm^{-2} in the MESFET shown in Fig. 8-12a.

8-9. In an enhancement GaAs MESFET, we have $a = 0.6\,\mu m$ and $N_d = 10^{15}$ cm^{-3}.

(a) Find the threshold voltage for different metal gates: Cu, Ti, Mg. Use Table 7-2 (I-V) data.

(b) Calculate the percentage change in threshold voltage if ϕ_b varies by ± 30 mV.

8-10. Estimate the gate capacitance C_G at zero bias and pinchoff in the JFET of the example in Sec. 8-2.

8-11. (a) Estimate the cutoff frequency of the JFET of Prob. 8-1.

(b) Repeat (a) if $L = 2\,\mu m$.

(c) Repeat (b) if n-type GaAs is used.

8-12. Calculate the drain resistance r_{ds} of the JFET of Prob. 8-1 at $V_D = V_P + 5$ V and $V_G = -1$ V.

8-13. Determine the pinchoff voltage, transconductance, and cutoff frequency for the permeable-base transistor shown in Fig. 8-18. Let $N_d = 10^{17}$ cm^{-3}, $L = 100$ nm, $W = 2\,\mu m$. The gate stripe width is $1\,\mu m$ and the spacing between stripes is 400 nm.

CHAPTER
9

MOS
TRANSISTOR
FUNDAMENTALS

In the last few years, the *metal-oxide-semiconductor* (MOS) transistor has emerged as the most important electronic device, superseding the bipolar junction transistor in sales volume and applications. Its strength derives from its simple structure and low fabrication cost. For these reasons, the MOS transistor should continue to be the most popular device for very large scale integration and mass-production products. This chapter deals with the basic principles of operation, with more advanced topics to be presented in following chapters.

The MOS transistor is a four-terminal device in which the lateral current flow is controlled by an externally applied vertical electric field. A typical MOS transistor is shown in Fig. 9-1, where the four terminals are designated as the *source, gate, drain,* and *substrate.*[1] With no voltage applied to the gate, the two back-to-back *pn* junctions between the drain and source prevent current flow in either direction. When a positive voltage is applied to the gate with respect to the substrate, mobile negative charge is induced in the semiconductor below the semiconductor-oxide interface. The negative carriers provide a conduction *channel* between the source and the drain. Since the current is controlled by the vertical as well as the lateral electric field, it is known as a *field-effect transistor* (FET). For the transistor shown in Fig. 9-1, the induced charge in the channel is *n* type, and the device is known as an *n-channel MOSFET*, or simply *NMOS*. A *p*-channel

---

[1] In most simplified analyses, the effect of the substrate is neglected, and the MOS transistor is considered as a three-terminal device.

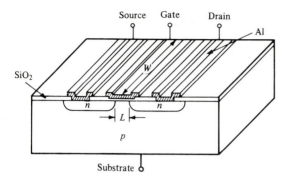

FIGURE 9-1
An *n*-channel MOS transistor.

transistor is obtained by interchanging the *n* and *p* regions. An important feature of this device is that the gate is insulated from the channel so that no current conducts through the oxide. For this reason, the device is also known as the *insulated-gate field-effect transistor* (IGFET).

Before we can understand the operation principle of the MOS transistor, we have to find out the mechanism of channel formation and the surface charge conditions. In this chapter, we begin our discussion on the formation of the surface space-charge region and energy-band diagram for an idealized MOS system. Subsequently, the topics of MOS capacitance and threshold voltage are presented for both ideal and practical structures. The basis device physics of the MOS transistor is then described.

9-1 SURFACE SPACE-CHARGE REGIONS OF AN IDEAL MOS STRUCTURE

Let us consider the metal, oxide, and semiconductor as three separate parts, as shown in Fig. 9-2a. Using the vacuum level as the energy reference, the energy-band diagrams for the three components are depicted in Fig. 9-2a. To simplify our discussion, we have assumed that the energy bands are flat and the work functions are the same for all three parts. As defined in Chap. 7, the work function specifies the work required to bring an electron from the Fermi level to the vacuum level. It should be noted that the band gap is 8 eV for SiO_2 and 1.1 eV for Si so that the vertical dimensions of the figure are not in proportion. But the diagram is correct qualitatively. As the three components are brought into contact, the Fermi levels align, as shown in Fig. 9-2b. This is an idealized situation, where the energy bands remain flat because the work functions are assumed to be the same so that there is no charge transfer upon contact. We are also assuming that there is no charge located inside the oxide or at the interface between the oxide and the semiconductor. With these two assumptions, we have eliminated localized space-charge regions and the built-in potential differences here. These assumptions will be removed later. The MOS structure shown in Fig. 9-2b may be considered as a vertically sliced section under the gate of the MOS transistor in Fig. 9-1.

The application of a voltage across the MOS capacitor establishes an electric

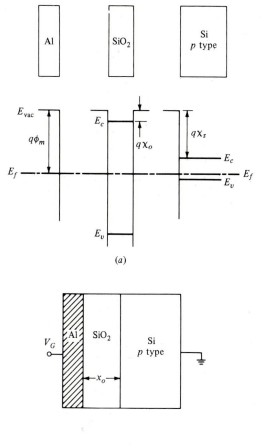

(a)

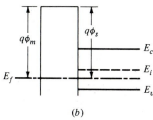

(b)

FIGURE 9-2
The MOS capacitor: structure and idealized energy-band diagram (a) before contact and (b) after contact.

field $\mathscr{E}_o$ between the plates. As a result, a displacement of mobile carriers near the surface of each plate takes place, giving rise to two space-charge regions. The density of the induced charge Q_s is given by Gauss' law

$$-Q_s = K_o \varepsilon_o \mathscr{E}_o = K_s \varepsilon_o \mathscr{E}_s \qquad (9\text{-}1)$$

where ε_o = free-space permittivity
K_o = dielectric constant of oxide
$\mathscr{E}_s$ = field at semiconductor surface
K_s = dielectric constant of semiconductor

The potential distribution is depicted in Fig. 9-3 for a p-type semiconductor for two different acceptor densities. To avoid the discontinuity of the field leading to an abrupt change of slope of the potential at the interface, the position coordinate x has been changed to x/K, with K as the appropriate dielectric constant in the material concerned. The penetration of the field into the semiconductor produces a potential barrier beneath the surface with the depth of penetration inversely proportional to the doping density. With negligible voltage drop in the metal plate, the applied voltage is shared by the voltage across the oxide V_o and *surface potential* ψ_s. Thus,

$$V_G = V_o + \psi_s \qquad (9\text{-}2)$$

The surface potential is the voltage across the semiconductor. Using the bulk semiconductor as the reference, ψ_s corresponds to the potential at the silicon surface, i.e., the silicon–silicon oxide interface. Therefore, it is called the surface potential. The electric field that exists in the semiconductor modifies the band diagram and establishes a space-charge region beneath the surface. Depending on the polarity of the applied voltage and its magnitude, it is possible to realize three different surface conditions: (1) carrier accumulation, (2) carrier depletion, and (3) carrier inversion.

The carrier densities under thermal equilibrium are given by Eqs. (1-25) and (1-26), repeated here for reference:

$$n = n_i e^{(E_f - E_i)/kT} \qquad (9\text{-}3a)$$

$$p = n_i e^{(E_i - E_f)/kT} \qquad (9\text{-}3b)$$

From these equations we find that $n > p$ for $E_f > E_i$ and the semiconductor is n the. Similarly, we have $n < p$ for $E_f < E_i$ in a p-type semiconductor. Equations (9-3) will be used to determine the carrier concentration at the surface under different biasing voltages on the MOS structure.

It should be noted that the oxide, being a good insulator, does not allow current conduction between the metal and the semiconductor even under bias. The condition of zero current corresponds to a constant Fermi level inside the bulk semiconductor. Physically, a flat Fermi level is equivalent to the state of thermal

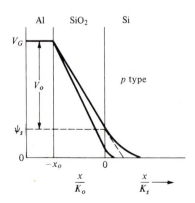

FIGURE 9-3
Potential distribution in an MOS structure with applied voltage V_G.

equilibrium where we have $np = n_i^2$. This condition prevails even under external bias in the MOS structure. We shall find this to be an important concept in examining the carrier population in the following discussion for a p-type substrate.

Carrier Accumulation

When the hole density just below the silicon surface is greater than the equilibrium-hole density in the bulk, we have the condition of carrier accumulation. This condition is realized by applying a negative voltage at the metal electrode. The resulting negative surface potential ψ_s produces an upward bending of the energy-band diagram, as shown in Fig. 9-4a. Since E_f remains constant, the band bending leads to a larger $E_i - E_f$ near the surface. According to Eq. (9-3), we have a higher hole density and a lower electron density at the surface compared with

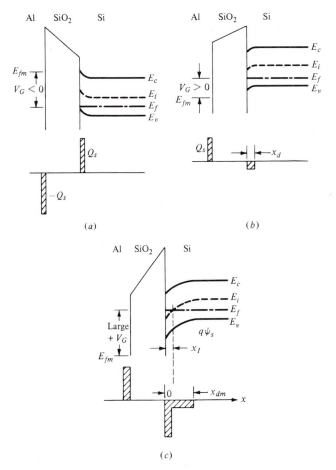

(a)

(b)

(c)

FIGURE 9-4
Energy band and charge distribution under (a) $-V_G$, (b) small $+V_G$, and (c) large $+V_G$.

that of the bulk. Consequently, holes are accumulated at the surface, and the surface conductivity is increased.

Carrier Depletion

When a small positive V_G is applied, the surface potential is positive and the energy bands bend downward, as shown in Fig. 9-4b. The Fermi level near the silicon surface is now farther away from the valence-band edge, indicating a smaller hole density. In other words, the value of $E_i - E_f$ is decreased, and holes are depleted from the vicinity of the oxide-silicon interface, establishing a space-charge region consisting of stationary acceptor ions. The total charge per unit area Q_s is given by

$$Q_s = Q_B = -qN_a x_d \tag{9-4}$$

where x_d is the width of the depletion layer, as shown in Fig. 9-4b. The symbol Q_B is defined as the *bulk charge* in the semiconductor, and the negative sign specifies the polarity of charge. The relationship between ψ_s and x_d can be obtained by solving Poisson's equation using the depletion approximation. The result is

$$\psi_s = \frac{qN_a x_d^2}{2K_s \varepsilon_o} \tag{9-5}$$

The potential distribution in the semiconductor is given by

$$\psi = \psi_s \left(1 - \frac{x}{x_d}\right)^2 \tag{9-6}$$

These equations are identical to those of the one-sided step junction having a lightly doped p side [compare with Eqs. (3-29) and (3-30)].

Carrier Inversion

If a large positive V_G is applied to the gate, the downward band bending would be more significant than that in carrier depletion. The large band bending may even cause the midgap energy E_i to cross over the constant Fermi level at or near the silicon surface. When this happens, as shown in Fig. 9-4c, an *inversion layer* is formed in which the electron density is greater than the hole density. Making use of Eqs. (9-3), we find that the right side of x_I in Fig. 9-4c remains p type but the left side of x_I becomes n type. Therefore, a pn junction is *induced* under the metal electrode.

The surface is inverted as soon as E_f becomes greater than E_i. However, the density of electrons remains small until E_f is considerably above E_i. This region of *weak inversion* will be discussed in the next chapter. For most MOSFET operation, it is desirable to define a condition after which the charge due to electrons in the inversion layer becomes moderately large. This condition, called the *onset of moderate inversion*, is reached when the electron density per unit volume at the surface is equal to the hole density in the bulk. From Eqs. (9-3), moderate inversion

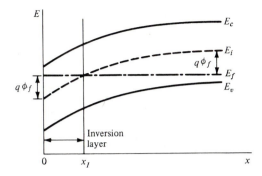

FIGURE 9-5
Energy-band diagram at onset of
moderate inversion.

demands that the midgap potential be below E_f at the surface by as much as it is above E_f in the bulk. The energy-band diagram in the semiconductor at the onset of moderate inversion is shown in Fig. 9-5. Since the bands are flat at zero bias, the required surface potential for the onset of moderate inversion is

$$\phi_{mi} = 2\phi_f \tag{9-7}$$

where $q\phi_f$ is defined as the difference between E_i and E_f in the bulk of the semiconductor.

In most MOS transistor operation, it is necessary to establish a higher inversion carrier density that corresponds to a surface potential of

$$\phi_{si} = 2\phi_f + 6\phi_T \tag{9-8}$$

This is known as the condition of the *onset of strong inversion*. Equations (9-7) and (9-8) clarify the difference between strong and moderate inversion, which follows Tsividis' treatment [1]. More will be said about the various inversion conditions in the next chapter.

The corresponding width of the surface depletion region of the induced junction is derived by using Eq. (9-5) and setting $\psi_s = \phi_{si}$:

$$x_{dm} = \sqrt{\frac{2K_s \varepsilon_o \phi_{si}}{qN_a}} \tag{9-9}$$

Let us examine what would happen if the applied voltage is so large that $\psi_s > \phi_{si}$. Since the increase of ψ_s is added to the difference of E_f and E_i, a small increase of ψ_s produces a large increase of electrons at the surface, according to the exponential nature of Eq. (9-3). Therefore, the surface inversion layer is acting like a narrow n^+ layer, and the induced junction resembles an n^+p junction for a large positive gate voltage. However, all induced charge will be in the inversion layer after strong inversion, and the charge inside the depletion layer will remain constant. Thus, the space-charge-layer width remains at x_{dm} as we further increase the gate voltage. With a depletion-layer width of x_{dm}, the bulk charge Q_B becomes

$$Q_B = -qN_a x_{dm} \tag{9-10}$$

In strong inversion, the charge condition in the surface region is given by

$$Q_s = Q_I + Q_B = Q_I - qN_a x_{dm} \tag{9-11}$$

where Q_I is the charge density per unit area in the inversion layer. It is important to note that Q_I is a function of the applied gate voltage, and these induced mobile charges become the current carriers in an MOSFET.

Example. Determine the surface band bending at strong inversion in a p-type silicon with $N_a = 3 \times 10^{14}$ cm³ at 300 K. Calculate the corresponding depletion-layer width and the total bulk charge.

Solution. Since

$$\phi_f = \phi_T \ln \frac{N_a}{n_i} = 0.255 \text{ V}$$

$$\phi_{mi} = 2\phi_f = 0.51 \text{ V} \tag{9-7}$$

Therefore,

$$\phi_{si} = 0.51 + 6 \times 0.026 = 0.67 \tag{9-8}$$

$$x_{dm} = \sqrt{\frac{2K_s \varepsilon_o \phi_{si}}{qN_a}} = 1.7 \times 10^{-4} \text{ cm} \tag{9-9}$$

$$Q_B = -qN_a x_{dm} = -8.3 \times 10^{-9} \text{ C/cm}^2 \tag{9-10}$$

9-2 THE MOS CAPACITOR

The MOS structure shown in Fig. 9-2 is basically a capacitor with the SiO_2 as the dielectric material. If the silicon were a perfect conductor, the parallel-plate capacitance per unit area would have been given by the oxide capacitance

$$C_o = \frac{K_o \varepsilon_o}{x_o} \tag{9-12}$$

where x_o is the oxide thickness. However, the MOS capacitor is more complicated because of the voltage dependence of the surface space-charge layer in silicon. The space charge of the depletion layer acts as another capacitor C_d in series with C_o, giving an overall capacitance of

$$C = \frac{C_o C_d}{C_o + C_d} \tag{9-13}$$

The space-charge-layer capacitance is

$$C_d = \frac{K_s \varepsilon_o}{x_d} \tag{9-14}$$

Under the condition of carrier accumulation, there is no depletion layer under the silicon surface, and the overall capacitance is equal to C_o. In strong inversion, the maximum space-charge width x_{dm} becomes a constant, and C_d is also a constant. With biasing voltage between the condition of carrier accumulation and strong inversion, the space-charge-layer width x_d is a function of the bias voltage V_G. Replacing $\mathscr{E}_o$ by V_o/x_o in Eq. (9-1) and solving it with Eq. (9-2), we obtain

$$V_G = -\frac{Q_s}{C_o} + \psi_s \qquad (9\text{-}15)$$

where Eq. (9-12) has been used. Substituting Eqs. (9-4) and (9-5) into Eq. (9-15) and solving for x_d gives

$$x_d = \frac{K_s \varepsilon_o}{C_o}\left(\sqrt{1 + \frac{2V_G}{qK_s\varepsilon_o N_a}\, C_o^2} - 1\right) \qquad (9\text{-}16)$$

Substitution of Eqs. (9-14) and (9-16) into Eq. (9-13) leads to

$$C = \frac{C_o}{[1 + (2C_o^2/qN_a K_s\varepsilon_o)V_G]^{1/2}} \qquad (9\text{-}17)$$

Equation (9-17) describes the MOS capacitor under the condition of carrier depletion since Eqs. (9-4) and (9-5) have been employed. This result is plotted as the solid curve in Fig. 9-6. The MOS capacitance follows this curve until carrier inversion sets in. As soon as the inversion-layer charge becomes significant, the capacitance starts to deviate from Eq. (9-17). In strong inversion, the inversion-layer charge is large enough so that the depletion-layer width x_d is essentially equal to x_{dm}. Consequently, the depletion capacitance given by Eq. (9-14) becomes a constant which leads to a constant MOS capacitance [Eq. (9-13)]. This result is plotted as the dashed line in Fig. 9-6. Near zero bias, the depletion approximation is also inaccurate because the transition between the depletion layer and the neutral region is not abrupt. This nonabruptness in the space-charge-layer edge is usually negligible, as described in Sec. 3-3 for a *pn* junction, but it can cause significant departure from the depletion approximation in an MOS capacitance. The actual depletion capacitance at zero bias is found to be $C_{FB} = (qK_s\varepsilon_o N_a/\phi_T)^{1/2}$, which will be derived in the next chapter.

The foregoing description satisfies the experimental capacitance-voltage (C–V) curve very well at high frequencies. At a very low frequency, for example, 10 Hz, additional complication sets in, and the C–V curve follows the dotted curve shown in Fig. 9-6. The mechanism responsible for the low-frequency C–V characteristics is carrier generation within the space-charge layer [2]. To understand this effect, let us examine the incremental change of charges upon the application of a positive voltage, shown in Fig. 9-7. The incremental positive gate voltage increases the silicon surface potential ψ_s so that holes are depleted and the depletion layer is widened. Therefore, more negative fixed charge is established at the edge of the neutral *p*-type semiconductor, as shown in Fig. 9-7a. This leads to two capacitors in series with an overall capacitance, expressed by Eq. (9-13). If we assume now that electron-hole pairs can be generated fast enough, the generated holes will replenish the depleted holes at the edge of the depletion region. At the same time, the generated electrons will be drawn by the field and accumulate at

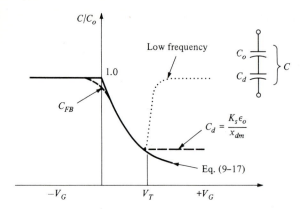

FIGURE 9-6
Capacitance-voltage characteristics
for high and low frequencies.

the silicon-oxide interface (Fig. 9-7b). The overall capacitance now is just C_o. As a result, the capacitance-voltage characteristics are frequency-dependent, and the generation rate determines the frequency range of transition.

In Fig. 9-6, the turning point of the low-frequency capacitance curve is designated as the *threshold voltage* V_T. This is the critical voltage at which the inversion layer is formed to a significant extent, giving rise to rapid increase of the inverse charge for higher gate voltages. In other words, the threshold voltage specifies the gate voltage at strong inversion. This condition is satisfied if the band bending is ϕ_{si}, and the bulk charge Q_B is given by Eq. (9-10). Since the gate voltage that induces a charge Q_B is $-Q_B/C_o$, the threshold voltage can be defined by

$$V_T \equiv -\frac{Q_B}{C_o} + \phi_{si} \qquad (9\text{-}18)$$

Physically, the threshold voltage supports a bulk charge Q_B and at the same time introduces a band bending at the surface to reach the strong inversion potential ϕ_{si}. Substituting Eqs. (9-15) and (9-18) into (9-11) yields

$$Q_I = -C_o(V_G - V_T) \qquad (9\text{-}19)$$

This equation relates the inversion-layer charge to the gate voltage above the threshold voltage. It is a simple linear relationship and is useful for MOS transistor analysis in a later section.

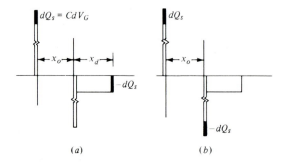

FIGURE 9-7
Charge distribution in an MOS
structure at (a) high and (b) low
frequencies.

Example. For the example in Sec. 9-1, an oxide of 100 nm is grown before a metal gate is deposited. Determine the threshold voltage.

Solution. Since

$$C_o = \frac{K_o \varepsilon_o}{x_o} = 3.5 \times 10^{-8} \text{ F/cm}^2 \qquad \text{Eq. (9-12)}$$

we have

$$V_T = -\frac{Q_B}{C_o} + 2\phi_f + 6\phi_T = 0.91 \text{ V} \qquad \text{Eq. (9-18)}$$

9-3 FLAT-BAND AND THRESHOLD VOLTAGES

In the previous discussions on the idealized MOS structure, we assumed that the energy-band diagram (Fig. 9-2b) is flat when the gate voltage is zero. In practice, this condition is not realized because of unavoidable work-function difference and charges in the oxide and surface states.

The work function of a material was defined in Chap. 7 as the energy required to bring an electron from the Fermi level to the vacuum level. In an MOS structure, the appropriate energies to be considered are the *modified* work functions from the respective Fermi level in the metal and semiconductor to the oxide conduction-band edge (Fig. 9-8a). In general, the modified work function of the metal is not

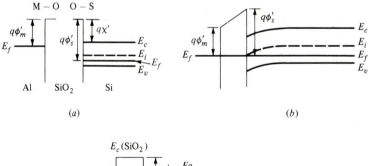

(a) (b)

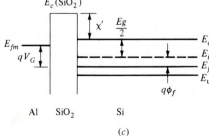

(c)

FIGURE 9-8
Energy-band diagrams for an Al–SiO$_2$–Si structure (a) before contact and (b) after contact at $V_G = 0$, (c) with $V_G = -V_{FB} = +q(\phi'_m - \phi'_s)$.

TABLE 9-1
Modified work function for metal–SiO$_2$ system

Metal	ϕ_m, V	ϕ'_m,† V
Al	4.1	3.2
Ag	5.1	4.2
Au	5.0	4.1
Cu	4.7	3.8
Mg	3.35	2.45
Ni	4.55	3.65

† $\phi'_m = \phi_m - 0.9$ V (SiO$_2$ electron affinity).
Source: After Sze [4].

equal to that of the semiconductor. Similar to the metal-semiconductor contact described in Chap. 7, when the metal, oxide, and semiconductor are joined together, a band bending exists in the semiconductor in order to satisfy the requirement of constant Fermi level under thermal equilibrium. This characteristic is illustrated in Fig. 9-8b for an Al–SiO$_2$–Si structure. The gate voltage required to eliminate the band bending so that no electric field exists in the silicon is clearly the difference of the modified work functions:

$$V_{G1} = \phi'_{ms} = \phi'_m - \phi'_s \tag{9-20}$$

The result is depicted in Fig. 9-8c. The modified work function for several metals is tabulated in Table 9-1. At room temperature, the experimental value of the

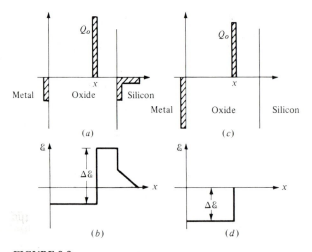

FIGURE 9-9
Effect of oxide charge on electric field distribution: (a) Q_o induces charges in metal and silicon at $V_G = 0$, (b) electric field for (a), (c) charge distribution for the flat-band condition, and (d) the electric field for (c).

modified electron affinity χ' for silicon is $3.25\,eV$ and $E_g = 1.1\,eV$. Therefore, the modified work function according to Fig. 9-8 is given by

$$\phi_s' = 3.25 + \frac{1.1}{2} + \phi_f = 3.8 + \phi_f \qquad V \qquad (9\text{-}21)$$

The gate voltage necessary to achieve the flat-band condition is further modified by charges residing within the oxide. Let us consider a thin layer of charge $+Q_o$ per unit area located at x from the metal as shown in Fig. 9-9a. With zero gate voltage, Q_o induces charges in both the metal and semiconductor, and the sum of the induced charges equals $-Q_o$. The corresponding electric field distribution is obtained by integrating the charge, and the result is plotted in Fig. 9-9b.. The induced charge in the silicon produces a nonzero field at the semiconductor surface so that the energy-band diagram is not flat. To achieve the flat-bend condition, we apply a negative gate voltage to lower the electric field distribution until the charge is reduced to zero at the silicon surface. The electric field increment is obtained by integrating the charge distribution, and the result is

$$\Delta\mathscr{E} = \int_0^{x+} \frac{\rho\,dx}{K_o\varepsilon_o} = \frac{Q_o}{K_o\varepsilon_o} \qquad (9\text{-}22)$$

Both the charge and field distributions are plotted in Fig. 9-9c and d. The gate voltage necessary to realize the flat-band condition (i.e., the electric field is zero at $x+$) is therefore

$$V_{G2} = -\int_0^{x+} \Delta\mathscr{E}\,dx = -\frac{xQ_o}{K_o\varepsilon_o} = -\frac{x}{x_o}\frac{Q_o}{C_o} \qquad (9\text{-}23)$$

This equation shows that the necessary gate voltage to reach the flat-band condition depends on the density and location of the oxide charge. If Q_o is located at $x = 0$, the corresponding flat-band voltage will be zero. On the other hand, the same charge residing at x_o requires the largest gate voltage to achieve the flat-band condition. In most cases, the charge at the silicon-oxide interface resulting from surface states dominates, and the flat-band voltage is

$$V_{G2} = -\frac{Q_o}{C_o} \qquad (9\text{-}24)$$

The effects of oxide charge and work-function difference can be combined, and the gate voltage required to achieve the flat-band condition, known as the *flat-band voltage*, is given by

$$V_{FB} = V_{G1} + V_{G2} = \phi_{ms}' - \frac{Q_o}{C_o} \qquad (9\text{-}25)$$

The flat-band voltage can be estimated from the C–V plot shown in Fig. 9-10. The shift of the C–V curve away from $V = 0$ is a reasonable measure of the flat-band voltage.

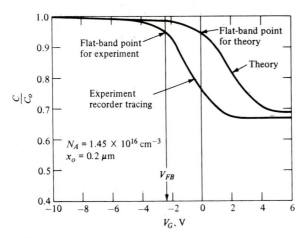

FIGURE 9-10
The combined effects of metal-semiconductor work-function difference and charges within the oxide on the capacitance-voltage curves. *(After Grove et al. [2].)*

The threshold voltage must also be modified to include the effects of work-function difference and oxide charge. The new threshold voltage is the sum of (1) the flat-band voltage, (2) the voltage to support a depletion-region charge Q_B, and (3) the voltage to produce band bending for strong inversion. Thus, Eq. (9-18) becomes

$$V_T = V_{FB} + \phi_{si} - \frac{Q_B}{C_o} = \phi'_{ms} + \phi_{si} - \frac{Q_o}{C_o} - \frac{Q_B}{C_o} \qquad (9\text{-}26)$$

If a voltage ψ is established in the channel by external voltages at the source and drain, Q_B under strong inversion can be rewritten as

$$Q_B = -[2qK_s\varepsilon_o N_a(\psi + \phi_{si})]^{1/2} \qquad (9\text{-}27)$$

Note that ψ is the voltage of the n-inversion-layer side of the junction with respect to the p-side substrate.

Example. An aluminum-gate MOS transistor is fabricated on an n-type $\langle 111 \rangle$ silicon substrate with $N_d = 10^{15}$ cm^{-3}. The thickness of the gate oxide is 120 nm, and the surface charge density at the oxide-silicon interface is 3×10^{11} cm^{-2}. Calculate the threshold voltage with the channel potential $\psi = 0$. Both Q_B and Q_o are positive charges.

Solution. We obtain first the oxide capacitance

$$C_o = \frac{K_o\varepsilon_o}{x_o} = 2.9 \times 10^{-8} \text{ F/cm}^2$$

Letting $n = N_d$, we obtain ϕ_f by using Eq. (9-3):

$$\phi_f = \phi_T \ln \frac{N_d}{n_i} = 0.29 \quad \text{V}$$

Therefore

$$\phi_{si} = 2\phi_f + 6\phi_T = 0.736 \text{ V}$$

Using Eq. (9-27),

$$Q_B = 1.6 \times 10^{-8} \text{ C/cm}^2$$

The work-function difference is calculated from Table 9-1 and Eqs. (9-20) and (9-21):

$$\phi'_{ms} = 3.2 - (3.8 - 0.29) = -0.31$$

The threshold voltage from Eq. (9-26) is

$$V_T = -0.31 - 2 \times 0.29 - 6 \times 0.026 - \frac{(1.6 \times 10^{-19})(3 \times 10^{11})}{2.9 \times 10^{-8}} - \frac{1.6 \times 10^{-8}}{2.9 \times 10^{-8}}$$

$$= -0.31 - 0.58 - 0.156 - 1.65 - 0.54 = -3.2 \text{ V}$$

In most recently fabricated MOS transistors, the metal gate has been replaced by polycrystalline silicon. An important advantage of this structure is that the work-function difference between the gate and the substrate can be eliminated since both are silicon. However, the doping of the polysilicon gate is usually very high. Since the electron affinity remains constant and the Fermi level shifts with the doping concentration, the effective work-function difference changes with the substrate doping density. For example, if the polysilicon gate is doped degenerately to 10^{19} or more boron atoms per cubic centimeter, the Fermi level is essentially pinned at E_v. The work-function difference for the MOS system is

$$\phi_{ms} = (E_{vac} - E_v) - (E_{vac} - E_{fs}) \tag{9-28}$$

where E_{fs} is the Fermi level of the substrate. Using Eq. (1-33) for an n-type substrate, we have

$$E_{fs} = E_c - kT \ln \frac{N_c}{N_d} \tag{9-29}$$

Substituting Eq. (9-29) into (9-28), we obtain

$$\phi_{ms} = E_g - kT \ln \frac{N_c}{N_d} \tag{9-30}$$

For a p-type substrate, the Fermi level and the work-function difference are

$$E_{fs} = E_v + kT \ln \frac{N_v}{N_a} \tag{9-31}$$

and

$$\phi_{ms} = kT \ln \frac{N_v}{N_a} \tag{9-32}$$

Equations (9-30) and (9-32) are plotted in Fig. 9-11 for n^+- and p^+-polysilicon gates.

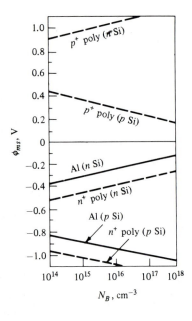

FIGURE 9-11
Work-function difference ϕ_{ms} vs. doping for degenerate polysilicon and Al electrodes. *(After Werner [3].)*

9-4 STATIC CHARACTERISTICS OF THE MOS TRANSISTOR

A schematic diagram of an *n*-channel MOSFET is shown in Fig. 9-12, in which bias voltages are included. To simplify our analyses, we have grounded the source and substrate terminals. The static characteristics of the MOSFET have two distinct regions of operation. At low drain voltages, the drain-to-source characteristics are basically ohmic, and the drain current is proportional to the drain voltage in a nearly linear fashion. This operation region is the linear region. At high drain voltages, the drain current remains almost constant with increasing drain voltage. This second region is the saturation region since the current is saturated. We derive first the current-voltage characteristics of the transistor in the linear region and then extend the results to the onset of saturation.

Linear Region of Operation

In the following analysis, we assume the vertical and lateral electric fields are independent of each other. In Fig. 9-12 the electric field in the *x* direction induces an inversion layer, whereas the electric field in the *y* direction produces a drain current flowing along the silicon surface.

Let us consider a small section at *y* of the transistor shown in Fig. 9-12 under the condition that the gate voltage is greater than the threshold voltage so that significant amounts of mobile carriers are induced in the inversion layer. The relatonship between the mobile charge Q_I and the gate voltage is given by Eq. (9-19) if we take the channel voltage as zero. As a result of the drain-to-source

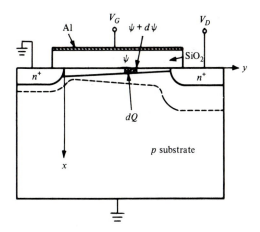

FIGURE 9-12
The n-channel MOS transistor.

bias, however, a potential ψ is established at y. Therefore, the induced channel charge should be modified to

$$Q_I = -C_o(V_G - V_T - \psi) \tag{9-33}$$

The channel current, being a majority-carrier current, can now be written as

$$I_D = W\mu_n Q_I \mathscr{E}_y \tag{9-34}$$

which is obtained by using the transport equation expressed in Eq. (2-69) and by neglecting the diffusion component. Substituting $\mathscr{E}_y = -d\psi/dy$ and Eq. (9-33) into Eq. (9-34), we obtain

$$I_D dy = W\mu_n C_o(V_G - V_T - \psi) \, d\psi \tag{9-35}$$

Integration of this equation from $y = 0$ to $y = L$ and from $\psi = 0$ to $\psi = V_D$ yields

$$I_D = C_o \mu_n \frac{W}{L}[(V_G - V_T)V_D - \tfrac{1}{2}V_D^2] \tag{9-36}$$

In the derivation of Eq. (9-36), we have assumed that V_T is independent of ψ. In other words, ψ is assumed to be zero in Eq. (9-27). This approximation can lead to erroneous results. In fact, V_T increases toward the drain due to the increase of Q_B, as shown in Eq. (9-27). If we use Eq. (9-27) without simplification, the drain-current equation becomes

$$I_D = C_o \mu_n \frac{W}{L} \left\{ \left(V_G - \phi'_{ms} - \phi_{si} + \frac{Q_o}{C_o} - \frac{V_D}{2} \right) V_D \right.$$

$$\left. - \frac{2}{3} \frac{\sqrt{2qK_s\varepsilon_o N_a}}{C_o} [(V_D + \phi_{si})^{3/2} - \phi_{si}^{3/2}] \right\} \tag{9-37}$$

Equations (9-36) and (9-37) are plotted in Fig. 9-13 for a p-channel device with typical device parameters. It is shown that the simplified expression overestimates the drain current and the difference is small here, but it often proves significant,

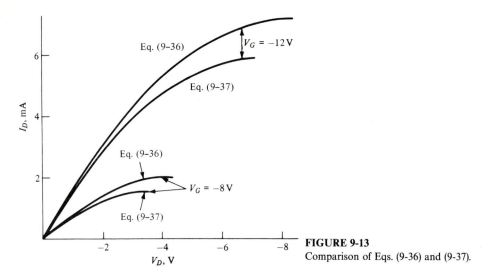

FIGURE 9-13
Comparison of Eqs. (9-36) and (9-37).

particularly if the substrate doping is high. Nevertheless, the simplicity of Eq. (9-36) provides better physical insight into the device operation, and it is useful in obtaining a first-order design, particularly in digital-circuit applications. For circuit design, a good empirical fit to actual device characteristics can be obtained by arbitrarily adjusting $C_o \mu_n$ so that Eq. (9-36) fits the observed characteristics in the I–V range of interest. We shall use the simplified expression in subsequent analyses; interested readers should consult the references at the end of the chapter.

Saturation Region

The foregoing analysis is based on the premise that an inversion layer is formed throughout the semiconductor surface between the source and the drain. If we increase the drain voltage so that the gate voltage is neutralized, the inversion layer will disappear at the drain side of the channel. The channel is now pinched off, and further increases of the drain voltage would not increase the drain current significantly. Therefore, the drain current is saturated, and the equations derived for the linear region are rendered inoperative.

The condition for the onset of the current saturation is given by setting Q_I equal to zero in Eq. (9-33). Therefore,

$$\psi = V_G - V_T = V_{DS} \tag{9-38}$$

where V_{DS} is the drain voltage, which is equal to ψ for the channel adjacent to the drain; it is known as the *drain saturation voltage*. Substitution of Eq. (9-38) into Eq. (9-36) yields

$$I_{DS} = \frac{\mu_n C_o W}{2L} (V_G - V_T)^2 \tag{9-39}$$

This equation is valid at the onset of saturation. Beyond this point the drain current can be considered as constant so that the foregoing equation is also applicable to the drain current for higher drain voltages. The complete current-voltage characteristics of the MOSFET are plotted in Fig. 9-14, in which the dashed curve specifies the onset of current saturation.

Example. The MOS structure of the example in Sec. 9-3 is used as an MOSFET. The following parameters are given: $L = 10\ \mu m$, $W = 300\ \mu m$, and $\mu_p = 230\ \text{cm}^2/\text{V} \cdot \text{s}$. Calculate I_{DS} for $V_G = -4\ \text{V}$ and $V_G = -8\ \text{V}$.

Solution. Since $C_o = 2.9 \times 10^{-8}\ \text{F/cm}^2$ in Sec. 9-3 we have

$$C_o \mu_p \frac{W}{L} = (2.9 \times 10^{-8})(230)(30) = 2 \times 10^{-4}\ \text{F/V} \cdot \text{s}$$

Substituting this value into Eq. (9-36) and setting $V_D = V_G - V_T$ yields

$$I_{DS} = \frac{2 \times 10^{-4}}{2} (V_G - V_T)^2$$

Since $V_T = -3.2\ \text{V}$,

$$I_{DS} = \begin{cases} 0.64 \times 10^{-4} = 64\ \mu A & \text{for } V_G = -4\ \text{V} \\ 23 \times 10^{-4} = 2.3\ \text{mA} & \text{for } V_G = -8\ \text{V} \end{cases}$$

Cutoff Region

If the gate voltage is smaller than the threshold voltage, no inversion layer is formed. As a result, the MOSFET behaves like two *pn* junctions connected back to back to prevent current flow in either direction. The transistor acts as an open circuit in this region of operation.

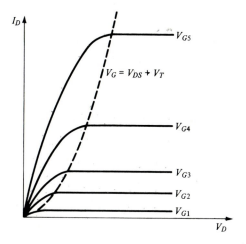

FIGURE 9-14
Current-voltage characteristics of an *n*-channel MOSFET.

9-5 SMALL-SIGNAL PARAMETERS AND EQUIVALENT CIRCUIT

The small-signal conductance is defined by

$$g_d \equiv \frac{\partial I_D}{\partial V_D}\bigg|_{V_G = \text{const}} \tag{9-40}$$

In the linear region where V_D is small, the conductance is obtained by differentiation of Eq. (9-36):

$$g_d = \mu_n C_o \frac{W}{L} (V_G - V_T - V_D) \approx \mu_n C_o \frac{W}{L} (V_G - V_T) \tag{9-41}$$

Note that the drain conductance is ohmic and linearly dependent on the gate voltage. Experimental data for an n-channel MOSFET are plotted in Fig. 9-15. We observe that g_d follows the gate voltage linearly except for high gate voltages, for which the decrease of mobility at high surface-carrier concentration is responsible. We also find that extrapolation of the data to the V_{GS} axis gives the experimental value of the threshold voltage. Besides the room-temperature data, two additional curves are included for the high- and low-temperature extremes. The decrease of slope with increasing temperature is caused by the reduction of mobility at high temperatures. The resistance in the linear region, frequently referred to as the ON resistance, is given by

$$R_{\text{on}} = \frac{1}{g_d} = \frac{L}{\mu_n C_o W (V_G - V_T)} \tag{9-42}$$

An important parameter in a MOSFET is the transconductance, defined by

$$g_m \equiv \frac{\partial I_D}{\partial V_G}\bigg|_{V_D = \text{const}} \tag{9-43}$$

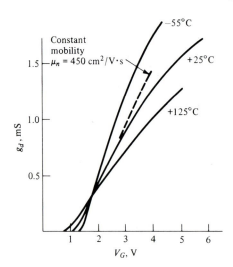

FIGURE 9-15
Channel conductance vs. V_G in an n-channel MOSFET.

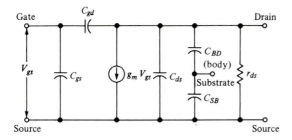

FIGURE 9-16
Small-signal equivalent circuit of an MOS transistor. C_{BD} and C_{SB} are the substrate-to-drain and substrate-to-source capacitances.

In the linear region, the transconductance is obtained by differentiation of Eq. (9-36), yielding

$$g_m = C_o \mu_n \frac{W}{L} V_D \tag{9-44}$$

The transconductance in the saturation region is derived by differentiating Eq. (9-39), which leads to

$$g_m = \frac{\mu_n C_o W}{L}(V_G - V_T) \tag{9-45}$$

Note that the expressions of the saturation g_m and the linear g_d are identical, which is valid only when Q_B is assumed to be constant.

According to the idealized theory, the drain current remains constant for any drain voltage beyond pinchoff. In other words, the drain resistance is infinite for $V_D > V_{DS}$. However, all experimental MOSFETs show a finite slope in their drain current-voltage characteristics beyond pinchoff. Therefore, we define a drain resistance in the saturation region as

$$r_{ds} = r_d(\text{sat}) \equiv \frac{\partial V_{DS}}{\partial I_{DS}} \bigg|_{V_G = \text{const}} \tag{9-46}$$

The drain resistance in the saturation region can be obtained graphically from the drain characteristics.

An equivalent circuit of the MOSFET is shown in Fig. 9-16. The gate-to-drain capacitance usually dominates the high-frequency characteristics because of the Miller effect.

9-6 CUTOFF FREQUENCY OF AN MOS TRANSISTOR

The maximum operating frequency or the cutoff frequency f_{co} is defined as the condition when the transistor no longer amplifies the input signal. Let us use the equivalent circuit shown in Fig. 9-16 with the output terminals short-circuited. The unity-gain condition is reached when the current through the input capacitance is equal to the output drain current. The input current is given by

$$I_i = 2\pi f_{co}(C_{gs} + C_{gd})V_{gs} = 2\pi f_{co} C_G V_{gs} \tag{9-47}$$

By neglecting the current passing through C_{gd}, the output current is

$$I_o = g_m V_{gs} \tag{9-48}$$

Equating Eqs. (9-47) and (9-48), we obtain the cutoff frequency as

$$f_{co} = \frac{g_m}{2\pi C_G} = \frac{\mu_n V_D}{2\pi L^2} \tag{9-49}$$

The last expression in Eq. (9-49) is obtained by using Eq. (9-44) and the relation[1]

$$C_G = WLC_o \tag{9-50}$$

In Eq. (9-49), it is noted that large mobility and short channel length are required to achieve the best high-frequency performance.

9-7   EFFECT OF SUBSTRATE (BODY) BIAS

In the previous analyses we have assumed that the source is connected to the substrate and both terminals are grounded. Let us keep the source at ground potential but apply a negative voltage $-V_{BS}$ to the p-type substrate (body) (Fig. 9-12). A reverse-bias voltage V_{SB} will be across the induced junction between the channel and body so that the space-charge layer is widened. As a result, the negative fixed charge Q_B in the space-charge layer is increased. When V_{SB} is zero, the charge in the space-charge layer is given by Eq. (9-4) and is rewritten as

$$Q_B = -qN_a x_d = -(2qK_s \varepsilon_o N_a \phi_{si})^{1/2} \tag{9-51}$$

where x_d is given by Eq. (3-31) with ψ_o and N_d replaced by ϕ_{si} and N_a, respectively. For an arbitrary reverse-bias voltage V_{SB} we have

$$Q_B = -[2qK_s \varepsilon_o N_a (V_{SB} + \phi_{si})]^{1/2} \tag{9-52}$$

Therefore, the incremental charge is

$$\Delta Q_B = -(2qK_s \varepsilon_o N_a)^{1/2}[(V_{SB} + \phi_{si})^{1/2} - \phi_{si}^{1/2}] \tag{9-53}$$

To reach the condition of strong inversion, the applied gate voltage must be increased to compensate for ΔQ_B. Therefore,

$$\Delta V_T = -\frac{\Delta Q_B}{C_o} = \frac{(2qK_s \varepsilon_o N_a)^{1/2}}{C_o}[(V_{SB} + \phi_{si})^{1/2} - \phi_{si}^{1/2}] \tag{9-54}$$

Figure 9-17 shows experimental data on the change of threshold voltage with V_{SB}, and it agrees with Eq. (9-54). Equations derived in Secs. 9-4 and 9-5 are valid even with substrate bias, provided that the change of threshold voltage is taken into account.

---

[1] Strictly speaking, $C_{gs} + C_{gd} \neq WLC_o$ in saturation. See Tsividis [1] for details.

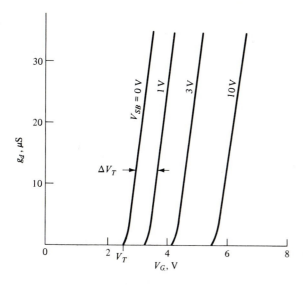

FIGURE 9-17
Effect of substrate bias on the threshold voltage.

9-8 MOS DEVICE FABRICATION

In this section, basic steps of the MOS transistor fabrication process will be presented. More details will be given in Chap. 12 on some of the latest developments.

One of the simplest processes in fabricating a p-channel metal-gate MOS transistor is shown in Fig. 9-18, where both the top view and the cross-sectional view are depicted, together with the major processing steps. The top view is the

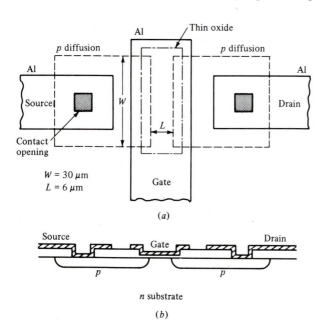

FIGURE 9-18
A p-channel MOS transistor.

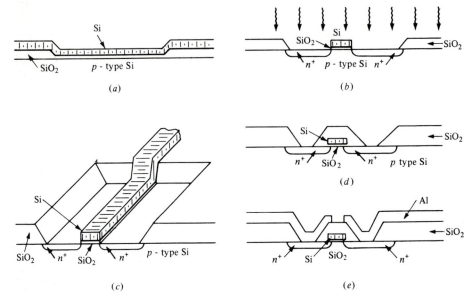

FIGURE 9-19
The silicon-gate technology.

layout diagram, in which outlines of four different masks are superimposed. The thin oxide under the metal gate is called the *gate oxide*, and its formation and protection are the most critical in the entire process. Frequently, charged ions contaminate the gate oxide and the threshold voltage is modified drastically. In the early development of the MOSFET, sodium ions were a serious problem because the ion migration causes the threshold voltage to change with time. In the past few years, the use of electron-beam evaporation and very clean processes enable manufacturers to produce stable MOS devices.

Reduction of the threshold voltage can be accomplished by employing polycrystalline silicon as the gate to replace the aluminum electrode. This process is called the *silicon-gate technology* The processing steps of an n-channel silicon-gate device are shown in Fig. 9-19. First, a thick oxide, approximately 1 μm thick, is thermally grown over the n substrate. Next, a window is opened by photomasking to define the source, channel, and drain regions of the final device. Subsequently, a thin layer of silicon dioxide is formed by thermal oxidation, followed by the deposition of a thin polycrystalline silicon layer. A second photomasking step is used to remove the polysilicon except in the specified gate area. The thin oxide is also removed by an oxide etch. Impurities are then diffused to produce the n^+-type drain and source regions as well as making the gate silicon highly n type. Next, a thick oxide layer is deposited over the entire wafer, and a mask is used to define the contact windows. Finally, aluminum is evaporated and etched to provide interconnections.

Self-alignment between the gate electrode and the source and drain regions can also be obtained by ion implantation with the gate as the mask. Figure 9-20

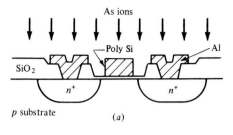

(a)

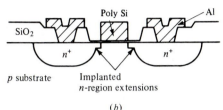

(b)

FIGURE 9-20
Self-alignment by ion implantation.

shows the device structure in which ion implantation is employed. The feature of self-aligned gate enables us to fabricate devices with small channel length which exhibit good high-frequency response.

REFERENCES

1. Tsividis, Y.: "The MOS Transistor," McGraw-Hill, New York, 1987.
2. Grove, A. S. et al.: Simple Physical Model for the Space-Charge Capacitance of MOS Structures, *J. Appl. Phys.*, **35**:2458 (1964).
3. Werner, W. M.: The Work Function Difference of the MOS System in Aluminum Field Plates and Polycrystalline Silicon Field Plates, *Solid-State Electron.*, **17**:769 (1974).
4. Sze, S. M.: "Physics of Semiconductor Devices," 2d ed., Wiley, New York, 1981.

Additional Readings

Brews, J.: Physics of the MOS Transistor, in D. Kahng (ed.), "Silicon Integrated Circuits," part A, Academic, New York, 1981.
Grove, A. S.: "Physics and Technology of Semiconductor Devices," Wiley, New York, 1967.
Muller, R. S., and T.I. Kamins: "Device Electronics for Integrated Circuits," 2d ed., Wiley, New York, 1986.
Pierret, R. F.: "Field Effect Devices," Addison-Wesley, Reading, Mass., 1983.

PROBLEMS

9-1. Sketch the energy-band diagrams and the charge distribution in an MOS structure under biasing conditions corresponding to carrier accumulation, depletion, and strong inversion. Use an n-type substrate and neglect the effects of surface states and work-function difference. (Specify the polarity of the gate voltage with respect to the substrate.)

9-2. Derive expressions of the bulk charge, surface potential, and surface field to show their dependence on the substrate doping density N_a at strong inversion. Plot the bulk charge, surface potential, and field as a function of N_a from 10^{14} to 10^{18} cm^{-3}.

9-3. An idealized MOS capacitor has an oxide thickness of 0.1 μm and $K_o = 4$ on a p-type silicon substrate with acceptor concentration of 10^{16} cm^{-3}. What is the capacitance at (a) $V_G = +2$ V and $f = 1$ Hz, (b) $V_G = +20$ V and $f = 1$ Hz, and (c) $V_G = +20$ V and $f = 1$ MHz?

9-4. Using superposition, show that the change in the flat-band voltage corresponding to a charge distribution $q\rho(x)$ in the oxide is given by

$$\Delta V_{FB} = -\frac{q}{C_o} \int_0^{x_o} \frac{x\rho(x)}{x_o} \, dx$$

9-5. Calculate the flat-band voltage of an MOS device with $x_o = 50$ nm, $q\phi_m = 4$ eV, $q\phi_s = 4.5$ eV, and a uniform positive oxide charge of 10^{16} cm^{-3}. Assume $K_o = 4$. Use the expression in Prob. 9-4.

9-6. Compare the following cases by using the result of Prob. 9-4:

(a) A density of 1.5×10^{12} charges/cm^2 is uniformly distributed in the oxide in an MOS structure. Calculate the flat-band voltage due to this charge if the oxide thickness is 150 nm.

(b) Repeat (a) if all the charge is located at the silicon–silicon oxide interface.

(c) Repeat (a) if the charge forms a triangular distribution which peaks at $x = 0$ and zero at $x = x_o$.

9-7. An aluminum-gate MOS transistor is fabricated on a p-type Si [111] substrate with $N_a = 10^{15}$ cm^{-3}. The thickness of the gate oxide is 120 nm, and the surface charge density is 3×10^{11} cm^{-2}. (a) Calculate the threshold voltage. (b) Repeat for an n^+-polysilicon gate and a p^+-polysilicon gate with $x_o = 50$ nm.

9-8. An MOS structure consists of an n-type substrate with $N_d = 5 \times 10^{15}$ cm^{-3}, an oxide of 100 nm, and an aluminum contact. The measured threshold voltage is -2.5 V. Calculate the surface charge density.

9-9. (a) Verify the following expression of the temperature dependence of the threshold voltage for an MOS structure with an n substrate, assuming $\phi_{Si} = 2\phi_f$ here and $\phi_f < 0$.

$$\frac{dV_T}{dT} = \frac{1}{T}\left(2 - \frac{Q_B}{2C_o\phi_f}\right)\left(\phi_f + \frac{E_g}{2q}\right)$$

(b) Does V_T increase or decrease with temperature? What is the effect of the variation of oxide thickness and substrate doping?

9-10. A p-channel aluminum-gate MOS transistor has the following parameters: $x_o = 100$ nm, $N_d = 2 \times 10^{15}$ cm^{-3}, $Q_{ss} = 10^{11}$ cm^{-2}, $L = 10$ μm, $W = 50$ μm, and $\mu_p = 230$ cm^2/V·s. Calculate I_{DS} for V_{GS} equal to -4 and -8 V, and plot the current-voltage characteristics.

9-11. Calculate the saturation current for $V_G = 4$ V of an n-channel MOS transistor with the following parameters: $K_o = 4$, $x_o = 100$ nm, $W/L = 10$, $\mu_n = 1000$ cm^2/V·s, and $V_T = +0.5$ V.

9-12. In the MOS transistor of Prob. 9-10, let $V_G - V_T = 1$ V and (a) calculate the oxide capacitance and the cutoff frequency. (b) Repeat (a) if $W = 10$ μm and $L = 50$ μm.

9-13. (a) Derive an equation describing the source-to-drain V–I characteristics for an n-channel MOS enhancement-mode FET if the source and substrate are grounded and the gate is short-circuited to the drain. Assume V_T is constant.

(b) Plot the V–I curves using data from Prob. 9-11.

(c) Calculate R_{on} at $V_G - V_T = 1$ V.

(d) Repeat (c) if $L/W = 1.0$.

9-14. (a) A p-channel MOS transistor is fabricated on an n substrate of 10^{15} cm^{-3} with a gate oxide thickness of 100 nm. Calculate the threshold voltage if $\phi_{ms} = -0.6$ eV and $Q_{ss} = 5 \times 10^{11}$ cm^{-2}.

(b) The threshold voltage of the MOSFET in part (a) is reduced by using boron ion implantation. What is the required boron doping density in order to obtain a threshold voltage of -1.5 V?

9-15 Sketch a set of masks for the fabrication of an MOS transistor using the silicon-gate technology.

CHAPTER
10

LONG-CHANNEL MOS TRANSISTORS†

In the last chapter, we introduced the basic concepts of the MOS capacitance and inversion-layer formation, yielding a simple model for the MOS transistor. The key approximations were that (1) the threshold voltage acts as a demarcation between zero and nonzero currents and (2) the carrier transport is by means of drift alone. The potential and charge distributions were not delineated clearly, and the carrier behavior below the threshold voltage was not described. Although the square law equation of Eq. (9-39) is useful in getting a first-order approximation for digital circuits, it is altogether inadequate for more exact analyses in both analog and digital applications. Furthermore, the equation is not applicable to the subthreshold regime. As the transistor's dimensions become smaller and the operating voltages and the oxide thickness are reduced, both diffusion and drift currents must be taken into account. It becomes increasingly necessary to develop more detailed device models capable of describing the characteristics of small-dimension devices.

In this chapter, we intend to develop a more general model based on the first principles of semiconductor physics. Unfortunately, the solution of exact equations requires numerical computation which does not provide physical insight of the important issues. Our approach will, therefore, use the *charge-sheet* model

---

† This chapter contains advanced topics that are more suitable for advanced undergraduates or first-year graduate students. The material in Secs. 10-3 to 10-5 follows closely the treatment of Tsividis [1], which gives a detailed account of accurate MOS transistor modeling without using the energy-band diagram.

[2] which is fairly involved but still manageable after some simplifications. We shall begin with the solution of Poisson's equation to obtain an expression of the inversion-layer charge as a function of the gate voltage, the oxide capacitance, and the depletion capacitance. In this derivation, it is necessary to use the concept of the Fermi level described in Chaps. 1 and 2. We shall assume that the vertical electric field and lateral electric field are independent of each other. The separation of the electric fields in the depletion layer and the inverted channel allows us to solve the one-dimensional Poisson equation in each region. This is the result of the gradual-channel approximation which was introduced in Chap. 8.

10-1 THE CHARGE-SHEET MODEL

The basic concept of the charge-sheet model is that the inversion layer is assumed to be extremely thin, typically in the order of 10 nm, as shown in Fig. 10-1. In other words, all the mobile charges are residing at the oxide-silicon interface as if the inversion layer were a two-dimensional sheet. Using this model, the total inversion-layer charge per unit area can be found accurately. The drain current can then be derived from the inversion-layer charge. The solution of the problem starts with Poisson's equation, Eq. (2-67):

$$\frac{d^2\psi}{dx^2} = \frac{q}{K_s \varepsilon_o} (N_a - N_d + n - p) \tag{10-1}$$

where

$$n = n_i e^{(\psi - \phi_f)/\phi_T} \tag{10-2}$$

$$p = n_i e^{(\phi_f - \psi)/\phi_T} \tag{10-3}$$

By integrating Eq. (10-1) once, we obtain the first derivative of the potential. This is equal to the negative of the electric field as expressed in Eq. (2-41). Since all the

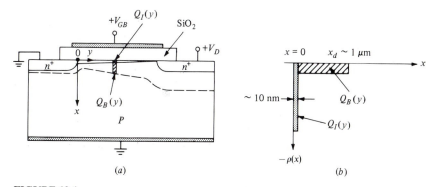

(a)

(b)

FIGURE 10-1
(a) Schematic diagram of an MOS transistor with bias. (b) Vertical distribution of inversion layer and bulk charges at y.

charge is confined to the layer next to the oxide-silicon interface, we may use Gauss' law to relate the charge to the electric field at the surface $\mathscr{E}_s$, that is,

$$Q_s = -K_s \varepsilon_o \mathscr{E}_s \tag{10-4}$$

Using this equation, the total surface charge is obtained if the surface electric field is known. The inversion-layer charge can then be calculated by subtracting the bulk charge. In this approach, we avoid calculating the spatial distribution of the charge and potential. Numerical integration has demonstrated that the electrons are indeed residing within 3 to 30 nm of the interface depending on the gate bias voltage. Thus, we can make use of the inversion-layer charge per unit area without introducing significant error in most applications.

Let us consider a p-type silicon substrate where the donor density is zero. The presence of holes in the depletion layer is neglected to simplify our derivation. Using the neutral p substrate as the potential reference, that is, $\psi = 0$, we may rewrite Eq. (10-3) as

$$\frac{n_i}{N_a} = e^{-\phi_f/\phi_T} \tag{10-5}$$

Equation (10-5) defines the Fermi level inside the p substrate by its doping concentration. Thus, Eq. (10-1) may be rewritten as

$$\frac{d^2\psi}{dx^2} = \frac{qN_a}{K_s \varepsilon_o} (1 + e^{(\psi - 2\phi_f)/\phi_T}) \tag{10-6}$$

By multiplying both sides of Eq. (10-6) with $d\psi/dx$ and using the identity

$$\frac{1}{2} \frac{d}{dx} \left(\frac{d\psi}{dx}\right)^2 = \frac{d\psi}{dx} \frac{d^2\psi}{dx^2} \tag{10-7}$$

we have

$$\frac{1}{2} \frac{d}{dx} \left(\frac{d\psi}{dx}\right)^2 = \phi_T \frac{qN_a}{K_s \varepsilon_o \phi_T} (1 + e^{(\psi - 2\phi_f)/\phi_T}) \frac{d\psi}{dx}$$

$$= \frac{\phi_T}{L_D^2} \frac{d}{dx} (\psi + \phi_T e^{(\psi - 2\phi_f)/\phi_T}) \tag{10-8}$$

where L_D is the extrinsic Debye length given by

$$L_D = \left(\frac{K_s \varepsilon_o \phi_T}{qN_a}\right)^{1/2} \tag{10-9}$$

The extrinsic Debye length is a measure of the abruptness of the depletion-layer edge. In a more precise sense, L_D represents the distance of falling off of the electric field originated from a charge by a factor of $1/e = 0.37$. For an acceptor density of 10^{16} cm^{-3}, L_D is found to be 3×10^{-6} cm. This corresponds to an effective width of the band-bending region under a flat-band condition at the silicon-oxide interface.

Integrating Eq. (10-8) once, we obtain

$$\frac{1}{2}\left(\frac{d\psi}{dx}\right)^2 \bigg|_0^{x_d} = \frac{\phi_T}{L_D^2}\left(\psi + \phi_T e^{(\psi - 2\phi_f)/\phi_T}\right)\bigg|_0^{x_d} \tag{10-10}$$

The boundary conditions in Fig. 10-1*b* are

$$\psi = \begin{cases} 0 \quad \text{and} \quad \dfrac{d\psi}{dx} = 0 \quad \text{at } x = x_d & (10\text{-}11a) \\[2mm] \psi_s \qquad\qquad\qquad\qquad \text{at } x = 0 & (10\text{-}11b) \end{cases}$$

where ψ_s is the surface potential. Equation (10-10) may be written as

$$-\frac{d\psi}{dx}\bigg|_0 = \mathscr{E}_s = \frac{\sqrt{2\phi_T}}{L_D}\left[\psi_s + \phi_T(e^{\psi_s/\phi_T} - 1)e^{-2\phi_f/\phi_T}\right]^{1/2} \tag{10-12}$$

Using Gauss' law and dropping the unity term in Eq. (10-12), we have

$$Q_s = -K_s\varepsilon_o\mathscr{E}_s = -(2\phi_T)^{1/2}C_{FB}(\psi_s + \phi_T e^{(\psi_s - 2\phi_f)/\phi_T})^{1/2} \tag{10-13}$$

where C_{FB} is known as the flat-band capacitance, defined by

$$C_{FB} = \frac{K_s\varepsilon_o}{L_D} \tag{10-14}$$

Physically, L_D represents the effective space-charge-layer width at flat band, and C_{FB} is the corresponding capacitance. For this reason, the capacitance at zero gate voltage in Fig. 9-6 is less than C_o.

Since the charge Q_s is the sum of the inversion layer and the bulk depletion-layer charges, we can obtain the inversion-layer charge as

$$Q_I = Q_s - Q_B = -(2\phi_T)^{1/2}C_{FB}[(\psi_s + \phi_T e^{(\psi_s - 2\phi_f)/\phi_T})^{1/2} - (\psi_s)^{1/2}] \tag{10-15}$$

In Eq. (10-15), the bulk charge is given by

$$Q_B = -(2qK_s\varepsilon_o N_a)^{1/2}(\psi_s)^{1/2} = -C_o\gamma(\psi_s)^{1/2} \tag{10-16}$$

where

$$\gamma \equiv \frac{(2qK_s\varepsilon_o N_a)^{1/2}}{C_o} \tag{10-17}$$

Equation (10-15) gives the relationship between the inversion-layer charge and the surface potential. It will be used in subsequent analyses to derive the current-voltage characteristics.

10-2 GATE-BIAS EFFECT ON INVERSION-LAYER CHARGE

When a gate voltage with respect to the substrate (body) V_{GB} is applied, the electric field across the oxide is given by

$$\mathscr{E}_o = -\frac{V_{GB} - \psi_s - V_{FB}}{x_o} \tag{10-18}$$

Making use of Gauss' law, we have

$$K_o \varepsilon_o \mathscr{E}_o = C_o (V_{GB} - \psi_s - V_{FB}) = -(Q_I + Q_B) \tag{10-19}$$

With the help of Eq. (10-13), the foregoing equation may be written as

$$V_{GB} = V_{FB} + \psi_s + \gamma(\psi_s + \phi_T e^{(\psi_s - 2\phi_f)/\phi_T})^{1/2} \tag{10-20}$$

In Eq. (10-20), the relationship between the gate bias and the surface potential is uniquely established. It is, however, not convenient to solve for the surface potential. Instead, we may use the surface potential as the independent variable and obtain the plot of V_{GB} vs. ψ_s, as shown in Fig. 10-2. Using the data points there and substituting them into Eq. (10-15), we obtain the results plotted in Fig. 10-3. In both figures, we have divided the characteristics into three regions of operation: *weak*, *moderate*, and *strong*-inversion regions. This division scheme adopts the treatment of Tsividis [1].

Weak Inversion

The onset of weak inversion is established at a surface band bending of ϕ_f, and it extends to $2\phi_f$. Within this region, the inversion-layer charge is small compared to the depletion-layer charge. Therefore,

$$|Q_I| \ll |Q_B|$$

Equation (10-19) may be written as

$$V_{GB} = V_{FB} + \psi_s - \frac{Q_B}{C_o} \tag{10-21}$$

By using Eq. (10-16), Eq. (10-21) becomes

$$V_{GB} = V_{FB} + \psi_s + \gamma(\psi_s)^{1/2} \tag{10-22}$$

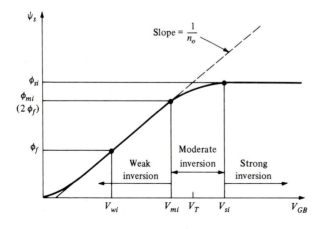

FIGURE 10-2
Surface potential as a function of gate-to-substrate bias.
(After Tsividis [1].)

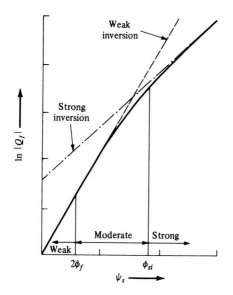

FIGURE 10-3
Dependence of inversion-layer charge on surface potential.

which can be solved for the surface potential to yield

$$\psi_s = \left[-\frac{\gamma}{2} + \left(\frac{\gamma^2}{4} + V_{GB} - V_{FB} \right)^{1/2} \right]^2 \qquad (10\text{-}23)$$

At the onset of weak inversion with a surface band bending of ϕ_f, we define the gate voltage $V_{GB} = V_{wi}$, where the subscript stands for weak inversion. Therefore, Eq. (10-22) becomes

$$V_{wi} = V_{FB} + \phi_f + \gamma(\phi_f)^{1/2} \qquad (10\text{-}24)$$

The voltage V_{wi} is shown in Fig. 10-2.

For a small ψ_s, Eq. (10-15) is simplified by assuming that the exponential term is small compared to ψ_s. This is accomplished by using the binomial expansion

$$(1 + y)^{1/2} = 1 + \frac{y}{2} + \cdots$$

to obtain

$$Q_I = -q N_a L_D \left(\frac{\phi_T}{2\psi_s} \right)^{1/2} e^{(\psi_s - 2\phi_f)/\phi_T} \qquad (10\text{-}25)$$

In the weak-inversion regime, the inversion-layer charge is essentially an exponential function of the surface potential. This is shown in Fig. 10-3, plotted as the dashed line. To relate the inversion-layer charge to the gate bias, Eq. (10-23) can be used to substitute in Eq. (10-25). It is, however, noted that the surface potential is linearly related to V_{GB} in Fig. 10-2. The slope n_o of the straight-line section may be obtained by differentiating Eq. (10-22) and evaluating it at $\psi_s = 1.5\phi_f$, the midpoint of the line. Therefore, we find

$$n_o = \frac{dV_{GB}}{d\psi_s} \bigg|_{\psi_s = 1.5\phi_f} = 1 + \frac{\gamma}{2\sqrt{1.5\phi_f}} \qquad (10\text{-}26)$$

Using Eq. (10-26) as the slope of the semilog plot of $\ln Q_I$ vs. V_{GB}, we may express the inversion-layer charge in the weak inversion as

$$Q_I = Q_{IO} e^{V_{GB}/n_o \phi_T} \tag{10-27}$$

This equation is plotted as the straight line in Fig. 10-4, where Q_{IO} can be determined from an experimental plot. It will be shown later that the subthreshold current is proportional to the inversion-layer charge, and it has the same relationship with the gate voltage.

Moderate Inversion

From the band bending of $2\phi_f$ to approximately $2\phi_f + 6\phi_T$, the incremental change of the inversion charge is comparable to that of the depletion-layer charge. As a result, no approximation can be made to simplify the expressions. The gate voltage at the onset of moderate inversion is defined by setting $V_{GB} = V_{mi}$ and $\psi_s = 2\phi_f$ in Eq. (10-22):

$$V_{mi} = V_{FB} + 2\phi_f + \gamma(2\phi_f)^{1/2} \tag{10-28}$$

This condition is used extensively in the literature to specify the threshold voltage. The definition of the threshold will be modified to achieve better correlation with experimental results.

Strong Inversion

Strong inversion is defined by the condition when $|Q_I| \gg |Q_B|$. This requires a large gate voltage that induces a large surface band bending. The inversion-layer charge given by Eq. (10-15) is now dominated by the exponential term. Thus, we have

$$Q_I = -\sqrt{2qL_D n_i} e^{\psi_s/2\phi_T} \tag{10-29}$$

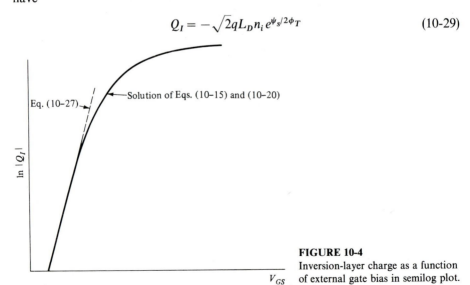

FIGURE 10-4
Inversion-layer charge as a function of external gate bias in semilog plot.

The inversion-layer charge is an exponential function of the surface potential with a slope of $1/2\phi_T$. Therefore, a small increment of the surface potential induces a large change in the inversion-layer charge. In other words, the surface potential remains almost the same for a large increment of Q_I or V_{GB}. In a realistic transistor, the onset of the strong inversion may be defined by

$$\phi_{si} = 2\phi_f + m\phi_T \tag{10-30}$$

where m is a parameter chosen to fit the experimental data. For $N_a = 10^{15} \text{ cm}^{-3}$ and $x_o = 50$ nm, the value of m is about 6 and the surface potential is about 150 meV above $2\phi_f$. Beyond this bias, the surface potential is pinned and the maximum depletion-layer width is given by

$$x_{dm} = \left(\frac{2K_s \varepsilon_o \phi_{si}}{qN_a} \right)^{1/2} \tag{10-31}$$

The corresponding gate voltage is given by

$$V_T = V_{FB} + \phi_{si} - \frac{Q_{BM}}{C_o} \tag{10-32}$$

where the maximum bulk charge is given by

$$Q_{BM} = -qN_a x_{dm} = -\gamma C_o \sqrt{\phi_{si}} \tag{10-33}$$

As an approximation, the inversion charge may be related to V_T by

$$Q_I = -C_o(V_{GB} - V_T) \tag{10-34}$$

This straight-line relationship is plotted in Fig. 10-5 to show that Eq. (10-34) is a good approximation for $V_{GB} > V_{si}$, where V_{si} is defined in Fig. 10-5 as the break point. But, between V_{mi} and V_{si}, there is no simple equation to describe the experimental results. Numerical methods may be required to obtain better modeling.

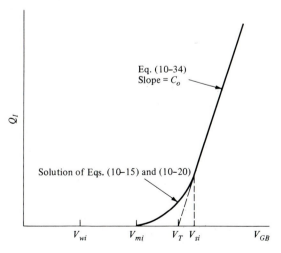

Eq. (10–34)
Slope = C_o

Solution of Eqs. (10–15) and (10–20)

V_{wi} V_{mi} V_T V_{si} V_{GB}

Q_I

FIGURE 10-5
Inversion-layer charge as a function of gate bias in linear plot.

Example. For an MOS transistor with $x_o = 50$ nm and $N_a = 10^{15}$ cm^{-3}, calculate V_{wi}, V_{mi}, and V_T. Assume $V_{FB} = -0.6$ V.

Solution. The oxide capacitance and the Fermi potential are

$$C_o = \frac{K_o \varepsilon_o}{x_o} = 7 \times 10^{-8} \text{ F/cm}^2$$

$$\phi_f = \phi_T \ln \frac{N_a}{n_i} = 0.29 \text{ V}$$

Also, $\gamma = 0.26 \text{ V}^{1/2}$ and $2\phi_f + 6\phi_T = 0.736 \text{ V}$

Thus,

$$V_{wi} = -0.6 + 0.29 + 0.26 \sqrt{0.29} = -0.17 \text{ V}$$

$$V_{mi} = -0.6 + 0.58 + 0.26 \sqrt{0.58} = 0.18 \text{ V}$$

$$V_T = -0.6 + 0.736 + 0.26 \sqrt{0.736} = 0.36 \text{ V}$$

10-3 CURRENT-VOLTAGE RELATION IN MOS TRANSISTORS

In Fig. 10-6, the MOS transistor is shown with appropriate biases. Using Eqs. (10-16) and (10-19), we obtain an expression for the inversion-layer charge as follows:

$$Q_I = -C_o(V_{GB} - V_{FB} - \psi_s - \gamma\sqrt{\psi_s}) \tag{10-35}$$

Theoretically, we may eliminate Q_I by equating Eqs. (10-15) and (10-35) and obtain a relation between the surface potential and the bias voltages. The result may then be substituted into the current-transport equation (2-69) to derive the current-voltage equation. This approach, however, does not produce an analytic solution. To avoid numerical computation which does not enhance our intuitive understanding, we use the following steps to find a solution:

1. Determine the current as a function of the surface potential.
2. Find the relation between the surface potential and the biases at the source and the drain.
3. Use the relation obtained in (2) as the boundary conditions for (1).
4. Separate the weak- and strong-inversion regimes with appropriate approximations.

We shall first develop a general expression of the I–V characteristics and then examine possible simplifications.

In the last chapter, the drift of carriers under an electric field was considered as the only important mechanism in deriving the current-voltage equations. In a

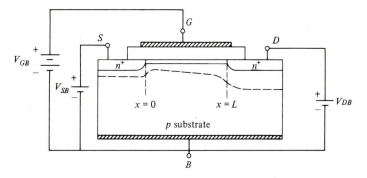

FIGURE 10-6
An *n*-channel MOS transistor with external biases.

more general formulation, both drift and diffusion have to be taken into account. We may rewrite Eq. (9-34) as

$$I_D \text{ (drift)} = \mu_n W Q_I \mathscr{E}_y = \mu_n W(-Q_I) \frac{d\psi_s}{dy} \tag{10-36}$$

where the electric field is expressed as the negative gradient of the surface potential. The diffusion current may be written as

$$I_D \text{ (diffusion)} = W D_n \frac{dQ_I}{dy} = \mu_n W \phi_T \frac{dQ_I}{dy} \tag{10-37}$$

where the Einstein relationship has been used. Since the total drain current is a constant given by

$$I_D = I_D(\text{drift}) + I_D(\text{diffusion}) \tag{10-38}$$

integration of the above equation gives

$$\int_0^L I_D \, dy = \int_0^L I_D \text{ (drift)} \, dy + \int_0^L I_D \text{ (diffusion)} \, dy \tag{10-39}$$

Using Eqs. (10-36) and (10-37), we have

$$I_D L = \mu_n W \int_{\psi_{so}}^{\psi_{sL}} (-Q_I) d\psi_s + \mu_n W \phi_T \int_{Q_I(0)}^{Q_I(L)} dQ_I \tag{10-40}$$

Using Eq. (10-35), the integration of the two right-hand terms yields

$$\int_0^L I_D \text{ (drift)} \, dy = W \mu_n C_o [(V_{GB} - V_{FB})(\psi_{sL} - \psi_{so}) - \tfrac{1}{2}(\psi_{sL}^2 - \psi_{so}^2) - \tfrac{2}{3}\gamma(\psi_{sL}^{3/2} - \psi_{so}^{3/2})]$$

and
$$\tag{10-41}$$

$$\int_0^L I_D \text{ (diffusion)} \, dy = \mu_n W \phi_T [Q_I(L) - Q_I(0)] \tag{10-42a}$$

$$= W \mu_n C_o \phi_T [(\psi_{sL} - \psi_{so}) + \gamma(\psi_{sL}^{1/2} - \psi_{so}^{1/2})] \tag{10-42b}$$

Equations (10-41) and (10-42) are the drift and diffusion components. What remains to be found are the boundary conditions of the surface potentials at the source and the drain.

The surface potential is given by Eq. (10-20), using the substrate as the reference. If a voltage V_{SB} is applied between the source and the substrate (body), the bias will shift the quasi-Fermi level at the source by V_{SB}. Therefore, the surface potential at the source ($y = 0$) is given by

$$\psi_{so} = V_{GB} - V_{FB} - \gamma \left[\psi_{so} + \phi_T \exp\left(\frac{\psi_{so} - 2\phi_f - V_{SB}}{\phi_T}\right) \right]^{1/2} \tag{10-43}$$

Similarly, the drain-to-body bias V_{DB} shifts the quasi-Fermi level at the drain, giving rise to

$$\psi_{sL} = V_{GB} - V_{FB} - \gamma \left[\psi_{sL} + \phi_T \exp\left(\frac{\psi_{sL} - 2\phi_f - V_{DB}}{\phi_T}\right) \right]^{1/2} \tag{10-44}$$

In either equation, it is not possible to obtain an expression of ψ_s as a function of the bias analytically, but it is rather simple to calculate the bias for a given surface potential. The total current may now be computed. The results agree with experiments showing that the current derived above is adequate to describe the MOS transistor. Since most calculations are done in a computer, this tedious procedure is not as formidable as it seems. It is, however, desirable to simplify the procedure to the extent that it can be accomplished by hand calculation. This can be done if we separate the different regimes of operation, in particular to consider the weak and strong inversions individually.

10-4 STRONG AND MODERATE INVERSION

In the region of strong inversion, the diffusion term is not important so that we may use Eq. (10-41) to represent the current of the transistor. The surface potential is the sum of ϕ_{si} and the appropriate bias. Thus,

$$\psi_{so} = \phi_{si} + V_{SB} \tag{10-45}$$

$$\psi_{sL} = \phi_{si} + V_{DB} \tag{10-46}$$

for the source and drain surface potentials. With these conditions, Eq. (10-41) becomes

$$I_D = \frac{W}{L} \mu_n C_o \left\{ (V_{GB} - V_{FB})(V_{DB} - V_{SB}) - \tfrac{1}{2}[(V_{DB} + \phi_{si})^2 - (V_{SB} + \phi_{si})^2] \right.$$

$$\left. - \tfrac{2}{3}\gamma[(\phi_{si} + V_{DB})^{3/2} - (\phi_{si} + V_{SB})^{3/2}] \right\} \tag{10-47}$$

In this equation, the current is expressed as a function of the bias voltages. Using $V_{GB} = V_{GS} + V_{SB}$ and $V_{DB} = V_{DS} + V_{SB}$, we obtain

$$I_D = \frac{W}{L}\,\mu_n C_o \{(V_{GS} - V_{FB} - \phi_{si})V_{DS} - \tfrac{1}{2}V_{DS}^2$$

$$- \tfrac{2}{3}\gamma[(\phi_{si} + V_{SB} + V_{DS})^{3/2} - (\phi_{si} + V_{SB})^{3/2}]\} \quad (10\text{-}48)$$

Equation (10-48) is in the same form as Eq. (9-37)

In the moderate-inversion regime, we are unable to simplify the equation since both the drift and diffusion components are important. The only useful point to bring out is that the onset of moderate inversion, Eq. (10-28), is now changed to

$$V_{mi} = V_{FB} + 2\phi_f + \gamma\sqrt{2\phi_f + V_{SB}} \quad (10\text{-}49)$$

This is the expression used extensively in the literature as the threshold voltage. The influence of body bias V_{SB} on V_{mi} gives rise to the well-known body effect, which was described in Sec. 9-7.

10-5 WEAK INVERSION AND SUBTHRESHOLD CURRENT

In weak inversion, the drift current is not important and the current is dominated by carrier diffusion. The surface potential in the channel in weak inversion, ψ_{sw}, is given by Eq. (10-23):

$$\psi_{sw} = \left[-\frac{\gamma}{2} + \left(\frac{\gamma^2}{4} + V_{GB} - V_{FB} \right)^{1/2} \right]^2 \quad (10\text{-}50)$$

The surface potential in weak inversion is the same along the channel and is independent of y. As a result, the electric field, which is the gradient of the surface potential, is zero, leading to zero drift current.

The inversion-layer charge depends not only on the surface potential but also on the quasi-Fermi level. Since a bias voltage shifts the quasi-Fermi level, the inversion charge expressed in Eq. (10-25) has to be modified to become

$$Q_I(0) = -qN_a L_D \left(\frac{\phi_T}{2\psi_{sw}} \right)^{1/2} e^{(\psi_{sw} - 2\phi_f - V_{SB})/\phi_T} \quad (10\text{-}51)$$

at the source end, and

$$Q_I(L) = -qN_a L_D \left(\frac{\phi_T}{2\psi_{sw}} \right)^{1/2} e^{(\psi_{sw} - 2\phi_f - V_{DB})/\phi_T} \quad (10\text{-}52)$$

at the drain end. Substituting Eqs. (10-51) and (10-52) into Eq. (10-42a), we find

$$I_{sub} = I_s (e^{-V_{SB}/\phi_T} - e^{-V_{DB}/\phi_T}) \quad (10\text{-}53)$$

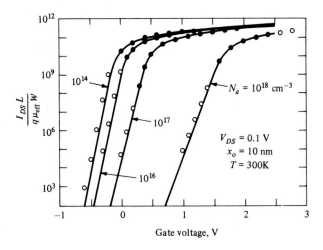

FIGURE 10-7
Normalized subthreshold current as a function of gate bias for transistors with a uniform substrate doping. *(After Wordeman [3].)*

where

$$I_s = \frac{\mu_n W C_o \gamma \phi_T^2}{2\sqrt{\psi_{sw}}\ L}\ e^{(\psi_{sw} - 2\phi_f)/\phi_T} \tag{10-54}$$

I_{sub} is known as the subthreshold current. Equation (10-53) has the same form as the Eberts-Moll equation in Chap. 5. This is indeed the case since the n^+(source)-p(channel)-n^+(drain) structure is an *npn* bipolar transistor. Under low bias at the gate, the source-channel junction is effectively forward-biased and the drain-channel junction is reverse-biased. The *npn* transistor is in its normal active mode. This is the condition for the surface condition of the MOS transistor in its weak-inversion regime, and the current is controlled by carrier diffusion.

Let us now consider the subthreshold current as a function of the gate bias. By using Eq. (10-27) to represent Q_I in Eq. (10-42a), the subthreshold current is plotted as a function of the gate voltage in Fig. 10-7. The slope of the curves is specified by $1/n_o \phi_T$, where

$$n_o = 1 + \frac{\gamma}{2\sqrt{1.5\phi_f + V_{SB}}} \tag{10-55}$$

This equation is in the same form as Eq. (10-26) except that the body bias is included. Because γ and ϕ_f are doping-dependent, the slope decreases with N_a. The shift of the curves is caused by the increase in threshold voltage introduced by the bulk charge.

10-6 ION-IMPLANTED MOS TRANSISTORS

In the previous analyses, the impurity concentration in the channel of an MOS transistor is assumed to be uniform without any spatial variation. This is a reasonable approximation for devices built on the technology described in Sec. 9-8, and its simplicity allows for straightforward analyses. Most modern practical MOS transistors have an ion-implanted channel which is not uniformly doped. The main reasons for the channel implantation are for the threshold-voltage control and prevention of punch-through between the source and the drain. Implantation provides a higher doping in the channel without a heavily doped substrate (or bulk) leading to lower parasitic capacitances. It introduces new ionized charge in the depletion layer and at the same time changes the depletion-layer width. Both of these modify the capacitance and threshold voltage, thus influencing the transistor behavior.

The basic concepts of ion implantation have been introduced in Chap. 3. The most important parameters are the dosage per unit area (10^{11} to 10^{12} cm^{-2}) and the implantation energy (between 30 and 300 keV) which determines the depth. Figure 10-8 shows the impurity profile of an implanted dosage of 8×10^{11} cm^{-2} at an energy of 30 keV. The impurity distribution is a gaussian function, and the total dosage is given by

$$D_I = \int_0^\infty (N_s - N_a)\, dx \qquad \text{cm}^{-2} \qquad (10\text{-}56)$$

After a post-implantation annealing, the distribution may be approximated by a box-shape function as shown. The density N_s is specified by

$$D_I = (N_s - N_a)x_s \qquad (10\text{-}57)$$

where N_s is the idealized uniform doping and x_s is the width of the idealized layer. If the substrate is p type and the implanted impurity is n type, an n channel is formed and the device will function as a normally on or depletion-mode transistor. On the other hand, if p-type impurity ions are implanted on a p-type substrate, the transistor will remain to operate in the enhancement mode or normally off.

Let us consider first the implantation of p impurities to a p substrate. Assuming a step junction of implanted profile as shown in Fig. 10-8, the depletion-layer width at the onset of strong inversion is

$$x_{dm} = \left[\frac{2K_s \varepsilon_o (\phi_{si} + V_{SB})}{q N_s}\right]^{1/2} \qquad (10\text{-}58)$$

where ϕ_{si} corresponds to the new doping N_s. The threshold voltage and the inversion-layer charge are, through the use of Eqs. (10-32) to (10-34),

$$V_T = V_{FB} + \phi_{si} + \gamma_1 \sqrt{\phi_{si} + V_{SB}} \qquad (10\text{-}59)$$

$$Q_I = -C_o(V_{GB} - V_T) \qquad (10\text{-}60)$$

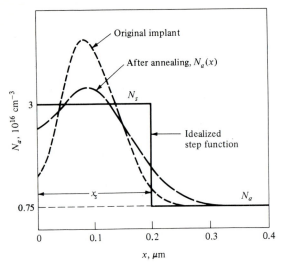

FIGURE 10-8
Doping profile of implanted region beneath the gate oxide. The original implant is broadened by thermal annealing. A step doping is used to approximate the actual doping. *(After Rideout, Gaensslen, and Le Blanc [4].)*

where

$$\gamma_1 = \frac{(2qK_s\varepsilon_o N_s)^{1/2}}{C_o} \tag{10-61}$$

The current-voltage equation remains the same as Eq. (10-48) except that N_a is replaced by N_s. There is an implicit assumption in the foregoing discussion; that is, x_{dm} is smaller than x_s. Let us assume that a large V_{SB} is applied so that x_{dm} is now greater than x_s. We may solve Poisson's equation (Prob. 10-12) to obtain

$$x_{dm} = \sqrt{\frac{2K_s\varepsilon_o}{qN_a}} \left[\phi_{si} + V_{SB} - \frac{qx_s^2}{2K_s\varepsilon_o}(N_s - N_a) \right]^{1/2} \tag{10-62}$$

The new threshold voltage for Eq. (10-60) becomes

$$V_T = V_{FB} + \phi_{si} + \frac{qD_I}{C_o} + \gamma_2 \left[\phi_{si} - \frac{q(N_s - N_a)x_s^2}{2K_s\varepsilon_o} + V_{SB} \right]^{1/2} \tag{10-63}$$

where

$$\gamma_2 = \frac{(2qK_s\varepsilon_o N_a)^{1/2}}{C_o}$$

Example. Calculate the threshold voltage for $N_a = 7.5 \times 10^{15}$ cm^{-3} and $x_o = 35$ nm (a) without implant and (b) with implant of $N_s = 3 \times 10^{16}$ cm^{-3} and $x_s = 0.2$ μm at $V_{SB} = 2$ V. Let V_{FB} be -1.22 V.

Solution

(a)
$$\phi_{si} = 6\phi_T + 2\phi_T \ln \frac{N_a}{n_i} = 0.83 \text{ V}$$

$$\gamma_2 = \frac{(2qK_s\varepsilon_o N_a)^{1/2}}{C_o} = 0.5 \text{ V}^{1/2}$$

We may now use Eq. (10-63) with $D_I = 0$ and $N_s - N_a = 0$:

$$V_T = -1.22 + 0.83 + 0.5(0.83 + 2)^{1/2} = 0.45 \text{ V}$$

(b) Using Eq. (10-58), we find that $x_{dm} = 0.35 \ \mu\text{m}$, which is greater than $0.2 \ \mu\text{m}$. Therefore, Eq. (10-63) should be used for the threshold calculaton. We have

$$\phi_{si} = 0.91 \text{ V}$$

$$D_I = 2.25 \times 10^{16} \times 0.2 \times 10^{-4} = 4.5 \times 10^{11} \text{ cm}^{-2}$$

$$\frac{qD_I}{C_o} = 0.71 \text{ V}$$

$$\frac{q(N_s - N_a)x_s^2}{2K_s\varepsilon_o} = 0.69$$

Therefore,

$$V_T = -1.22 + 0.91 + 0.71 + 0.5(0.91 - 0.69 + 2)^{\frac{1}{2}}$$

$$= 1.14 \text{ V}$$

Additional calculation using the same procedure was made, and the results are plotted in Fig. 10-9. In this figure, we have included the example of an extremely shallow implant to show that the implantation shifts the threshold voltage by qD_I/C_o, but the substrate bias sensitivity remains low.

We have described the two cases where the depletion layer is either greater than or smaller than the implanted-layer width along the entire channel. The reader is referred to Tsividis [1] for the more general case where part of the channel has a depletion layer greater than x_s and the other part less than x_s.

If the donor impurities are implanted in a p substrate, an n-type channel is formed between the source and the drain. Without any bias at the gate, there is a conducting path allowing a drain current to flow. The structure, shown in Fig. 10-10, is similar to the JFET in Chap. 8 except that the upper gate junction is replaced by an MOS capacitor. If a negative voltage is applied to the gate, the energy band at the silicon-oxide interface will bend upward, depleting electrons as the depletion layer extends into the bulk. As a result, the conducting channel contracts until pinchoff of the channel takes place. This situation is identical to that of the pinchoff condition in the JFET giving rise to current saturation. For this reason, the structure shown is known as the *depletion-MOS* transistor. Because the conducting channel is inside the bulk, it is also called the *buried-channel* transistor. The current-voltage characteristics of a typical depletion-MOS transistor are shown in Fig. 10-11. Notice that the current at zero gate voltage is large. An

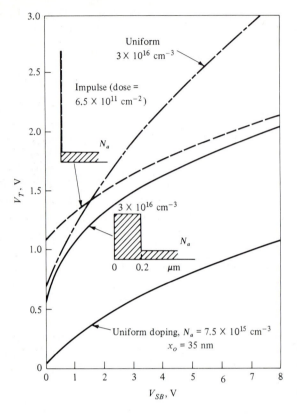

FIGURE 10-9
Calculated substrate sensitivity for various doping profiles. *(After Rideout, Gaensslen, and LeBlanc [4].)*

important application of the depletion transistor is the load device of an inverter, a subject to be discussed in a later chapter. The theory of the depletion transistor is slightly more involved, but the basic principle is the same. It can be developed from studying the MOS capacitor having a buried np junction in the substrate. The subsequent steps follow the derivation of that of the enhancement transistor.

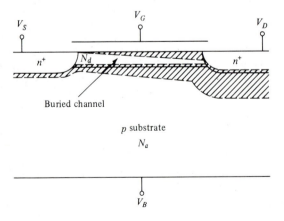

FIGURE 10-10
A depletion-mode MOS transistor.

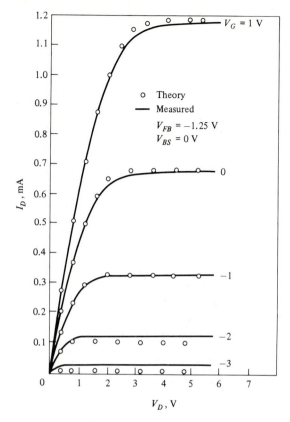

FIGURE 10-11
Measured and calculated drain characteristics for a normally on MOSFET. (*After Huang and Taylor [5].*)

10-7 EFFECTIVE MOBILITY IN THE INVERSION LAYER

So far, we have assumed that the carrier mobility is a constant, independent of impurity concentration and gate voltage. This is not a good approximation because the surface mobility varies with the doping density and with the electric field in both the vertical and horizontal directions. The effect of doping is of secondary importance to the device operation,[1] but the field dependence cannot be ignored.

Let us first consider the influence of the vertical electric field. By applying a positive gate voltage, electrons in the inversion layer are attracted toward the surface, leading to increased surface scattering and reduced mobility. At the same time, the interaction between the electrons and the fixed oxide charge slows down the electrons, which also lowers the electron mobility. The relationship between the surface mobility and the electric field is usually obtained experimentally. The surface electric field is related to the charges through Gauss' law:

$$\mathscr{E}_{sv} = \frac{-1}{K_s \varepsilon_o} (Q_I + Q_B) \tag{10-64}$$

---

[1] Recent results show that the surface mobility is essentially independent of surface doping [5].

Since the inversion-layer charge is a strong function of vertical dimension, the electric field is spatial-dependent in the vertical direction. For the calculation of the MOS transistor drain current, we define an effective field

$$\mathscr{E}_{\text{eff}} = \frac{-1}{K_s \varepsilon_o}\left(\frac{Q_I}{2} + Q_B\right) \tag{10-65}$$

This equation specifies a field corresponding to the point that covers half the inversion-layer charge and is found to be a good approximation of the average field in the channel. Using the effective field as a parameter, we can measure the channel mobility by means of the channel conductance as a function of the gate voltage for small drain voltages. The result is plotted in Fig. 10-12, where the mobility may be represented by

$$\mu_{\text{eff}} = \mu_o \left(\frac{\mathscr{E}_{\text{eff}}}{\mathscr{E}_c}\right)^{-1/3} \tag{10-66}$$

where μ_o and $\mathscr{E}_c$ are constants obtained experimentally. (The power law of $-\frac{1}{3}$ appears to be useful in the range of 2×10^4 to 5×10^5 V/cm.) These data were taken from samples with high-quality SiO_2–Si interfaces where the oxide charge is less than 10^{10} cm^{-2} and thus negligible. Various oxide thickness and substrate doping were used to demonstrate that the effective electric field is the most important parameter in determining the mobility. It is seen that the saturation of mobility for 3×10^{17} cm^{-3} is caused by the bulk mobility limit.

The reduction of mobility by the gate voltage bias lowers the transconductance of the transistor. This is observed in devices operated above 200 K. The effect is less significant at 77 K. For this reason, research in low-temperature operation of MOS transistors may be important in enhancing the device performance.

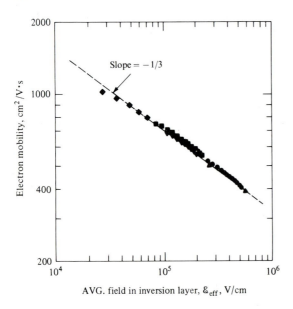

FIGURE 10-12

Measured inversion-layer electron mobility at $T = 30°C$ for different doping and oxide thickness (*After Wordeman [3].*)

REFERENCES

1. Tsividis, Y.: "Operation and Modeling of the MOS Transistor," McGraw-Hill, New York, 1987.
2. Brews, J.: Physics of the MOS Transistor, in D. Kahng (ed.), "Silicon Integrated Circuits," part A, Academic, New York, 1981.
3. Wordeman, M. R.: Doctoral dissertation, Columbia University, New York, 1985.
4. Rideout, V. L., F. H. Gaensslen, and A. LeBlanc: Device Design Consideration for Ion-Implanted *n*-Channel MOSFETS, *IBM J. Res. Dev.*, **50** (1975).
5. Huang, J. S. T., and G. W. Taylor: Modeling of an Ion-Implanted Silicon-Gate Depletion Mode IGFET, *IEEE Trans. Electron Devices*, **ED-22**:995 (1975).
6. Sabnis, A. G., and J. T. Clemens: Characterization of the Electron Mobility in the Inverted (100) Si Surface, *Tech. Dig., Int. Electron Device Meet*, 18 (1979).

PROBLEMS

10-1. Calculate the extrinsic Debye length and the flat-band capacitance for $N_a = 10^{14}$, 10^{15}, 10^{16}, 10^{17}, and 10^{18} at 300 K.

10-2. Plot the inversion-layer charge as a function of the surface potential; that is, $\ln Q_I$ vs. ψ_s/ϕ_T for $15 < \psi_s/\phi_T < 35$ with $N_a = 10^{14}$, 10^{15}, and 10^{16} cm^{-3} in the form of Fig. 10-3.

10-3. In an MOS capacitor, the oxide thickness is 50 nm and the substrate doping is 10^{15} cm^{-3}. Assuming $V_{FB} = 0$, calculate and plot the surface potential as a function of the gate voltage. Indicate the points of ϕ_f, $2\phi_f$, and $2\phi_f + 6\phi_T$. Relate this plot to the physical meaning of strong inversion.

10-4. Repeat Prob. 10-3 for $x_o = 20$ nm and $N_a = 5 \times 10^{16}$ cm^{-3}. Find the *m* value for the condition of strong inversion, Eq. (10-30).

10-5. The small-signal MOS and semiconductor capacitances are defined as $-dQ_s/dV_{GB}$ and $-dQ_s/d\psi_s$, respectively. Show that

$$C_{\text{MOS}} = \frac{C_s C_o}{C_s + C_o}$$

where C_o is the oxide capacitance. What is the C_{MOS} at the flat-band condition?

10-6. Using the definition in Prob. 10-5, derive an expression for C_s. Then calculate and plot C_{MOS} as a function of the gate voltage for $x_o = 50$ nm and $N_a = 10^{15}$ cm^{-3}. Specify the points of the surface potential at ϕ_f, $2\phi_f$, and $2\phi_f + 6\phi_T$. Let $V_{FB} = 0$.

10-7. In an MOS capacitor with $x_o = 50$ nm and $N_a = 10^{15}$ cm^{-3}, use the result of Prob. 10-3 to obtain n_o and Q_{IO} under weak inversion.

10-8. For an MOS transistor with $x_o = 20$ nm, $N_a = 5 \times 10^{16}$ cm^{-3}, and $V_{FB} = -0.5$ V, calculate V_{wi}, V_{mi}, and V_T.

10-9. For the transistor in the example of Sec. 10-2, $L = 2 \mu$m, $W = 10 \mu$m, and $\mu_n = 500$ cm^2/V · s. Plot I_D vs. V_{DS} under strong inversion with V_{GS} as a parameter for 0, 1, 2, 3, and 4 V. Let $V_{SB} = 0$ at 300 K.

10-10. Using the parameters in Prob. 10-9, plot I_D vs. V_{GS} with $V_{DS} \approx 0$ for $V_{SB} = 0$, 1, 2, and 4 V. Find V_T, and compare the result with Eq. (10-49).

10-11. (*a*) Plot the subthreshold current as a function of the gate voltage for the transistor in Prob. 10-9 with $V_{DS} = 0.1$ V, $V_{SB} = 0$.
(*b*) Repeat (*a*) for $N_a = 10^{17}$ and 10^{18} cm^{-3}.

(c) Repeat (a) for $V_{SB} = 1$, 2, and 4 V.

(d) Calculate n_o in parts (a) and (c).

10-12. The implanted profile of an MOS transistor may be represented by the idealized step function shown in Fig. 10-8. Derive Eq. (10-62) for the condition $x_{dm} > x_s$.

10-13. Calculate the threshold voltage for $N_a = 10^{15}$ cm^{-3} and $x_o = 50$ nm in the following cases:

(a) Without implant. Let $V_{FB} = -0.1$ V.

(b) With implant of $N_s = 10^{16}$ cm^{-3} and $x_s = 1.0$ μm at $V_{SB} = 3$ V.

(c) Repeat (b) with $x_s = 0.4$ μm.

10-14. Derive the drain I–V characteristics for the depletion-mode transistor shown in Fig. 10-10.

10-15. In a depletion-mode MOS transistor, the implant dosage of As in a boron doped substrate (10^{16} cm^{-3}) is $N_s = 10^{17}$ cm^{-3} and $x_s = 0.2$ μm. Let $x_o = 30$ nm, $L = 3$ μm, $W = 12$ μm, and $V_{FB} = +1$ V. Calculate the drain I–V curves for $V_{GB} = -1$, 0, and 1 V with $V_{SB} = 0$.

CHAPTER
11

SMALL-DIMENSION MOS TRANSISTORS

The purpose of microminiaturization is to increase the packing density and to improve the circuit performance. Ever since the birth of the integrated circuit, researchers have attempted to reduce the size of devices. However, simply reducing device dimensions without paying attention to other processing parameters gives rise to a variety of nonideal characteristics, e.g., threshold-voltage variation and nonsaturated drain current. Consequently, scaling laws have been developed to preserve the ideal characteristics of a long-channel device while decreasing its size. These laws predict the behavior of an MOS transistor when it is scaled down. Its performance is improved over the larger counterpart. In this chapter, we shall first present some experimentally observed effects when device dimensions are reduced. For each of these effects, the physical mechanism is examined to understand the origin of the nonideal behavior. The scaling laws are then discussed to explore possibilities for device optimization. If the electric field is not kept small, velocity saturation can become important. In the last section, the lightly doped drain transistor is used as an example to illustrate how to design a short-channel MOS transistor.

11-1 MINIATURIZATION AND CIRCUIT SPEED

It is obvious that as device dimensions are decreased we can put more transistors on a chip with the same area. It is, however, not obvious what it will do to the circuit

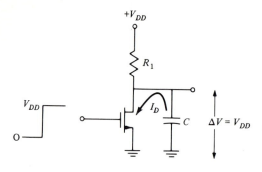

FIGURE 11-1
Load capacitance discharged by output current in an FET circuit.

speed. For this reason, we shall make a simple estimate of the effect of miniaturization on the delay time per gate. In a field-effect transistor circuit, the output current is used to discharge its load capacitance, as illustrated in Fig. 11-1. The incremental voltage change is ΔV, and the charging time is Δt. Neglecting the current through R_1, we have

$$I_D = \frac{C \, \Delta V}{\Delta t} \tag{11-1}$$

where $C \, \Delta V$ is the charge stored in the capacitor. Therefore, the delay time is given by

$$\tau_d \approx \Delta t = \frac{C \, \Delta V}{I_D} \tag{11-2}$$

for a fixed I_D. Let us assume that the logic swing is equal to the drain supply voltage V_{DD}, and the output load is an identical transistor gate. By using the simplified form of Eq. (9-39) for I_D with $V_G - V_T = V_{DD}$, we have

$$I_D = \frac{\mu W C_o V_{DD}^2}{2L} \tag{11-3}$$

The load capacitance C is the gate capacitance of the following stage. Therefore,

$$C = C_o W L \tag{11-4}$$

Substituting Eqs. (11-3) and (11-4) into (11-2) and simplifying, we find

$$\tau_d = \frac{2L^2}{\mu V_{DD}} \tag{11-5}$$

In the above equation, the delay time is proportional to the square of the channel length. Therefore, significant improvement in speed is realized by downsizing the transistor's channel length.

11-2 EXPERIMENTAL SHORT-CHANNEL CHARACTERISTICS

When the dimensions of an MOS transistor are reduced, three distinct features are seen in the device's characteristics. First, the drain current is found to increase with the drain voltage beyond pinchoff. This is in contrast with the $I–V$ curves of a long-channel transistor, where the drain current becomes constant after the pinchoff condition is reached. As shown in Fig. 11-2, the output current of a short-channel transistor does not saturate. It also exhibits a soft breakdown that is not seen in long-channel devices. Furthermore, the drain current is not zero at zero gate voltage and large drain voltages, indicating that the gate has lost control of shutting off the device.

The second distinct short-channel characteristic is seen in the subthreshold regime. To illustrate this point, the current-voltage curves are plotted in a semilog graph, as shown in Fig. 11-3. The device with a 5-μm channel length exhibits clearly the long-channel ideal characteristic. As the length L is reduced, the $I–V$ curve shifts to the left as a result of a drop of the threshold voltage, but the basic shape of the plot is of a long-channel device. The slope of the plot remains essentially unchanged. When the length is below 1.5 μm, the slope of the curve starts to change, and the drain current cannot be reduced to zero. In the extreme case with $L = 0.8$ μm, the gate voltage does not control the drain current. In other words, the output current cannot be turned off and the transistor can no longer function as a switch.

The third feature is the shift of the threshold voltage with the channel length, as plotted in Fig. 11-4. In the long-channel theory described in the last chapter, the threshold voltage is not a function of the channel length. But in the figure shown, the threshold voltage decreases with the channel length with a sharp drop for

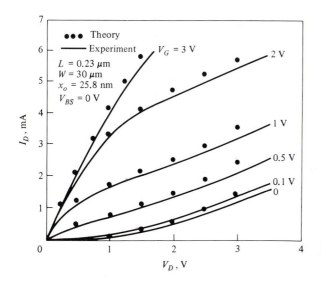

FIGURE 11-2

Drain characteristics of a short-channel MOSFET having a channel length $L = 0.23$ μm. *(After Fichtner [1].)*

$L < 2\ \mu m$. In addition, the threshold is very sensitive to the drain bias voltage. This behavior will be explained by examining the space charge in the depletion layer in the next section.

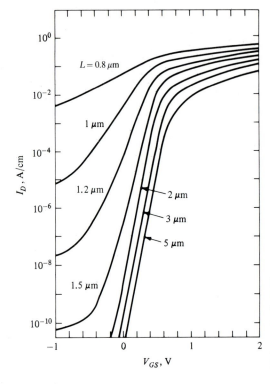

FIGURE 11-3
Log current vs. gate bias characteristics for various channel lengths, L (from full two-dimensional computer calculation), $N_a = 10^{15}\ cm^{-3}$, $x_o = 0.5\ \mu m$, $r_j = 0.33\ \mu m$, $V_{DS} = 2.0\ V$, $V_{BS} = 0\ V$. (*After Kotani and Kawazu* [2].)

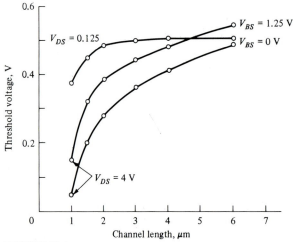

FIGURE 11-4
Experimental threshold vs. channel length for implanted devices ($6 \times 10^{11}\ cm^{-2}$ at 35 keV annealed at 1000°C for 10 min). $N_a = 8 \times 10^{15}\ cm^{-3}$, $x_o = 0.028\ \mu m$, $r_j = 1\ \mu m$. (*After Fichtner and Pötzl* [3].)

11-3 SPACE CHARGE IN THE DEPLETION LAYER

The depletion-layer width and its associated space charge in an MOS transistor with a p-type substrate are given by

$$x_d = \left[\frac{2K_s \varepsilon_o (\phi_o + V)}{qN_a} \right]^{1/2} \tag{11-6}$$

$$Q_B = -qx_d N_a = -[2qK_s \varepsilon_o N_a (\phi_o + V)]^{1/2} \tag{11-7}$$

where ϕ_o represents the built-in potential for a pn junction or the surface band bending ϕ_{si} in the MOS transistor and V is the reverse-bias voltage. A schematic diagram of a long-channel transistor is shown in Fig. 11-5. The depletion layer has space charge that can be divided into vertical and lateral components controlled by the vertical and lateral electric field, respectively. In the previous analyses, the space-charge extension in the vertical direction has been taken into account as the bulk charge in determining the threshold- and current-voltage characteristics. The lateral extension of the depletion layer into the channel ΔL has been ignored. This is a reasonable approximation for long-channel devices since the depletion-layer width is much smaller than the channel length. As the channel length is reduced, ΔL may become comparable to the channel length. Therefore, the lateral dimension of the space charge must be included in device characterization.

Channel-Length Modulation

The lateral extension of the depletion layer into the channel region reduces the effective channel length. Since this depletion layer is bias-dependent, it changes with the drain voltage and thus modulates the effective channel length. This is known as *channel-length modulation*, which has been discussed in the junction field-effect transistor. At the onset of current saturation, zero inversion-layer charge is situated at the drain end of the channel at a bias voltage V_{DS}. If the drain voltage is

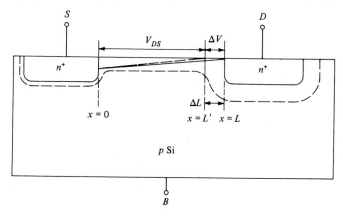

FIGURE 11-5
Channel-length modulation in an MOS transistor; the gate metal and oxide are not shown.

increased to $V_{DS} + \Delta V$, the depletion layer is widened so that the pinchoff location is moved into the channel, as shown in Fig. 11-5. The incremental change of the depletion layer is given by

$$\Delta L = \sqrt{\frac{2K_s \varepsilon_o}{qN_a}} \left(\sqrt{\phi_p + V_{DS} + \Delta V} - \sqrt{\phi_p + V_{DS}} \right) \qquad (11\text{-}8)$$

where ϕ_p specifies the surface potential at pinchoff and is a parameter to be determined experimentally. As a first-order approximation, ϕ_p is sometimes considered to be zero to represent a zero vertical field at pinchoff. This is not true in most cases, and ϕ_p is likely to be between 0.5 and 1 V depending on the doping and biases. Notice that the space charge extends only toward the channel because of the heavy doping in the drain. Since the drain current is inversely proportional to the channel length, the change of the drain current may be written as

$$\frac{I'_D}{I_D} = \frac{L}{L - \Delta L} \qquad (11\text{-}9)$$

where I'_D represents the current at $V_{DS} + \Delta V$. The drain current can now be calculated by using Eqs. (11-8) and (11-9). In a short-channel device, ΔL can be comparable to L so that the modulation effect could be very significant. This gives rise to the poor I-V characteristics observed in Fig. 11-2.

Subthreshold Current

As mentioned previously, the subthreshold current is caused by carrier diffusion from the source to the drain. Consequently, it behaves like a bipolar transistor which is very sensitive to the base width, i.e., the channel length in this case. The subthreshold current is given by Eq. (10-53) and will not be reproduced here. We should just note that in Eq. (10.54) the channel length L in the denominator is now replaced by L'. Because of the channel-length modulation, $L' < L$. Therefore, the subthreshold current is increased for a smaller L'. The increase in current is moving the I–V curve upward in Fig. 11-3, which appears as a shift toward the left.

As the channel length continues to decrease, the depletion layer of the drain starts to interact with the source-channel junction to lower the source junction potential barrier. This is known as *drain-induced barrier lowering* (DIBL). As illustrated in Fig. 11-6, this effect is specially strong for a large V_{DD} and a short channel. The lowering of the source barrier allows electrons to be injected into the channel regardless of the gate voltage. As a result, the gate voltage loses control of the drain current in the subthreshold regime. As seen in Fig. 11-3, a large subthreshold current is observed when the channel length is below 1.5 μm. It should be pointed out that the DIBL is initiated before the punch-through is reached. The punch-through condition is defined when the source and the drain depletion layers meet, assuming the depletion approximation is used to calculate the space-charge-layer width. As it turns out, the extrinsic Debye length is about 100 nm so that the boundary layer between the depletion and neutral regions is

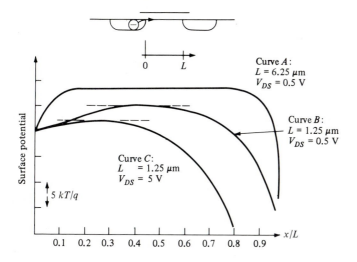

FIGURE 11-6
Surface potential in the channel for devices with different channel lengths. *(After Troutman [4].)*

approximately 300 nm for a substrate doping of 10^{15} cm^{-3}. Thus, the two junctions start to interact when the depletion layers are $0.6\,\mu$m apart, long before the punch-through condition is established. The situation is worsened if a light doping is used for the substrate. A heavily doped channel by ion implantation can reduce this short-channel effect.

The large subthreshold current is really a leakage current preventing the turning off of the transistor. It limits the transistor's ability to isolate capacitive nodes in a dynamic circuit and allows excess current in static inverters. Therefore, care must be taken to minimize its magnitude.

11-4 GEOMETRY EFFECT ON THRESHOLD VOLTAGE

It has been observed experimentally that the threshold voltage does not remain the same if the dimensions of L and W are reduced. While such a phenomenon can be modeled by using two-dimensional numerical calculations to solve the Poisson and transport equations, it is desirable to obtain a simple analytic solution. In general, a simple physical model can give better insight and enhance our understanding of the device operation. The *charge-sharing* model developed by Yau [5] serves this function very well by presenting the essential features of the geometry effect on the threshold voltage.

The cross section of a short-channel transistor is shown in Fig. 11-7, illustrating the depletion-layer charge distribution. Using the method described previously, the threshold voltage is given by

$$V_T = V_{FB} + \phi_{si} - \frac{Q_B}{C_o} \tag{11-10}$$

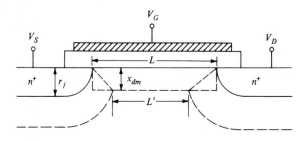

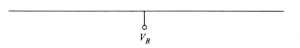

p substrate

FIGURE 11-7
Yau's model of charge sharing.
(After Yau [5].)

where Q_B is given by Eq. (11-7). The total charge contributing to the threshold under the gate contact is WLQ_B, which is represented by the rectangle with a width of x_{dm} and a length of L. However, if we assume part of this charge is shared by the source and the drain such that only the area inside the trapezoid is controlled by the gate, the bulk charge becomes

$$Q'_B L = qN_a x_{dm} \frac{L + L'}{2} \tag{11-11}$$

where $Q'_B < Q_B$ because $L' < L$. In fact, it can be readily proved (Prob. 11-4) that the following relationship holds:

$$\frac{L + L'}{2L} = 1 - \left(\sqrt{1 + \frac{2x_{dm}}{r_j}} - 1 \right) \frac{r_j}{L} \tag{11-12}$$

where r_j is the junction depth. Let us now assume that the charge Q'_B is uniformly spread out under the gate. Taking the average value of the total bulk charge, we obtain

$$V_T = V_{FB} + \phi_{si} + \frac{Q_B}{C_o} \left[1 - \left(\sqrt{1 + \frac{2x_{dm}}{r_j}} - 1 \right) \frac{r_j}{L} \right] \tag{11-13}$$

In the above equation, the threshold is found to be a function of L, r_j, and N_a. For a given junction depth and doping, Yau's model predicts accurately the drop in experimental threshold voltage as shown in Fig. 11-8. The same concept can be applied to narrow-channel devices where the threshold voltage has to increase as the channel width is narrowed (Prob. 11-5).

In the charge-sharing model, we have assumed that the drain-to-source voltage is zero. This is not a necessary condition. It can be verified that the trapezoid concept will work even under different values of V_{DS}. In general, we find that the threshold variation is smaller for longer devices with a shallower junction. In addition, reducing the Q_B/C_o term in Eq. (11-13) minimizes the short-channel effect.

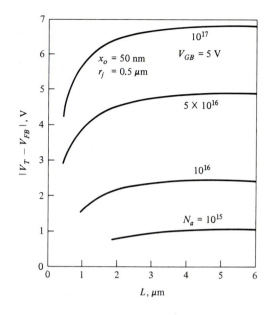

FIGURE 11-8
Theoretical threshold voltage as a function of channel length for various substrate dopings. *(After Yau [5].)*

11-5   HOT-CARRIER EFFECTS

In the reverse-bias drain-to-substrate junction, the electric field may be quite high in short-channel devices. Carriers that are injected into the depletion layer are accelerated by the high field, and some of them may gain enough energy to cause impact ionization. These carriers have higher energy than the thermal energy and are called *hot carriers*. As shown in Fig. 11-9, the holes generated by multiplication can flow to the substrate, giving rise to a large substrate current. Some of the holes may find their way to the source, effectively lowering the source barrier to induce electron injection. The drain-channel-source structure now acts as an *npn* transistor with a floating base and with its collector under avalanche multiplication. Thus, the injected electrons from the source will reach the drain depletion layer, leading to more carrier multiplicaton.

The electrons generated in the drain depletion layer are attracted to the positive gate voltage, as shown in Fig. 11-9. If these electrons have an energy greater than 1.5 eV, they may be able to tunnel into the oxide or to surmount the silicon-oxide potential barrier to produce a gate current. In either case, electrons can be trapped inside the gate oxide, thus changing the threshold voltage and the current-voltage characteristics. This is not desirable and should be avoided. The hot-carrier effects can be minimized if the electric field of the junction can be reduced. This can be accomplished by using the lightly doped drain structure to be described in a later section.

Because the substrate current can be easily measured, it has been employed to study the hot-carrier effect. An experimental plot of the substrate current is given in Fig. 11-10. It is seen that the substrate current increases rapidly with the gate voltage at first, but it starts to decrease at high gate voltages after peaking.

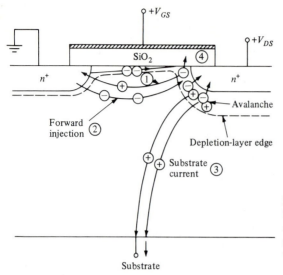

FIGURE 11-9
Hot-carrier generation and current components. ① Holes reaching the source. ② Electron injection from the source. ③ Substrate hole current. ④ Electron injection into the oxide.

The increase of the substrate current is caused by the increase of electrons entering the depletion layer. The electrons are induced by the gate voltage, and they experience multiplication, giving rise to a large number of holes which exit through the substrate. The surface electric field along the channel is illustrated in Fig. 11-11, where the point of pinchoff separates the high and low field regions. The high electric field is created by the bias voltage across the drain junction. When the gate voltage is increased, the pinchoff point moves toward the drain. The surface potential is lowered, and the field pattern is illustrated by the broken line. The net voltage in the lateral direction from the drain to the pinchoff location is decreased, and the avalanche multiplication and the substrate current are reduced.

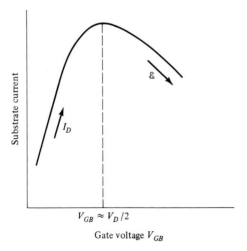

FIGURE 11-10
Substrate current as a function of gate bias.

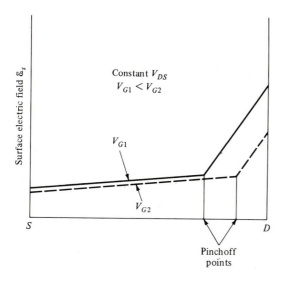

FIGURE 11-11
Surface electric field along the channel.
The sharp rise in field beyond pinchoff
indicates carrier depletion in this region.

11-6 SCALING LAWS OF THE MOS TRANSISTOR

The fundamental issue of downsizing the MOS transistor is to preserve the long-channel characteristics after miniaturization. The first scaling proposal is known as the *constant electric field* (CE) *scaling law* [6]. According to this proposal, the device dimensions, voltages, and doping profiles must be adjusted to keep a constant electric field. This is accomplished by *scaling* (dividing) all dimensions by λ, voltages by λ, and impurity concentrations by $1/\lambda$, where λ is greater than unity. In Fig. 11-12, a transistor is shown before and after scaling by a factor of 5.

The threshold voltage and the drain current after scaling become

$$V_T' = \phi_{ms} - \frac{Q_o}{\lambda K_o \varepsilon_o / x_o} + \phi_{si} + \frac{[2qK_s \varepsilon_o \lambda N_a(\phi_{si} + V_{SB}/\lambda)]^{1/2}}{\lambda K_o \varepsilon_o / x_o}$$

$$\approx \frac{V_T}{\lambda} \qquad \text{if} \quad \phi_{si} + \phi_{ms} \text{ is small} \tag{11-14}$$

$$I_D' = \frac{W/\lambda}{L/\lambda} \mu_n \frac{\lambda K_o \varepsilon_o}{x_o} \left(\frac{V_{GS}}{\lambda} - \frac{V_T}{\lambda} - \frac{V_{DS}}{2\lambda} \right) \frac{V_D}{\lambda} \approx \frac{I_D}{\lambda} \tag{11-15}$$

where the primed terms represent scaled parameters. In the approximation used for the threshold voltage, we have neglected ϕ_{ms} and ϕ_{si}. This is valid only for large V_T. With the scaled-down threshold, these nonscalable terms cannot be ignored. In an NMOS transistor with an n^+-polysilicon gate, the work-function difference and surface-potential terms roughly cancel out in the threshold-voltage equation, and they can be neglected.

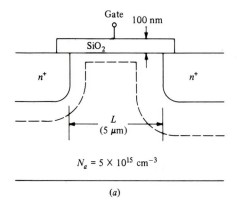

(a)

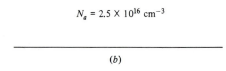

(b)

FIGURE 11-12
An MOS transistor (a) before and (b) after
scaling by $\lambda = 5$.

The scaled gate capacitance is

$$C'_G = \frac{W}{\lambda} \frac{L}{\lambda} \frac{K_o \varepsilon_o}{x_o/\lambda} = \frac{C_G}{\lambda} \qquad (11\text{-}16)$$

The gate delay is

$$\tau' = \frac{C_G}{\lambda} \frac{V_D/\lambda}{I_D/\lambda} = \frac{\tau}{\lambda} \qquad (11\text{-}17)$$

where V_D/I_D represents the effective resistance charging or discharging the gate capacitance. The power dissipation density and power-speed product after scaling are given by

$$P' = \frac{I_D}{\lambda} \frac{V_D}{\lambda} \frac{1}{A/\lambda^2} = P \qquad (11\text{-}18)$$

and

$$P'\tau' = \frac{P\tau}{\lambda} \tag{11-19}$$

These results are summarized in Table 11-1. Notice that, besides the reduction in size and the increase in packing density, the performance in terms of power-speed product is also improved.

It is noted that the built-in potential, work-function difference, and sub-threshold current do not scale. In fact,the subthreshold current is a diffusion current which increases with smaller dimensions. As pointed out earlier, the nonscaled potentials will impact the threshold and bias voltages, making the constant electric field approximation difficult to be realized. The subthreshold current in scaled devices becomes larger while current above the threshold is reduced. This will impose a lower limit on the minimum voltage that can turn off the transistor.

As a result of these considerations, most realistic designs have not reduced the voltage as sharply as required by the CE scaling. Instead, most designers maintain a constant voltage or power supply. This is known as *constant voltage* (CV) *scaling*, and its consequences are outlined in Table 11-1. Notice that the gate delay actually improved by λ^2. An implicit assumption has been made in that the electric field is low enough to avoid velocity saturation.

Under velocity saturation, the drain current is limited to

$$I_D = W(-Q_I)v_{\text{sat}} \tag{11-20}$$

It can be shown that the gate delay will improve by λ. The most serious problem of CV scaling is the increase in the electric field. It induces oxide breakdown and

TABLE 11-1
Comparison of scaling laws [7]

Physical parameter	Expression	1/Scaling factor			
		CE	CV	QCV	GS
Linear dimensions	W, L, x_o, x_j	$1/\lambda$	$1/\lambda$	$1/\lambda$	$1/\lambda$
Potentials	ψ_G, ψ_s, ψ_D	$1/\lambda$	1	$1/\sqrt{\lambda}$	$1/\kappa$
Impurity conc.	$N_a, N_d,$	λ	λ	λ	λ^2/κ
Electric field	$\mathscr{E}$	1	λ	$\sqrt{\lambda}$	λ/κ
Capacitance	AC_o, AC_j	$1/\lambda$	$1/\lambda$	$1/\lambda$	$1/\lambda$
Current	$\dfrac{W}{L}\mu C_o V_D(V_G - V_T)$	$1/\lambda$	λ	1	λ/κ^2
Power density	$I_D V_D/A$	1	λ^3	$\lambda^{1.5}$	λ^3/κ^3
Gate delay	$C_G V_D/I_D$	$1/\lambda$	$1/\lambda^2$	$1/\lambda^{1.5}$	κ/λ^2
Electric field pattern preserved		Yes	Yes	No	Yes

hot carriers, causing concerns for device reliability. To reduce the oxide field, one may scale the oxide thickness with $1/\lambda^{1/2}$ rather than $1/\lambda$. This approach changes the field pattern and exaggerates the short-channel effects.

The *quasi-constant voltage* (QCV) *scaling* is the same as the CV scaling except that the voltages are scaled by $1/\lambda^{1/2}$. The result is a slower increase in the electric field and the power density. Velocity saturation and hot electrons remain to be important in QCV scaling.

By using different scaling factors for the dimensions and the potentials, Wordeman et al. [7] have arrived at a more *general scaling (GS) law*. This method introduces additional flexibility so that the scaled devices will perform somewhere between CE and CV scaling. Practical implementation depends on the specific applications. For example, the high-power density in CV scaling is not suitable for NMOS circuits which consumes dc power, but it may be acceptable in CMOS designs. A comparison of these scaling laws is given in Table 11-1.

After scaling all the transistor's active regions, a natural extension is to reduce the size of the contacts and interconnects. If the CE scaling law is employed, the area of the contact windows is decreased by $1/\lambda^2$, and the contact resistance will increase by λ^2. Since the current is scaled by $1/\lambda$, the voltage drop across the contact will increase by a factor of λ. This is undesirable since all other voltages are scaled by $1/\lambda$. Scaling of the interconnect is left as a homework problem. In general, scaling of interconnects and contact windows degrades the performance and should be avoided.

11-7 VELOCITY SATURATION

The mobility dependence on the horizontal electric field appears in the form of velocity saturation. As carriers are accelerated in an electric field, they gain velocity and kinetic energy. The increase in energy increases the scattering probability between the carriers and the lattice. Imparting of the carrier energy to the lattice generates phonons which limit the carrier velocity at high electric fields. Velocity saturation affects the performance of short-channel devices by reducing the transconductance in the saturation mode.

Current saturation is defined as the condition when the inversion-layer charge becomes zero. This corresponds to the pinchoff condition of

$$V_{DS} = V_G - V_T \tag{11-21}$$

which is satisfied at $x = L'$ in Fig. 11-5. The average electric field inside the channel is given by

$$\mathscr{E}_{av} = \frac{V_G - V_T}{L'} \tag{11-22}$$

As an example, let us assume a gate voltage of 5 V and a threshold of 1 V for a transistor with a channel length of 1 μm. The average electric field is found to be 4×10^4 V/cm, clearly above the critical field for velocity saturation (Fig. 11-13).

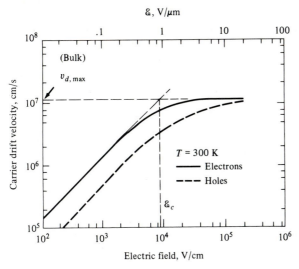

FIGURE 11-13
Carrier velocity becomes constant for fields above a critical value.

Under this condition, the drain current is limited by carrier velocity so that

$$I_D = W\bar{Q}_I v_{d,\,\text{max}} = WC_o\left(V_G - V_T - \frac{V_{DS}}{2}\right)v_{d,\,\text{max}} \tag{11-23}$$

The critical field is defined as the intersecting point between the linear and horizontal lines, as shown in Fig. 11-13. Thus,

$$v_d = \begin{cases} \mu\mathscr{E} & \text{for } \mathscr{E} \ll \mathscr{E}_c \tag{11-24} \\ v_{d,\,\text{max}} & \text{for } \mathscr{E} \gg \mathscr{E}_c \tag{11-25} \end{cases}$$

A better approximation to match the experimental data is given by

$$v_d = v_{d,\,\text{max}}\frac{\mathscr{E}/\mathscr{E}_c}{1 + \mathscr{E}/\mathscr{E}_c} \tag{11-26}$$

Frequently, an alternative form is used based on the mobility:

$$\mu = \frac{\mu_{\text{eff}}}{[1 + (\mu_{\text{eff}}\mathscr{E}/v_{d,\,\text{max}})^2]^{1/2}} \tag{11-27}$$

For most purposes, Eq. (11-26) is sufficient. Differentiating Eq. (11-23), we obtain

$$g_m = \frac{\partial I_D}{\partial V_G} = \tfrac{1}{2}WC_o v_{d,\,\text{max}} \tag{11-28}$$

It is noted that the transconductance is independent of the gate and drain voltages and the channel length. This tends to compress the drain current, as shown in Fig. 11-14. The drain current is saturated not by pinchoff but by velocity saturation, and it leads to a constant transconductance. This is likely to happen in short-channel devices where the dimensions have been reduced without lowering the bias voltages. It is less important in scaled-down devices following the constant electric field approach.

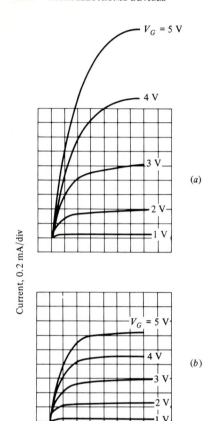

Current, 0.2 mA/div

$V_G = 5$ V

4 V

3 V

(a)

2 V

1 V

$V_G = 5$ V

4 V

(b)

3 V

2 V

1 V

Drain voltage, 1 V/div

FIGURE 11-14
Theoretical comparison of current vs. drain bias curves for *(a)* a device with constant mobility (no velocity saturation) and *(b)* a device exhibiting velocity saturation. $N_a = 10^{15}$ cm^{-3}, $x_o = 0.05\,\mu$m, $r_j = 0.4\,\mu$m, $L = 2.7\,\mu$m, implant dose 3×10^{11} cm^{-2}. *(After Yamaguchi [8].)*

11-8 THE LIGHTLY DOPED DRAIN TRANSISTOR

As discussed in the scaling of the MOS transistor, the constant electric field law may not be adhered to strictly. In fact, a common practice in the industry is to keep a high bias voltage using either CV or QCV scaling rules. In these cases, the internal electric field becomes higher, and the hot-carrier effects are more serious in small-dimension devices. One approach to maintain relatively high bias voltages with minimal hot-electron injection is to tailor the impurity profile near the drain to reduce the junction electric field. The *lightly doped drain* (LDD) design shown in Fig. 11-15 is such a structure. In the same figure, we have included a conventional device along with its doping profile. By introducing a lightly doped section between the drain and the channel, the depletion layer's peak field is shifted toward the drain, and the field is lowered, as illustrated in Fig. 11-16. Obviously, the LDD structure is more complicated and takes more steps in fabricaton, but the few added processing steps produce significant improvements in performance. It has a higher breakdown voltage, and the substrate current is reduced by a factor of 30. The

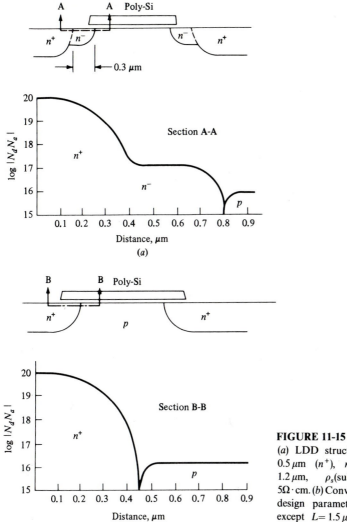

FIGURE 11-15

(a) LDD structure, $x_o = 45$ nm, $n_j = 0.5\,\mu$m (n^+), $n_j = 0.3\,\mu$m (n^-), $L = 1.2\,\mu$m, ρ_s(substrate resistivity = $5\,\Omega \cdot$cm. (b) Conventional structure with design parameters the same' as (a) except $L = 1.5\,\mu$m and $\rho_s = 15\,\Omega \cdot$cm. (*After Ogura [9].*)

peak field is shifted away from the oxide so that hot-electron injection into the oxide is much less. Furthermore, the lightly doped drain means a thinner depletion layer, thus less charge sharing and less threshold reduction in short-channel devices. There is also less overlapping capacitances, resulting in a faster circuit.

The improvements of the LDD structure are not achieved without a price. Besides adding process complexity, the lightly doped regions increase the series resistance, so the available drain current is reduced. In addition, the higher voltage means higher power dissipation. For these reasons, one must consider the power-delay tradeoff as well as heat generation of a particular design. The LDD configuration is a popular structure, and it has been employed extensively in both NMOS and CMOS circuits.

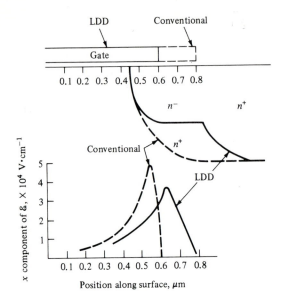

FIGURE 11-16
Magnitude of the electric field at the Si–SiO$_2$ interface as a function of distance; $L = 1.2\,\mu m$, $V_{DS} = 8.5\,V$, $V_{GS} = V_T$. The physical geometries for both devices are shown above the plot. *(After Ogura [9].)*

REFERENCES

1. Fichtner, W.: Scaling Calculation for MOSFETs, IEEE Solid S. Circuits and Tech. Workshop on Scaling and Microlithography, New York, 4/22/1980.
2. Kotani, N., and S. Kawazu: Computer Analysis of Punch-Through in MOSFETs, *Solid State Electron.*, **22**:63 (1979).
3. Fichtner, W., and H. W. Pötzl: MOS Modeling by Analytical Approximations, I Subthreshold Current and Threshold Voltage, *Int. J. Electron.*, **46**:33 (1979).
4. Troutman, R. R.: VLSI Limitations from Drain Induced Barrier Lowering, *IEEE Trans. Electron Devices*, **ED-26**:461 (1979).
5. Yau, L. D.: A Simple Theory to Predict the Threshold Voltage of Short-Channel IGFETs, *Solid State Electron.*, **17**:1059 (1974).
6. Dennard, R. H., et al.: Design of Ion-Implanted MOSFETs with Very Small Physical Dimensions, *IEEE J. Solid-State Circuits,* **SC-9**:256 (1974).
7. Baccarani, G., M. R. Worderman, and R. H. Dennard: Generalized Scaling Theory and Its Application to a $\frac{1}{4}$ micrometer MOSFET Design, *IEEE Trans. Electron Devices*, **ED-31**:452 (1984).
8. Yamaguchi, K.: Field-Dependent Mobility Model for Two-Dimensional Numerical Analysis of MOSFETs, *IEEE Trans. Electron Devices*, **ED-26**:1068 (1979).
9. Ogura, S., et al.: Design and Characterization of the Lightly Doped Drain-Source (LDD) Insulated Gate Field-Effect Transistor, *IEEE Trans. Electron Devices*, **ED-27**:1359 (1980).

PROBLEMS

11-1. (*a*) An MOS transistor is used to drive an identical transistor. The channel length is 10 μm, $\mu_n = 1000\,cm^2/V \cdot s$, and the drain supply is 5 V. Estimate the delay time.

(*b*) The MOS transistor is used to drive six identical transistors. If the delay per gate is to remain the same as in (*a*), what is the necessary change in device dimension?

11-2. Derive Eq. (11-8), using the one-sided step junction for the drain-to-channel depletion layer.

11-3. (a) Calculate I'_D/I_D for a transistor with a channel length of 5 μm, $N_a = 10^{15}$ cm^{-3}, and $V_{DS} = 3$ V. Assume $\phi_p = 1$ V and $\Delta V = 0.1$ V.
   (b) Repeat (a) for $L = 0.5$ μm.
   (c) Calculate the drain resistance r_{ds}, defined by $\Delta V_{DS}/\Delta I_D$ at $I_D = 1$ mA for (a) and (b)

11-4. (a) From geometrical consideration, show that the effective channel length is related to the physical channel length by Eq. (11-12).
   (b) Derive Eq. (11-13) by following the procedure outlined in the text.

11-5. If the width W of an MOS channel is narrow as shown in Fig. P11-5, the threshold voltage will be affected by the channel width because of the additional space charge. Show that

$$V_T = V_{FB} + \phi_{si} + \frac{Q_B}{C_o}\left[1 + \frac{\alpha x_{dm}}{W} - \frac{r_j}{L}\left(1 + \frac{4}{3}\frac{\alpha x_{dm}}{W}\right)\left(\sqrt{1 + \frac{2x_{dm}}{r_j}} - 1\right)\right]$$

where α is a parameter that represents the effective channel depletion under the thick oxide. (See Fig. P11-5.)

11-6. In Fig. 11-7 the depletion layers on the drain and source junctions are symmetrical. This is not true if $V_S = 0$ and $V_D = V_{DD}$. Redraw the space-charge diagram to show the effective L and L'. Explain the large drop in V_T for $V_{DS} = 4$ V and $L < 2$ μm in Fig. 11-4.

11-7. An MOS transistor has the following parameters: $N_a = 10^{16}$ cm^{-3}, $x_o = 50$ nm, $r_j = 0.3$ μm, $V_{FB} = -1$ V, $L = 1.6$ μm, and $W = 25$ μm.
   (a) Calculate V_T without considering charge sharing at $V_{DS} = 0$.
   (b) Repeat (a) with charge sharing.
   (c) Repeat (b) with $V_{DS} = 5$ V.

11-8. Using constant electric field scaling, the transistor of the example in Sec. 9-3 is scaled by a factor of (a) 2 and (b) 10. Calculate the scaled V_T. Comment on the results.

11-9. Using the CE scaling, calculate the current in the transistor of the example in Sec. 9-4, for $V_G = -8$ V if it is scaled by a factor of (a) 2 and (b) 10.

11-10 Verify the scaled electric field and current in Table 11-1 for CV, QCV, and GS scalings.

11-11 Repeat Probs. 11-8 and 11-9 for CV scaling.

11-12 In an MOS transistor circuit having a 5-V power supply, estimate the channel length L that will give rise to velocity saturation. Assume a threshold voltage of 1 V and a surface mobility of 500 $cm^2/V \cdot s$.

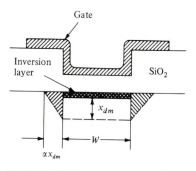

FIGURE P11-5

11-13. The channel length of the transistor in Prob. 11-12 is changed from 8 μm to 2 μm and then to 0.5 μm in CV scaling. Determine the ratio of the drain saturation current.

11-14. Calculate the effective channel length (i.e., subtracting the depletion-layer widths) for the LDD and conventional structures shown in Fig. 11-15 for (*a*) $V_{DS} = 0$ and $V_{SB} = 5$ V and (*b*) $V_{DS} = 5$ and $V_{SB} = 5$ V. To simplify the calculation, assume the doping level in the n^+ region to be 10^{20} cm^{-3} and in the n^- region to be 10^{17} cm^{-3}. The substrate doping is 5×10^{16} cm^{-3} and $L = 1.5$ μm for both devices. Assume ϕ_p is zero.

11-15. Estimate the threshold-voltage drop in Prob. 11-14*a* by charge sharing for both type of devices.

CHAPTER 12

MOS
INTEGRATED
CIRCUITS
AND
TECHNOLOGY

As pointed out in Chap. 9, the MOS transistor has emerged as the most important electronic device in the last few years because of its simple structure and low fabrication costs. Since the source, channel, and drain are surrounded by a depletion region, there is no need to isolate individual components. This elimination of isolation regions makes possible the high packing density of MOS transistors on a chip.

The most basic circuit is the MOS inverter. There are a number of arrangements for this circuit depending on the type of load device. Of these circuits, the most widely used structures are the NMOS with a depletion load and the complementary MOS (CMOS) circuit. In the latest technology, the CMOS has emerged as the standard. The CMOS inverter makes use of a *p*-channel and *n*-channel pair of transistors. Its virtue lies with the fact that it consumes no power under a steady-state dc condition. Because of the low overall power dissipation, CMOS is becoming the dominant technology for very large scale integration, and it is likely to surpass the NMOS and bipolar processes in the near future. From its appearance, CMOS is a circuit-design topic not suitable for a device textbook. However, the key issues in CMOS are essentially device- and technology-oriented. As device dimensions are scaled down, the interaction between devices gives rise to the problem of latch-up. Modification of device structure and technology to eliminate the latch-up is a major device problem.

In this chapter, we shall present the various forms of the inverter. The emphasis is on the CMOS design. Topics to be taken up include the threshold matching, latch-up and its prevention, the CMOS with Schottky source and drain contacts, and trench isolation.

12-1 MOS INVERTERS

The most common applications of MOS transistors are in integrated digital logic gates and memory arrays, in which the MOS inverter is the basic circuit. We consider several types of inverters used in IC designs. Each inverter is characterized by the arrangement of its load. By examining the load line drawn in the output characteristics, we can understand the basic operation of each circuit. Design consideration is given from two points of view: (1) circuit performance including power dissipation, maximum voltage swing, and speed and (2) fabrication factors, including isolation, device area, and cost.

Linear Resistive Load

An MOS inverter circuit with a linear resistive load is shown in Fig. 12-1a, and the load line is plotted in Fig. 12-1b. This circuit has large output-voltage swing, and its switching speed is limited by the product $R_L C_L$. Because C_L is fixed by parasitic capacitances, high speed is achieved with a smaller R_L, which leads to high power dissipation and low voltage swing (see Fig. 12-1b). The more detrimental factor of the circuit is the large area required for a linear diffused resistor, which also needs electrical isolation. At present, the MOS inverter with a linear resistor is not competitive with other inverter circuits described below. However, it is conceivable that in the future we can use an ion-implanted region or doped polycrystalline silicon embedded in the oxide as the load resistance. The advantages of a resistive load include low temperature dependence and reasonable power-speed product.

Saturated MOS Load

Most MOS logic circuits use an MOS transistor load, as shown in Fig. 12-2a. The MOS load is used in ICs because it takes up a much smaller area than a resistive load. The operation of the circuit can be understood through the current-voltage characteristics shown in Fig. 12-2b, where the curve under the condition $V_G = V_D$ is plotted to represent the load characteristics with the gate tied to the drain. If the inverter device and the load device are identical, the load line is obtained by using the $V_G = V_D$ curve and plotting it as shown by the dotted line. This arrangement, however, produces an output swing significantly less than V_{DD} and is not desirable. The output swing can be increased if the transconductance of the load device is reduced to one-tenth of the inverter device. This change leads

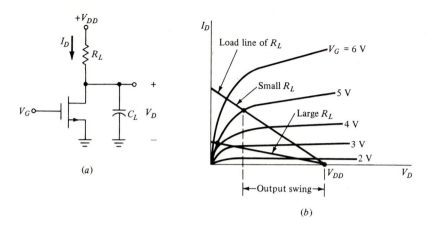

(a)

(b)

FIGURE 12-1
An MOS inverter with linear resistive load: (a) circuit and (b) output characteristics and load line.

to an ON resistance of the load device that is 10 times that of the inverter transistor. In practical design, we make use of Eq. (9-42) to obtain

$$R_{\mathrm{on},l} = \frac{L_l}{\mu_n C_o W_l (V_{Gl} - V_T)} \tag{12-1}$$

for the load device and

$$R_{\mathrm{on},i} = \frac{L_i}{\mu_n C_o W_i (V_{Gi} - V_T)} \tag{12-2}$$

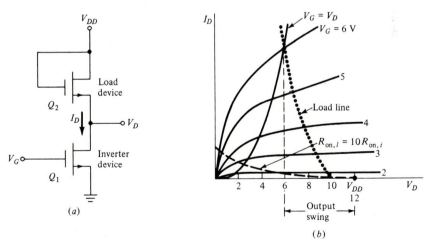

(a)

(b)

FIGURE 12-2
An MOS inverter with a saturated MOS load: (a) circuit and (b) I-V characteristics with load curves.

for the inverter device. Since $V_{GI} = V_{DS} \approx V_{DD}$ and $V_{Gi} \approx V_{DD}$, the resistance ratio becomes

$$\frac{R_{on,l}}{R_{on,i}} = \frac{L_l/W_l}{L_i/W_i} \tag{12-3}$$

For example, a resistance ratio of approximately 10 can be obtained by setting

$$\frac{L_l}{W_l} = 3 \quad \text{and} \quad \frac{L_i}{W_i} = \frac{1}{3}$$

As we can see, the length-to-width ratio of the MOS transistor is the most important design parameter in a saturated MOS load inverter. The layout diagram shown in Fig. 12-3a illustrates the required area for the complete inverter. The new load line is sketched as the dashed curve in Fig. 12-2b. The output swing is now substantially improved, and the transfer characteristic is displayed in Fig. 12-3b. Note in Fig. 12-2b that the load resistance is not linear and has a high value at V_{DD} and a low value at $V_D \to 0$ V. With the same parasitic capacitance, we find that

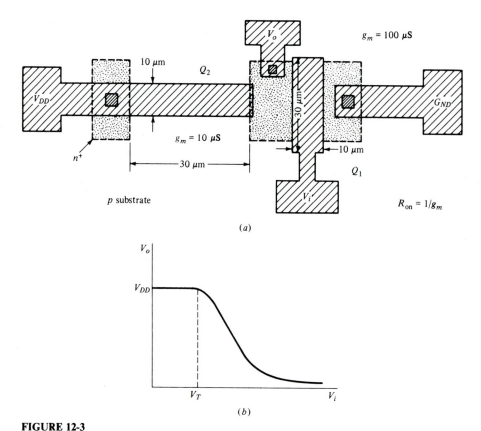

(a)

(b)

FIGURE 12-3
(a) Layout diagram and (b) transfer characteristic of an MOS inverter with saturated MOS load.

the charging time through the load device is much slower than the discharging time through the inverter device.

Nonsaturated MOS Load

If an additional power supply is available, we can bias the gate of the load device instead of having it shorted to the drain. This bias arrangement is shown in Fig. 12-4, in which the load device changes from a highly nonlinear resistor to a linear one and approaches a linear resistor for high gate bias voltage. The charging time through the load device is significantly reduced because of the lower load impedance. In addition, the voltage swing is increased by V_T compared with the saturated load.

Depletion MOS Load

When the gate is short-circuited to the source, a depletion MOS transistor has a current-voltage characteristic shown in Fig. 12-5. With the inverter circuit shown in Fig. 12-6a, the output characteristics and load line are shown in Fig. 12-6b. An important feature of the depletion-load circuit is its essentially symmetrical charging and discharging time constants. The switching speed is about 3 times faster than the all-enhancement MOS inverter.

Complementary MOS (CMOS) Inverter

The standard dc power dissipation of an MOS inverter can be reduced to very small (10-nW) levels by using a complementary p-channel and n-channel pair connected as shown in Fig. 12-7a. Transistor Q_1 is an n-channel device, and Q_2 is a p-channel device. When the input voltage is high (V_{DD}), Q_1 is turned on and Q_2 is turned off. When the input voltage is low (0 V), Q_1 is turned off and Q_2 is turned on. The operating points are shown by the curves in Fig. 12-7b. Note that under either input condition, very little current is drawn in the steady state. This current is the leakage current of the OFF device. By examining the load lines, it is seen that the turn-on and turn-off time constants are about the same.

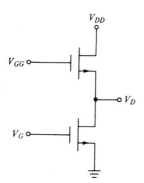

FIGURE 12-4
MOS-inverter circuit with nonsaturated load.

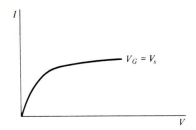

FIGURE 12-5
I-V characteristic of the depletion load.

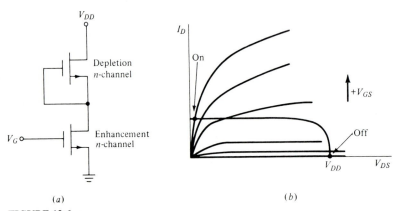

(a)

(b)

FIGURE 12-6
The depletion-load MOS inverter with load curve.

The total power dissipation of a CMOS inverter is the sum of the standby power and the transient power loss during switching. The latter is given by fCV^2, where f is the switching frequency, C is the total output capacitance, and V is the output voltage.

12-2   THE n-CHANNEL (NMOS) TECHNOLOGY

The n-channel MOS transistor is preferred over the p-channel transistor in ICs because of the higher electron mobility. A two-input NAND gate using NMOS transistors with a depletion load is shown in Fig. 12-8. In this figure, the top view of the circuit layout is also shown. The fabrication process of this structure is illustrated in Fig. 12-9. The starting p substrate is lightly doped, upon which an oxide is first grown and is then covered by a silicon-nitride deposition. An isolation mask is used to define the active device areas (i.e., covered by Si_3N_4–SiO_2) and the isolation or field areas (etched by plasma or reactive ion etching). In Fig. 12-9a, boron ions are implanted as the channel stop to prevent inversion under the field oxide. After photoresist stripping and cleaning, the wafer is put in an oxidation furnance to grow a thick LOCOS or FOX which surrounds the active devices.

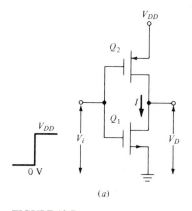
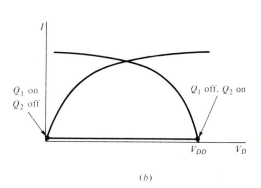

(a)

(b)

FIGURE 12-7
The complementary MOS inverter and its load curves.

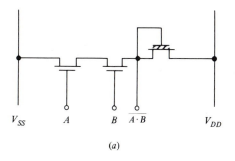

(a)

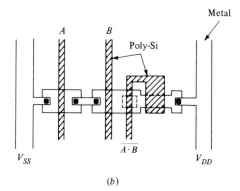

(b)

FIGURE 12-8
(a) A two-input NAND gate circuit and (b) top view of layout.

This step also drives in the channel-stop implant (Fig. 12-9b) and provides a contact to the substrate (not shown in the figure). The nitride/oxide layers are stripped and the surface is cleaned to prepare for the critical step of growing a thin gate oxide (about 20 nm in thickness). Using the photoresist to mask the enhancement-mode device (EMD), an n-channel implant is made to form the depletion-mode device (DMD) (Fig. 12-9c). The polysilicon is deposited and patterned as the gates which are also used as the self-aligned mask for source and drain arsenic

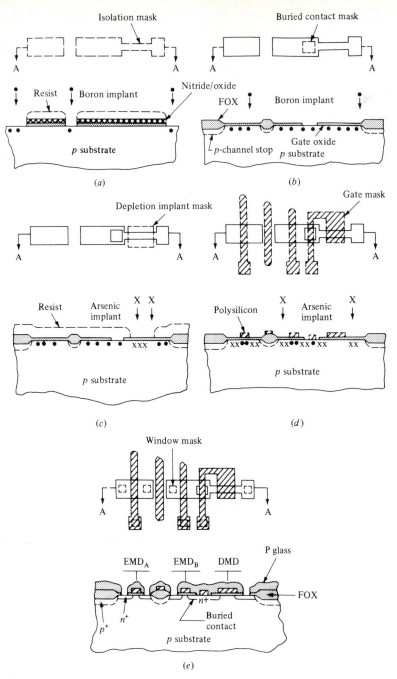

FIGURE 12-9
Top and cross-section views of NMOS logic gate fabrication. (*a*) Isolation mask and cross section after nitride/oxide etch and boron channel-stop implant. (*b*) Buried contact mask and cross section after field oxidation (FOX), gate oxidation, buried contact window etch, and boron-enhancement threshold-adjustment implant. (*c*) Depletion implant mask and cross section after resist-masked arsenic-depletion implant. (*d*) Gate mask and cross section after polysilicon gate definition and arsenic source-drain implant. (*e*) Window mask and cross section after P-glass flow and window etch. (*After Parrillo* [1].)

implantation (Fig. 12-9d). Contact openings are made and metal films are evaporated and etched to produce the final structure in Fig. 12-9e.

The higher electron mobility means a higher speed and transconductance in NMOS devices over the p-channel MOS (PMOS) devices. Since CMOS makes use of both NMOS and PMOS devices, it is inherently slower than the NMOS circuit. However, the lower power-dissipation requirement in scaled devices makes the CMOS technology more attractive. In some CMOS designs, the NMOS circuit is incorporated, e.g., in the domino-CMOS, to take advantage of the NMOS's speed and CMOS's low power, producing a special family of integrated circuits. Interested readers are referred to the literature for this unusual design concept.

12-3 THE CMOS TECHNOLOGY [2, 3]

Since there are two types of devices in a CMOS inverter, it is necessary to have two electrically isolated regions to accommodate the n- and p-channel transistors. The standard technique is to form a p well in an n substrate or an n well in a p substrate. A well, also called a tub, is produced by an extra diffusion step. The cross section of a finished p-well CMOS structure is shown in Fig. 12-10a. The p well is formed by a deep diffusion after the removal of the oxide by a masking step. Then, a field oxide is grown by local oxidation, followed by gate oxidation and polysilicon-gate deposition. The self-aligned source and drain are produced by ion implantation, and the metal interconnects are deposited and patterned after opening of contact holes. What has been described is a simplified procedure for the CMOS process. In practice, additional steps are needed to control the threshold voltage, to form a channel stop to avoid inversion under the field oxide, and to reduce the short-channel effects. These will be discussed in a realistic example in the next section. Modification of the above procedure is required to contact the substrate and the p well so that they would not be floating electrically.

The p-well process had been the commonly used technology for earlier CMOS circuits before the technique of ion implantation was widely applied for threshold control. With a high-resistivity n substrate (for example, $10 \, \Omega \cdot cm$), the threshold of the p-channel transistor is typically between -1 and -2 V because of the positive oxide charge and the work-function difference with the n^+-polysilicon gate. At the same time, the diffused p well has a high boron surface concentration preventing its inversion to an n channel without a gate bias. Consequently, the threshold voltages of the p- and n-channel transistors are well-matched. This consideration is no longer important today because the oxide charge is now negligible and the threshold can be controlled easily with ion implantation.

The n-well process has the advantage of starting with an existing NMOS technology which produces excellent n-channel transistors. Since n-channel devices have a higher current and speed capability, and they are already optimized in the p substrate, the n well is added to form the p-channel devices with less stringent requirements. The structure is shown in Fig. 12-10b. An additional advantage of the n-well design is that the high substrate current caused by impact ionization in the n-channel transistor can pass through the substrate with less feedback action.

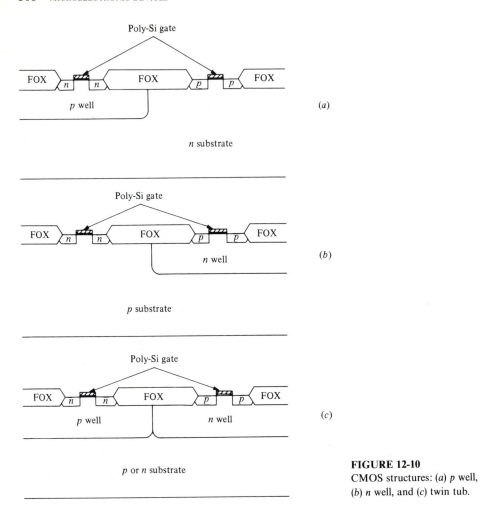

FIGURE 12-10
CMOS structures: (*a*) *p* well,
(*b*) *n* well, and (*c*) twin tub.

(On the other hand, the *p*-well process has a better *p*-channel transistor. Therefore, the performance of both *n*- and *p*-channel transistors are about the same, which is advantageous in static circuit applications.) More will be said about the *n*-well technology in the next section.

The process of having both *n* and *p* wells together is known as the *twin-tub* or *twin-well* CMOS. This technology provides separate optimization for the *n*- and *p*-channel devices. Therefore, one can control the threshold, transconductance, and body effect independently. The two wells are formed by separate implantation and drive-in. Other steps are similar to those described earlier for the *p*-well process. The cross section of the twin-tub structure is shown in Fig. 12-10*c*. Frequently, an epitaxial layer is grown in which the wells are formed. The epitaxial layer is used to prevent latch-up, a subject that will be discussed in Sec. 12-6. The twin-tub process also allows tighter packing density because of the butted wells and self-aligned channel stops. It is attractive for submicron devices where the two device types have about the same performance because of velocity saturation. Symmetrical n- and p-channel transistors can be obtained with either an epitaxial substrate or trench isolation.

Traditional p- and n-well structures make use of a deep impurity diffusion to form the well. Since impurity atoms diffuse vertically and laterally, significant lateral areas are used up, resulting in a lower packing density. Recently, high-energy ion implantation has been employed to obtain deep penetration of impurity. Since the annealing temperature is lower than the diffusion temperature, the implanted impurities stay put with minimal lateral spread. The range of the implanted ions is deep so that the impurity concentration decreases toward the surface, forming a retarding field. This is known as a *retrograde* well. The impurity profile has a peak deep in the well, and it has a high conductivity and a low ohmic drop. The vertical punch-through voltage is improved in comparison with a shallow well. Furthermore, it reduces the junction capacitance and body effect since the highly doped layer is away from the channel of the inversion layer. The retarding field also decreases the vertical bipolar current gain, resulting in better latch-up immunity.

12-4 THRESHOLD CONTROL AND MATCHING

Since the gates of the n- and p-channel transistors are tied together and the applied gate voltage keeps one device on and the other off, the threshold voltages of the two devices should be closely matched. This is achieved in the p-well technology because of the positive oxide charge and the work-function difference with an n^+-polysilicon gate. In most recent designs, the control of the threshold is accomplished by ion implantation. This is necessary for the n-well technology. In this section, a short-channel n-well process is presented with emphasis on the channel implantation and threshold control.

The n well is formed by diffusion of implanted phosphorus at 1100°C on a $10\,\Omega\cdot$cm p-epitaxial layer on a p^+ substrate (Fig. 12-11a). A thin layer of oxide is grown, followed by the deposition of a silicon-nitride film. Borons are implanted into the field-oxide region of the n-channel devices as the channel stops (Fig. 12-11b). After photoresist and masking steps, a field oxide is grown by LOCOS, and a gate oxide is formed subsequently. The channels are now ready for threshold-control implantations in the structure shown in Fig. 12-11c. There are different approaches to get the same result. For example, the whole structure may be exposed to both 150- and 30-keV implantations. The deep implantation controls the subthreshold current of the n-channel device, and the shallow implantation adjusts the threshold voltages of both n- and p-channel transistors. This situation is depicted in Fig. 12-11c. Alternatively, the p-channel device may be masked to avoid the deep implantation which degrades the short-channel V_T falloff resulting from the lighter doping due to charge compensation in the n well (Fig. 12-11d). Subsequent steps include the n^+-polysilicon-gate deposition and etching as well as the self-aligned source and drain implantations (Fig. 12-11e). The final structure is shown in Fig. 12-11f. Note that the double-implanted p-channel device (DIP)

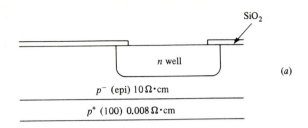

(a)

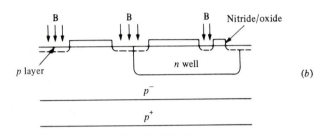

(b)

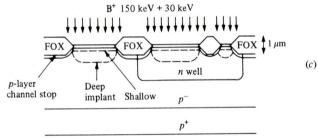

(c)

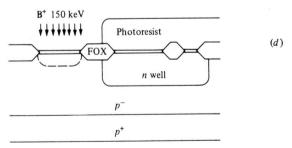

(d)

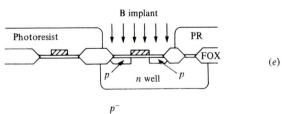

(e)

FIGURE 12-11
Threshold control of *n* well
technology.

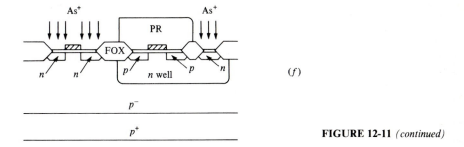

(f)

FIGURE 12-11 (continued)

and the single-implanted p-channel device (SIP) are obtained by using the processes illustrated in Fig. 12-11c and d, respectively.

The threshold voltage of the n-channel transistor is controlled by the shallow boron-implantation dosage D_I. Its magnitude is given by either Eq. (10-59) or (10-63), depending on the doping and biases. The threshold voltage of the p-channel transistor is a function not only of the shallow boron implant but also of the n-well's surface concentration. Exact modeling can be made to calculate the actual threshold voltage shift (Probs. 12-5 and 12-6). The result is plotted graphically in Fig. 12-12 to show how the correct implantation dosage is selected. The NMOS's threshold depends only on the shallow boron implant, and it is seen to increase linearly with the dosage.

The deep implant does not affect the threshold since it is far away from the channel. In the p-channel device, the threshold voltage is determined by first selecting the surface impurity concentration of the n well (C_{SN}). This is specified by the phosphorus n-well implant dose and the subsequent drive-in diffusion. Then, the shallow boron implant is used to match the n-channel threshold. For

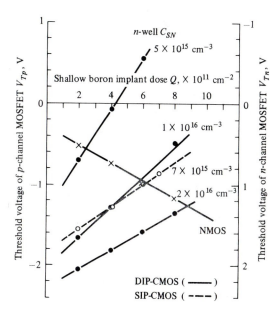

FIGURE 12-12
Extrapolated threshold voltages of n- and p-channel MOSFETs as a function of shallow boron-implantation dose. $V_{DS} = 0.1$ V. Deep boron-implantation dose is 6×10^{11} cm^{-2}. (After Yamaguchi [2].)

example, if a shallow boron-implant dosage of $6 \times 10^{11}\,\text{cm}^{-2}$ is used for the n-channel device, the threshold is 0.95 V from Fig. 12-12. Selecting an n-well surface concentration of $10^{16}\,\text{cm}^{-3}$ and using the same boron implant produces a threshold of -0.95 V for the p-channel device. By employing an additional masking step, the threshold of the p-channel device may be controlled separately. In this way, the n- and p-channel devices could be optimized individually to achieve better short-channel behavior.

12-5 CMOS LATCH-UP

The major problem in CMOS circuits is device latch-up, an internal feedback mechanism that gives rise to temporary or permanent loss of circuit function. Let us consider a p-well structure as shown in the cross-sectional diagram in Fig. 12-13. The n^+ source, p well, and the n substrate constitute a vertical npn bipolar transistor, and the p well, n substrate and p^+ source from a lateral pnp bipolar transistor. The base of each transistor is driven by the collector of the other to form a positive feedback loop. This is the structure of a $pnpn$ switch which may be turned on or off as described in Chap. 5. When the loop gain is greater than 1, the $pnpn$ device is switched to a low impedance state with large current conduction. This parasitic $pnpn$ action interferes with the circuit function of the CMOS circuit and must be avoided. In large-area devices, the pnp and npn transistors can be placed far apart so that they don't interact. This is not possible in small-dimension structures where devices are tightly packed to save space. To prevent latch-up from happening, we have to understand the basic switching operation.

By examining the physical structure of the parasitic devices, one may arrive at the equivalent circuit shown in Fig. 12-14. In addition to the two bipolar transistors, we have included two resistors which represent a key departure from the simple $pnpn$ switch described in Sec. 5-10. The resistor R_w arises from the series resistance in the p well for the current flowing into the p^+ contact. The resistor R_s represents the resistance to reach the substrate n^+ contact. Let us first consider the case when these resistances are very large. The circuit acts as a $pnpn$ diode, and it has a current-voltage characteristic shown in Fig. 12-15. If a large reverse bias is applied across the p-well–to–n-substrate junction, the avalanche condition sets in. The current is increased which enhances the current gains. It can be shown that the loop gain is given by the product $\beta_n \beta_p$. The switch is turned from its OFF (blocking) state to its ON state when the loop gain is greater than unity. The onset of its switching is specified by the switching current and voltage in Fig. 12-15. The current in the ON state can be very large, and it could destroy the device. The device returns to the OFF state if the current is reduced to below the holding current. When the two resistors are included, a portion of the base current is siphoned off so that the effective current gains of the bipolar transistors are reduced. The latch-up condition with finite R_w and R_s is (Prob. 12-8)

$$\beta_n \beta_p > 1 + \frac{(\beta_p + 1)(I_{RS} + I_{RW}/\beta_p)}{I - I_{RW}(1 + 1/\beta_p)} \tag{12-4}$$

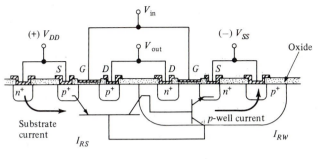

FIGURE 12-13
Cross-sectional view of a p-well CMOS inverter with parasitic bipolar transistors and lateral currents schematically shown. *(After Chen [3].)*

Since the emitter forward-bias voltage is approximately 0.7 V, the currents through the resistors can be estimated if the resistances are known. The substrate resistance can be estimated from the structure, but the well resistance is more difficult to obtain because of both nonuniform doping (implanted or diffused) and the depletion-layer width.

The obvious solution to the latch-up problem is to reduce the current gains or the series resistances. In either case, the loop gain will be lowered; the latch-up condition is harder to satisfy. If R_w and R_s are zero, the two emitters are short-circuited to prevent the turning on of the bipolar transistors. In practice, this approach is not realistic. For example, device scaling using the CE rules reduces the base width of the transistors and increases the current gains with the resistances unchanged. Frequently, the impurity concentration of the scaled devices may not be increased as aggressively as the CE law, so the resistance of R_w and R_s is actually higher. As a result, the scaled devices have a more serious problem in latch-up.

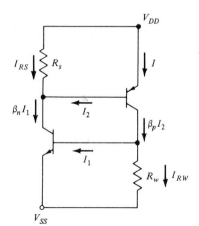

FIGURE 12-14
Equivalent circuit for latch-up currents of a CMOS structure shown in Fig. 12-13.

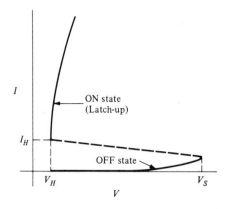

FIGURE 12-15
Current-voltage characteristic of a *pnpn* latch-up.

12-6 LATCH-UP PREVENTION [4]

The latch-up in CMOS can be prevented if $\beta_n\beta_p$ is less than unity at all time. This is not always possible because of other device design constraints. Practical methods to avoid latch-up follow two approaches known as *bipolar spoiling* and *bipolar decoupling*. The first approach is to spoil the transistor action by decreasing the injection efficiency or the transport factor, e.g., using a base retarding field, lifetime

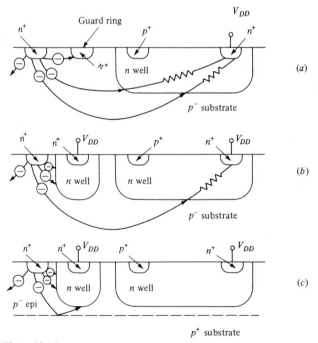

Figure 12-16
Minority-carrier guard in substrate. (*a*) n^+-diffusion guard. (*b*) *n*-well guard. (*c*) *n* well in epitaxial CMOS. *(After Troutman [4].)*

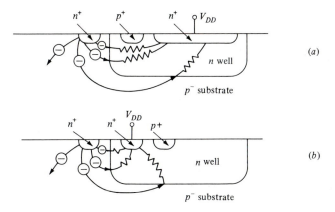

FIGURE 12-17
Majority-carrier guard in well. (a) n^+-diffusion guard to reduce n well sheet resistance. (b) n^+-diffusion guard to steer current away from vertical pnp emitter. *(After Toutman [4].)*

killing with gold doping or neutron irradiation. Alternatively, Schottky-barrier source and drain contacts are used to eliminate minority-carrier injection. In the second approach, layout techniques such as butted contacts and guard rings are employed. Decoupling can also be achieved by using a lightly doped thin epitaxial layer on a heavily doped substrate or by trench isolation. Gold doping or neutron irradiation introduces leakage current and is difficult to control. The retrograde well is effective in reducing the vertical bipolar gain by providing a high base Gummel number as well as a retarding field for minority carriers. But it does not increase the switching current sufficiently to provide latch-up immunity. The Schottky-barrier contacts increase the series resistance of the MOS transistor and degrade the transconductance. Thus, all the bipolar spoiling techniques are interesting device concepts but are inadequate to eliminate latch-up.

The more useful methods combine the guard rings, butted contacts, and thin epitaxial layer structures. Let us first consider the guard rings. There are two type of guard rings. The minority-carrier ring acts as a pseudo collector to collect minority carriers, thus lowering the bipolar gain. The majority-carrier ring is used to reduce the local sheet resistance, thus decoupling the feedback between the two bipolar transistors. In Fig. 12-16a, a minority-carrier guard ring in the substrate is inserted between the parasitic emitter and the n well. It decreases the carrier collection by the n well, which acts as the collector. If a thin epitaxial layer is used (Fig. 12-16c) the path of minority-carrier diffusion is narrowed, forcing the carriers closer to the guard ring and thus be collected.

The majority-carrier guard ring in the well is shown in Fig. 12-17. The injected carriers from the n^+ in the substrate are collected by the n well. The collected current is, however, diverted to the n^+ ring inside the well (Fig. 12-17b), effectively decoupling the feedback by lowering R_w. A similar guard ring of the p^+ region can be used in the substrate (Fig. 12-18a). However, the same function can be performed by the p^+ substrate as shown in Fig. 12-18b. When the epitaxial layer

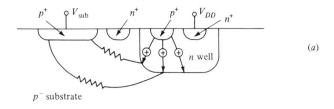

(a)

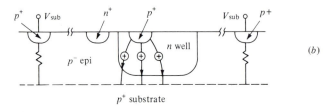

(b)

Figure 12-18

Majority-carrier guard in substrate. (a) p^+ diffusion guard to reduce substrate sheet resistance. (b) Contact ring is preferable to p^+ diffusion guard in epi-CMOS. *(After Troutman [4].)*

is thin, the p^+ substrate is extremely effective in taking away the current from the vertical bipolar transistor. This is the result of the low resistivity giving rise to a small R_s. It is an effective ground plane to drain away a large substrate current since the area of the substrate contact ring can be significantly larger than a local ring. This is an ideal structure for latch-up prevention since it does not occupy surface area. In fact, for a very thin epitaxial layer, the guard rings in the well are not needed.

One of the critical dimensions in CMOS designs is the n^+-to-p^+ separation d. The smaller distance indicates a better packing density but higher probability of latch-up. This dimension d is illustrated in Fig. 12-19a, and its relation to the triggering current is plotted in Fig. 12-19b for different epitaxial-layer thicknesses. The triggering current is assumed to be caused by a surge in V_{DD}. Notice that for a 12-μm epitaxial layer the triggering current required for latch-up is less than 1 mA for $d = 10\ \mu$m. This current becomes 18 mA for a 5-μm epitaxial layer. When the epitaxial-layer thickness is reduced to 3 μm, the n^+-to-p^+ spacing can be reduced to 5 μm without latch-up at a triggering current of 80 mA. For practical purposes, such large triggering current is not available in the circuit, and the CMOS is considered latch-up-free. In general, the epitaxial layer thickness should be smaller than the device separation in order to steer the current away from the vertical bipolar transistor. The minimum thickness of the epitaxial layer is limited by out-diffusion of the p^+ substrate.

In the device shown in Fig. 12-19a, a p^+ region is added abutting the n^+ source of the n-channel MOS transistor, and the contiguous n^+ and p^+ regions are connected to the same potential. This is known as the *butted contact*. A similar butted contact is found for the p-channel device. These contacts minimize the shunt resistances R_w and R_s so that the current gain of the lateral transistor is reduced

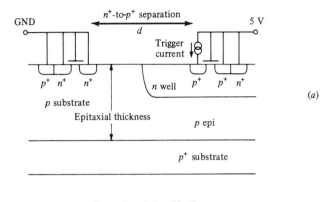

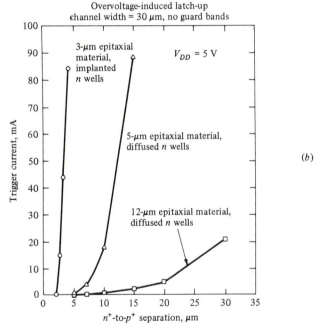

FIGURE 12-19
An n-well CMOS (a) structure with butted contacts and (b) triggering current vs. n^+-to-p^+ separation. *(After Chen [3].)*

at low forward bias. As a result, the holding current and voltage are increased. This is particularly effective in eliminating latch-up when it is combined with a thin epitaxial layer. However, these contacts can only be used with the grounded-source configuration.

12-7   SCHOTTKY-MOS TRANSISTOR

A Schottky barrier is a majority-carrier device whose minority-carrier current is typically 5 orders of magnitude lower than the majority current. With such a low emitter injection efficiency, the bipolar transistor action is eliminated when the Schottky barrier is used for the source and drain contacts of an MOS transistor. A Schottky-MOS transistor is shown in Fig. 12-20, where platinum silicide is used as the contacts to the n substrate. Since the Schottky-barrier height of PtSi–nSi is 0.85 eV, the contacts have high resistance to the channel at zero gate voltage (Fig. 12-20b). When a gate voltage is applied, the channel is inverted to a p-type material. Assuming that $\phi_n + \phi_p = E_g/q$, the barrier height of the PtSi–p channel becomes 0.25 eV. This low barrier provides a reasonably good ohmic contact under the condition of carrier inversion. The energy-band diagram of the Schottky-MOS transistor with inverted channel is shown in Fig. 12-20c. The difficulty with the Schottky-MOS device comes from the gap between the source Schottky barrier and the inverted channel. It increases the source resistance and degrades the transconductance. Furthermore, the leakage current of the junctions are higher than the pn junctions. An improved version of the Schottky-MOS transistor makes use of a recessed or trenched structure with boron-implanted p as self-aligned source and drain.

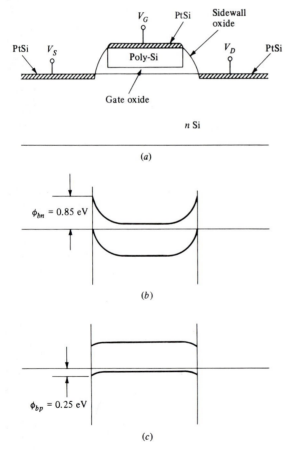

FIGURE 12-20
A platinum-silicide Schottky-MOS transistor. (a) Structure with sidewall oxide spacers obtained from plasma etching of CVD oxide. (b) Energy-band diagram without inversion layer. (c) Energy-band diagram with inversion layer.

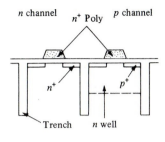

n channel   *n*⁺ Poly   *p* channel

n⁺   *p*⁺

Trench   *n* well

p substrate

FIGURE 12-21

Trench-isolated *n*-well CMOS technology. *(After Rung et al. [6].)*

For an *n*-channel Schottky-MOS transistor, a low metal–*n*Si Schottky barrier is required. Since the lowest M–*n*Si Schottky barrier is about 0.5 eV, the resistance associated with such a contact is too high to be usable. For this reason, only *p*-channel Schottky-MOS transistors can be fabricated at the present.

12-8 TRENCH ISOLATION

As device size is scaled downward, the area occupied by LOCOS is proportionally larger because of both electrical isolation and latch-up requirements. A recently proposed alternative of isolation makes use of trenches. By using dry anisotropic etching techniques such as reactive ion etching (RIE) or plasma etching, a narrow and deep trench can be produced which can then be filled with oxide or undoped polysilicon. A schematic diagram for this structure is shown in Fig. 12-21. When the trench is deeper than the *n* well as shown, the *p*-channel transistor is isolated from the *n*-channel device. Since the LOCOS is not needed, the problems associated with the "bird's beak" are eliminated. In addition, the filled trench gives rise to a flat surface so that there is no difficulty with step coverage.

The disadvantages of the trench isolation are, first of all, a much more complicated process than the LOCOS. Major issues include trench etching, conformal oxide, or poly deposition to fill the trench. Defects induced during dry etching could produce extra leakage current and interface charge. But the most difficult problem is caused by the sidewall inversion. Since the isolation trench is typically 1 μm wide, the bias of 5 V on the *n* well could induce negative charge on the *p* side of the trench. This sidewall inversion effectively shorts all n^+ regions abutted to it. Solution of this problem includes a thicker trench or forming a channel stop vertically. A practical approach may be to oxidize the trench first, and then to fill it with doped polysilicon to control the induced charge. These problems are apparently solved since a 256K static RAM with trench isolation has been successfully fabricated.

REFERENCES

1. Parrillo, L. C: VLSI Process Integration, in S. M. Sze (ed.), "VLSI Technology," McGraw-Hill, New York, 1983.

2. Yamaguchi, T., et al.: Process and Device Performance of 1-Micron Channel *n*-Well CMOS Technology, *IEEE Trans. Electron Devices*, **ED-31**:205 (1984).

3. Chen, J. Y.: CMOS—The Emerging VLSI Technology, *IEEE Circuits & Device* Magazine, **2**:16 (March 1986).

4. Troutman, R. R.: "Latchup in CMOS Technology," Kluwer, Boston, 1986.

5. Koeneke, C. J., et al.: Schottky MOSFET for VLSI, *Tech. Dig., Int. Electron Device Meeting*, 367 (1981)

6. Rung, R. D. et al.: Deep Trench Isolated CMOS Devices, *Tech. Dig., Int. Electron Device Meeting*, 237 (1982).

PROBLEMS

12-1. Draw the first four masks for the standard *p*-channel MOS transistor. Specify the device dimensions.

12-2. Plot the transfer characteristics of an *n*-MOS inverter with an MOS load where the load device is biased with (*a*) $V_G = V_{DD}$ and (*b*) $V_G - V_{DD} = V_T = 4$ V. The *V-I* characteristic of the inverter device is shown in Fig. P12-2. Assume that the transconductance of the load device is one-tenth that of the inverter device.

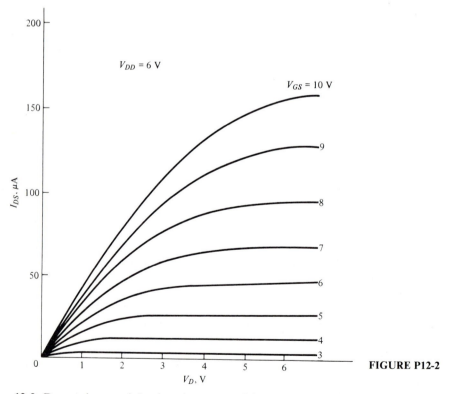

FIGURE P12-2

12-3. Draw a layout of the three-input monolithic MOS NOR/NAND gate shown in Fig. P12-3.

12-4. Sketch a set of masks for the *n*-well CMOS technology shown in Fig. 12-11.

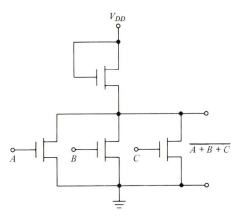

V_{DD}

$\overline{A + B + C}$

A

B

C

FIGURE P12-3

12-5. Show how to obtain the plot of V_T as a function of shallow boron implant for the NMOS transistor in Fig. 12-12.

12-6. Repeat Prob. 12-5 for the PMOS for an n-well surface concentration of $10^{16}\,\mathrm{cm}^{-3}$.

12-7. (a) In Fig. P12-7, find the condition of latch-up in terms of β_n, β_p and R_w, R_e.

  (b) Show that the condition obtained in (a) can be written as $K\alpha_n + \alpha_p = 1$, where $K = R_w(R_e + R_w)$.

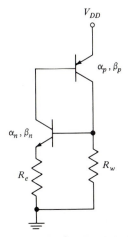

V_{DD}

α_p, β_p

α_n, β_n

R_e

R_w

FIGURE P12-7

12-8. Derive Eq. (12-4), using the equivalent circuit shown in Fig. 12-14.

12-9. Draw an equivalent circuit to illustrate the function of the minority-carrier guard ring in Fig. 12-16a.

12-10. In Fig. 12-16a, assume the carrier injection efficiency at the effective n^+ emitter is 0.99. Without the guard ring, the n^+-to-n-well spacing is $6\,\mu\mathrm{m}$. The minority-diffusion length is $20\,\mu\mathrm{m}$.

  (a) Estimate the lateral npn transistor gain α without the guard ring.

  (b) Tepeat (a) with the guard ring. If 70% of the minority-carrier current is collected by the guarding.

12-11 (a) Draw an equivalent circuit of the majority-carrier guard ring structure in Fig. 12-17a.

  (b) Repeat (a) for Fig. 12-17b.

CHAPTER

13

CHARGE-COUPLED AND NONVOLATILE MEMORY DEVICES

This chapter is devoted to two distinctly different MIS device families: the charge-coupled device (CCD) and the nonvolatile memory, more commonly known as the *erasable programmable read-only memory* (EPROM) or the *electrically erasable programmable read-only memory* (EEPROM). These two device families are not directly related, and the reader may skip either one without losing continuity.

The charge-coupled device consists of a chain of closely spaced MOS capacitors. In these capacitors, charge packets are stored and are transferred from one to the other to perform useful functions. The basic operating principle is presented along with various device structures and technology. The most important application of the CCD is in a television camera where an optical image is converted into electric signals. Other potentially useful functions include digital and analog memories, signal processing, and analog computation.

The nonvolatile memory device can keep data for as long as 10 years even when the power supply is removed. There are two basic structures: the double-layer dielectric MIS transistor and the floating-gate transistor. Of the first type, the *metal-nitride-oxide-semiconductor* (MNOS) sandwich is used as an example. As for the floating-gate variety, the *floating-gate avalanche-injection MOS device* (FAMOS) is presented to represent the EPROM where an ultraviolet light is needed for erasure. In the last few years, the EEPROM has begun to appear in the market, with the *floting-gate tunnel oxide* (FLOTOX) as the most widely adopted structure. Operating principles and characteristics of the FLOTOX will be discussed.

13-1 THE CONCEPT OF CHARGE TRANSFER

The concept of charge transfer can be explained by using a chain of amplifiers with unity gain and infinite input impedance connected as shown in Fig. 13-1*a*. Upon the closing of the switch S_1, the input signal is stored in the form of a charge packet in the capacitor C_1. Now, let us open S_1 and then close S_2; the stored charge will be transferred to the capacitor C_2. Following the same procedure, the charge will eventually reach the output terminal. It is obvious that this system can be used as a digital shift register or an analog delay line. If we replace each amplifier-and-switch pair by an MOS transistor, we obtain the circuit shown in Fig. 13-1*b*. The transistors can be turned on and off sequentially by applying voltage pulses at the respective gate electrode, and the charge is stored and transferred just as in Fig. 13-1*a*. In practical systems, gates 1 and 3 are connected and pulsed together and gates 2 and 4 are similarly joined. The transfer of charge is analogous to filling and emptying water buckets, as illustrated in Fig. 13-1*c*. The circuit shown

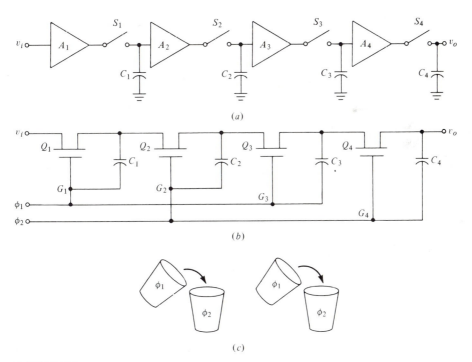

FIGURE 13-1
Charge-transfer systems using (*a*) operational amplifiers, (*b*) MOS transistors, and (*c*) water buckets.

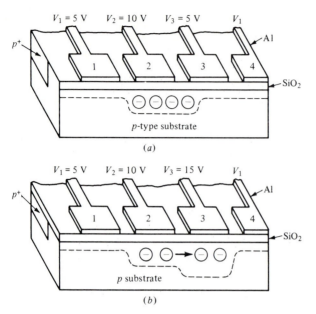

FIGURE 13-2
The basic operation of a three-phase CCD: (*a*) charge storage and (*b*) charge transfer. The p^+ diffusion is used for channel confinement; its function will be explained later.

in Fig. 13-1*b* is known as the *bucket-brigade device*. It is a *two-phase* system because two separate clock pulses ϕ_1 and ϕ_2 are needed.

In the bucket-brigade device, the charge transfer is implemented on the *circuit* level by using either discrete or integrated components. Implementation of charge transfer on the *device* level is realized by the *charge-coupled device* (CCD). In the CCD, minority carriers are stored in potential wells created at the surface of a semiconductor. These carriers are transported along the surface by filling and emptying a series of potential wells sequentially. In its simplest form, the CCD is a string of closely spaced MOS capacitors, as illustrated in Fig. 13-2. If electrode 2 is biased at 10 V, more positively than its two adjacent electrodes (at 5 V), a potential-energy well is set up, as depicted by the dashed line, and charge is stored under this electrode, as in Fig. 13-2*a*. Now let us bias electrode 3 with 15 V; a deeper potential well is established under electrode 3 (Fig. 13-2*b*). The stored charge seeks the lowest potential and therefore travels along the surface when the potential wells are moved. Note that we need three electrodes in this structure to facilitate charge storage and transfer in one direction only. The three electrodes will be referred to as one *stage* or *cell* of the device.

13-2   TRANSIENT RESPONSE OF THE MOS CAPACITOR

The charge-coupled device stores and transports minority carriers between potential wells created by voltage pulses on closely spaced MOS capacitors. Therefore, it is important to understand an MOS capacitor under pulsed conditions. Figure 13-3 shows the structure and energy-band diagrams of an MOS capacitor on a p-type substrate. Upon the application of a positive gate voltage, a depletion layer is formed under the gate. The applied voltage is large enough to induce an inversion layer. However, no inversion layer is formed at $t = 0+$ because no minority carriers are available. As a result, a *deep depletion* layer exists, as seen in the energy-band diagram in Fig. 13-3b. Under the condition of deep depletion, a substantial portion of the external voltage is across the depletion layer, and the surface potential is large. As time elapses, electron-hole pairs are generated in the depletion region. The electrons accumulate at the Si–SiO$_2$ interface, and they influence the charge distribution and energy-band diagram. With a positive gate voltage, holes are driven to the substrate to reduce the depletion-layer width, and at the same time electrons are attracted to the surface to form an inversion layer. When a sufficient

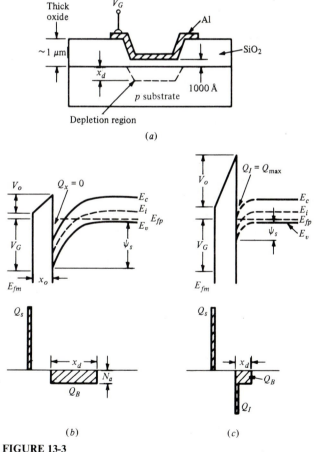

FIGURE 13-3
The MOS capacitor: (a) structure, (b) energy-band diagram and charge distribution under deep depletion at $t = 0+$, and (c) energy-band diagram and charge distribution at thermal equilibrium ($t = \infty$).

number of electrons has been collected under the surface, a saturation condition is reached in which the electron diffusion current away from the surface is exactly balanced by the electron drift current toward the surface. This condition is illustrated by the energy-band diagram in Fig. 13-3c. If the accumulated electrons are less than the saturation value, the net flow will be toward the surface; but if the saturation value is exceeded, there will be a net flow into the undepleted bulk, where the minority carriers recombine. The time required to reach the saturation condition is known as the *thermal-relaxation time*. The thermal-relaxation time at room temperature has been measured from 1 s to several minutes, depending on the structure and fabrication processes of the CCD. Since a useful potential well does not exist in the saturation condition, the CCD is basically a *dynamic* device in which charges can be stored for a time *much shorter* than the thermal-relaxation time.

Let us now derive the surface potential ψ_s as a function of the applied voltage V_G at the gate electrode. The surface potential of an MOS capacitor is given by Eq. (9-15), repeated here,

$$V_G = -\frac{Q_s}{C_o} + \psi_s \tag{13-1}$$

where Q_s is the total charge per unit area in the semiconductor surface region and C_o is the oxide capacitance, given by

$$C_o = \frac{K_o \varepsilon_o}{x_o} \tag{13-2}$$

where x_o is the oxide thickness. Equation (13-1) was derived under the assumption that the energy bands in the semiconductor are flat at $V_G = 0$. Therefore, it must be modified to include the flat-band voltage V_{FB}:

$$V_G - V_{FB} = -\frac{Q_s}{C_o} + \psi_s \tag{13-3}$$

For a time much smaller than the thermal-relaxation time, there is no inversion layer in a CCD. Therefore, the charge Q_s is the sum of the depletion-layer charge and an externally introduced *signal charge* Q_{sig}. The depletion-layer charge is given by Eq. (9-4), so that the total surface charge is

$$Q_s = -qN_a x_d - Q_{sig} \tag{13-4}$$

where x_d, the depth of the depletion layer, is given by Eq. (9-5):

$$x_d = \left(\frac{2K_s \varepsilon_o}{qN_a}\psi_s\right)^{1/2} \tag{13-5}$$

Substituting Eqs. (13-4) and (13-5) into Eq. (13-3) yields

$$V_G - V_{FB} - \frac{Q_{sig}}{C_o} = \psi_s + \frac{1}{C_o}(2qK_s \varepsilon_o N_a \psi_s)^{1/2} \tag{13-6}$$

Solving for the surface potential from the foregoing equation leads to

$$\psi_s = V - B\left[\left(1 + \frac{2V}{B}\right)^{1/2} - 1\right] \tag{13-7}$$

where

$$V = V_G - V_{FB} - \frac{Q_{sig}}{C_o} \tag{13-8}$$

and

$$B = \frac{qK_s\varepsilon_o N_a}{C_o^2} \tag{13-9}$$

Equation (13-7) is very important in the design of a CCD because the gradient of this potential governs the motion of minority carriers. The value of ψ_s also specifies the depth of the potential well. In Eqs. (13-7) to (13-9), we find that the surface potential is controlled by the substrate doping concentration N_a and the oxide thickness x_o, which determines C_o. If we set V to be constant, then ψ_s increases as N_a and x_o are reduced. Equation (13-7) is plotted as a function of V in Fig. 13-4 with N_a and x_o as parameters. Since Eq. (13-8) indicates that V decreases with an increase of Q_{sig}, the surface potential is also a function of the magnitude of the signal charge.

Example. A CCD electrode has an area of 5 by $10\,\mu m$ with $x_o = 50$ nm, $N_a = 2 \times 10^{15}\,cm^{-3}$, and $V_{FB} = 1$ V. Calculate the surface potential and depletion-layer charge at $V_G = 5$ V.

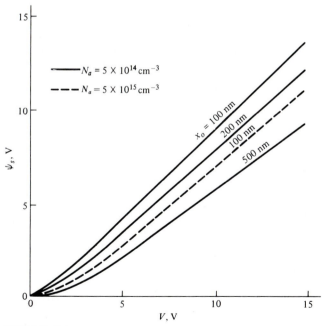

FIGURE 13-4
Surface potential as a function of voltage in Eq. (13-7). *(After Amelio et al. [1].)*

Solution. The oxide capacitance is

$$C_o = \frac{K_o \varepsilon_o}{x_o} = 6.8 \times 10^{-8} \text{ F/cm}^2$$

Since $Q_{sig} = 0$, we have

$$V = V_G - V_{FB} = 4 \text{ V} \qquad \text{Eq. (13-8)}$$

In addition,

$$B = \frac{q K_s \varepsilon_o N_a}{C_o^2} = 0.072 \text{ V}$$

Therefore

$$\psi_s = 4 - 0.072 \left[\left(1 + \frac{2 \times 4}{0.072} \right)^{1/2} - 1 \right]$$

$$= 3.3 \text{ V}$$

$$x_d = \left(\frac{2 K_s \varepsilon_o \psi_s}{q N_a} \right)^{1/2} = 1.47 \ \mu m$$

Thus

$$Q_d = q N_a x_d = 4.7 \times 10^{-8} \text{ C/cm}^2$$

Since $A = 5 \times 10 \ \mu m^2 = 5 \times 10^{-7} \text{ cm}^2$, we find

$$\frac{Q_d A}{q} = 1.47 \times 10^5 \qquad \text{charge}$$

The capacitance C_{GS} between the gate electrode and substrate is the series combination of the oxide and depletion capacitances. By making use of Eqs. (9-13), (9-14), (13-2), and (13-5), we derive

$$C_{GS} = \frac{C_o}{1 + (2\psi_s/B)^{1/2}} \tag{13-10}$$

from which we can calculate ψ_s if C_{GS} is measured. The signal charge can then be calculated from Eqs. (13-7) and (13-8). Alternatively, the signal charge can be estimated from the consideration of charging up the oxide and depletion-layer capacitances. Usually, the depletion charge is much smaller than the oxide charge, and it can be neglected. For an oxide thickness of 100 nm and electrode area of 10 by 20 μm, the oxide capacitance is calculated to be 0.068 pF [Eq. (13-2)]. Assuming that half the gate voltage (10 V) is across the oxide capacitor, we find each charge packet to be 0.34 C. Since an electron has a charge of 1.6×10^{-19} C, there are 2×10^6 electrons in each packet, giving an electron density of 10^{12} cm^{-2}.

For time intervals that are short compared with the thermal-relaxation time, the MOS capacitor serves as a storage element for analog information represented by the amount of charge in the well.

13-3 CHARGE-COUPLED DEVICE STRUCTURES

The transfer of charge can be implemented by different MOS structures and electrode arrangements. The approach in designing a CCD depends on consideration of electrical performance, fabrication difficulty, and cell size. One question is how many phases we should have in the system, as it is practical to build a two-, three-, or four-phase system. In this section, we present the basic principles of two- and three-phase CCDs and discuss some merits of each system along with representative fabrication technology.

The Three-Phase CCD

A three-phase CCD is a linear array of closely spaced MOS capacitors with three electrodes per stage or cell. As shown in Fig. 13-5a, every third electrode is connected to the same clock voltage so that three separate clock generators are required. The basic principle was discussed in connection with Fig. 13-2. In practice, the driving click pulses display the special features shown in Fig. 13-5e. These waveforms are designed to achieve better efficiency in charge transfer, as explained in the following paragraph.

If a positive voltage applied to ϕ_1 is higher than that applied to ϕ_2 and ϕ_3, surface potential wells will be formed under the ϕ_1 electrodes. Charge packets, which have been introduced either optically or electrically, are accumulated in these wells at $t = t_1$. These charge packets may be of different magnitudes, as depicted in Fig. 13-5a. To facilitate charge transfer to the right, a positive voltage step is applied to ϕ_2 so that the potential wells under the ϕ_1 and ϕ_2 electrodes are the same in depth. Thus, the stored charge packets spread out, as seen at $t = t_2$. Almost immediately after the application of the positive pulse at ϕ_2, the voltage at ϕ_1 starts to decrease linearly so that the potential wells under the ϕ_1 electrodes rise slowly rather than abruptly. The charge packets tend to spill over to the potential wells under gates 2 and 5, as shown at $t = t_3$. The slow rise of the potential wells under gates 1 and 4 provides a more favorable potential distribution for the complete transfer of charge. When we reach the condition at $t = t_4$, the charge has been transferred over to the wells under the ϕ_2 electrodes. Note that the charge is prevented from moving to the left by the barriers under the ϕ_3 electrodes. Repeating the same procedure, we can move the charge from ϕ_2 to ϕ_3 and then from ϕ_3 to ϕ_1. As a full cycle of clock voltages is completed, the charge packets advance one stage to the right.

In designing a CCD, the capacitors must be physically close together so that the depletion layers overlap strongly and the surface potential has a smooth transition at the boundaries between neighboring electrodes. The first realization of the CCD is by means of a *single metal gate*, shown in Fig. 13-5. Typically, the oxide thickness is between 100 and 200 nm, and the gap or spacing between aluminum gates is about 2.5 μm. This gap size is smaller than the minimum size of the standard photolithographic process, which is approximately 5 μm. Therefore, etching the gaps between metal electrodes is difficult, and slight flaws in masking

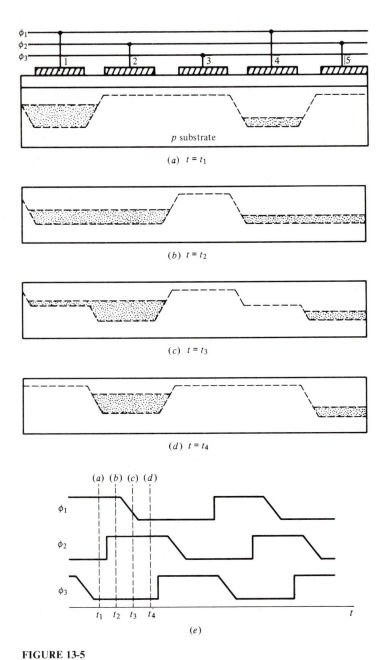

FIGURE 13-5
Potential wells and timing diagram at different time intervals during the transfer of charge in a three-phase CCD.

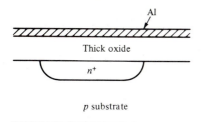

FIGURE 13-6
A diffused cross-under used as a conductor normal to the paper.

or photoresist steps could short-circuit the electrodes or (at the other extreme) could cut down the charge-transfer efficiency significantly in electrodes that are far apart. In addition, the channel oxide in the gaps is exposed, and electrostatic charge residing there leads to device instabilities. Furthermore, the need for three separate interconnections for three phases means that two conducting bus lines must cross over to address all electrodes. To avoid making electrical contact between conducting bus lines, a *cross-under* diffusion (Fig. 13-6) or two-level metalization is required in the area away from the active-device region. An alternative to the single-metal-gate structure is the *doped-polysilicon-gate* CCD, depicted in Fig. 13-7a where the bare gaps are covered with high-resistivity polysilicon to eliminate device instabilities. The difficulty with this metod is that the doping of polysilicon is not precisely localized, resulting in large cell dimensions. Another sealed-channel structure uses a *triple-polysilicon gate*, as illustrated in Fig. 13-7b. Each polysilicon

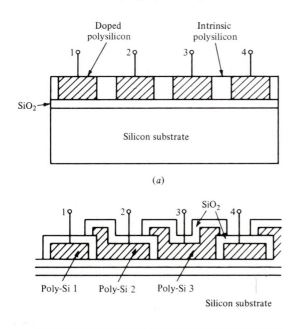

FIGURE 13-7
Three-phase CCD structures: (a) doped-polysilicon gate *(after Kim and Snow [2])* and (b) triple-polysilicon gate *(after Bertram et al. [3])*.

gate is covered with oxide and is isolated from others so that shorts between gates are unlikely. With this technology it is possible to fabricate a small CCD. The drawback is the complexity of the process.

The Two-Phase CCD

In a three-phase CCD, the potential well is symmetrical so that charge can flow to the right or left. Directionality of the signal flow is provided by blocking the charge transfer in one directon with appropriate *external* gate voltages. If the potential well is constructed to provide *built-in* directionality, we obtain a *two-phase* CCD system, as shown in Fig. 13-8. Note that the oxide thickness is stepped so that a different potential appears beneath each individual electrode. To facilitate charge transfer, the potentials on adjacent electrodes are alternated between $V_o + V$ and $V_o - V$ to obtain unsymmetrical potential distribution. In both potential diagrams shown in Fig. 13-8a and c, the signal is always directed toward the right. Because of the stepped oxide, the two-phase CCD can be operated satisfactorily without overlapping clock-voltage pulses, as in a three-phase system. Interconnections to the electrodes can be made easily since no cross-under is needed.

A popular fabrication technology for building a two-phase CCD is the *polysilicon-aluminum gate* using the silicon-gate process described in Sec. 9-8. After the formation of the silicon gates, a thermal oxidation is performed to cover the entire wafer. Subsequently, aluminum gates are produced by metalization and etching in areas between the polysilicon gates leading to the structure shown in

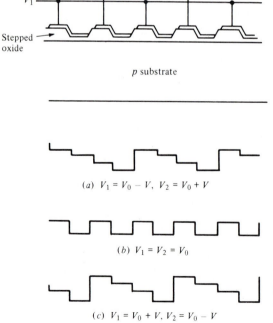

(a) $V_1 = V_0 - V$, $V_2 = V_0 + V$

(b) $V_1 = V_2 = V_0$

(c) $V_1 = V_0 + V$, $V_2 = V_0 - V$

FIGURE 13-8
Potential-well diagrams of a two-phase charge-coupled device.

Fig. 13-9. Commonly used layout rules are 2.5-μm gate overlap and 5-μm silicon gate. The main disadvantage of this structure is its large RC time-constant delay in charging the electrodes due to the high resistivity of the polysilicon gates.

13-4 TRANSFER EFFICIENCY

When a charge packet is traveling along the CCD, a small fraction of charge is left behind at each transfer. The fraction of charge transferred from one potential well to the next is known as the *transfer efficiency* η. The fraction left behind is known as *transfer inefficiency* ε so that $\eta + \varepsilon = 1$. It is obvious that the longer we allow the transfer to take place, the more charge is moved over to the next potential well. Therefore, the fraction of charge left behind is a function of time. Experimentally, it has been observed that most of the charge appears to transfer rapidly, whereas a small fraction b of the total charge packet transfers more slowly with an exponential time constant τ. Thus, the slower charge transfer limits the frequency response of the device, and the transfer efficiency conforms to

$$\eta = 1 - be^{-t/\tau} \tag{13-11}$$

Let us examine the carrier behavior during the process of transfer carefully. In the beginning, the charge packet is very dense and localized with large density gradient at the edges of the well. For a short time, a large fraction of the packet is transferred over due to the strong repulsive force between the electrons. As time evolves, this repulsive force decreases, and electrons are transported by means of thermal diffusion and/or drift caused by the fringing field. Usually, it is the last two mechanisms that are responsible for the loss of charge, although the first mechanism can also be used to improve the transfer efficiency.

For a small amount of signal charge, the transfer mode is governed by thermal diffusion. This mechanism leads to an exponential decay of charge under the transferring electrode with a time constant given by [6]

$$\tau = \frac{L^2}{2.5D} \tag{13-12}$$

where D is the carrier diffusivity and L is the center-to-center electrode spacing.

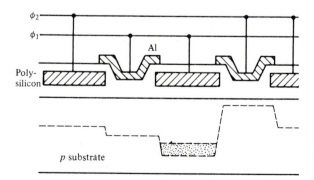

FIGURE 13-9
Two-phase polysilicon-aluminum-gate structure. *(After Kosonocky and Carnes [4].)*

By means of thermal diffusion alone in a p-channel CCD, 99.99 percent of the charge is removed each cycle at a clock frequency of

$$f_c = \frac{5.6 \times 10^7}{L^2} \quad \text{Hz} \qquad (13\text{-}13)$$

where $D = 6.75 \text{ cm}^2/\text{s}$ for holes at the surface and L is in micrometers.

The charge-transfer process can be speeded up by the fringing field established between the electrodes in the direction of charge propagation along the channel. This field has its maxima at the boundaries between electrodes and minima at the centers of the transferring electrodes. The magnitude of the fringing field increases with gate voltage and oxide thickness and decreases with gate length and substrate doping density. Computer simulation has been performed for this process, and the results are plotted in Fig. 13-10 for transfer time at 99.99 percent of charge transfer vs. gate length with the substrate doping as a variable. The thermal-diffusion time is also plotted. Above the dashed line, it takes longer to remove the signal charge by means of the fringing field, so that the effect of thermal diffusion dominates. Below the dashed line, the charge is transferred by the fringing field. Thus, we obtain 99.99 percent charge transfer at a clock frequency of 10 MHz if the substrate doping is 10^{15} cm^{-3} and $L = 7 \mu m$.

The self-induced drift in the beginning of transfer is caused by the repulsive force between the carriers with the same sign of charge. For this reason, it is important only when the signal density is large (typically greater than 10^{10} cm^{-2}), and in most cases the transfer of the first 99 percent of charge obeys this mechanism. In some CCDs, the entire channel is filled with a large background charge, known as *fat zero*, to improve the transfer efficiency. Self-induced drift is significant under this mode of operation.

In practical devices, the above consideration overestimates the transfer efficiency because charges can be trapped in surface states. While the capture rate of carriers in these states is proportional to the free-carrier density, the empty rate depends only on the energy level of the surface states. Consequently, the filling rate can be much faster than the empty rate, and some carriers are trapped there to be released later as noise. This type of loss can be reduced by propagating a small background charge throughout the channel to fill these traps.

If a charge packet with an initial amplitude of A_o is traveling down a three-phase CCD shift register, the charge amplitude after m stages is

$$A_n = A_o(1 - 3m\varepsilon) = A_o(1 - n\varepsilon) \qquad (13\text{-}14)$$

where $n = 3m$ is the number of transfers. For a sinusoidal input signal with a frequency f, the output amplitude has been derived and is given here without proof [8]:

$$\frac{A_n}{A_o} = \exp\left[-n\varepsilon\left(1 - \cos\frac{2\pi f}{f_c}\right)\right] \qquad (13\text{-}15)$$

where $f_c = 1/T$ is the operating clock frequency. This result is plotted in Fig. 13-11.

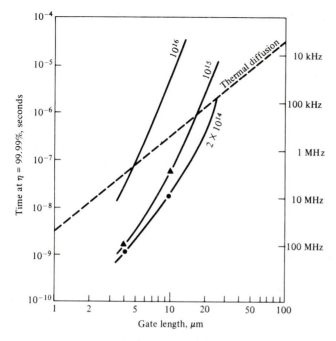

FIGURE 13-10
The time required to achieve 99.99 percent transfer efficiency as a function of gate length for various substrate doping densities: $\mu_p = 250\,\text{cm}^2/\text{V}\cdot\text{s}$, $x_o = 200\,\text{nm}$, $V_{\text{substrate}} = 7\,\text{V}$, and $V_G = 10\,\text{V}$ (pulse). *(After Carnes et al. [7].)*

We observe that the attenuation in amplitude is significant for a large $n\varepsilon$ product and for $f/f_c > 0.1$. An additional phase delay with respect to the ideal case is

$$\Delta\phi = n\varepsilon \sin \frac{2\pi f}{f_c} \qquad (13\text{-}16)$$

Equations (13-15) and (13-16) can be used to give an estimate of performance degradation in a particular application due to transfer inefficiency.

13-5 CHARGE INJECTION, DETECTION, AND REGENERATION

Charge packets are injected either electrically or optically. While optical injection is necessary in an image sensor, electrical injection is preferred in a shift register or delay line. Figure 13-12a illustrates the method using a *pn* junction to introduce minority carriers. The *n*-type diffusion, also known as the *source*, is short-circuited to the substrate. When a positive pulse is applied to the input gate, it allows the electrons to flow from the source to the potential well under the ϕ_1 electrode. The current source keeps filling the first potential for the duration of the pulse Δt of the input. This is similar to the carrier flow from source to drain in the inversion

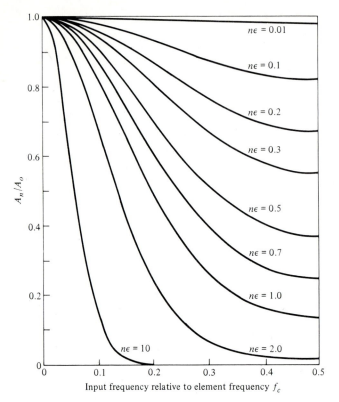

FIGURE 13-11

The frequency response of a CCD plotted for different values of the transfer inefficiency $n\varepsilon$. *(After Joyce and Bertram [8].)*

layer of an MOS transistor. By biasing the source and input gate, the pn junction can be made more efficient. A possible optical-injection technique is shown in Fig. 13-12b, where optically generated minority carriers are attracted by the electrodes and accumulate in the potential wells.

Methods of charge detection are illustrated in Fig. 13-13. A simple approach of sensing the charge in a CCD is by means of a pn junction diode at the end of the line, as shown in Fig. 13-13a. The diode is reverse-biased, so it acts as a drain. When the signal charge reaches the drain, a current spike is detected at the output as a capacitive charging current. This is known as the *current-sensing method*. The most popular scheme in charge detection is the *charge-sensing* method shown in Fig. 13-13b, in which the floating diffusion region is periodically reset to a reference potential V_D through the reset gate. When the signal charge of the CCD reaches the floating diffusion region, the voltage there becomes a function of the signal charge. The variation of voltage at the floating region is detected via an MOS transistor amplifier, as shown.

As a charge packet is traveling along a CCD, its magnitude is changed because of both transfer inefficiency and thermal generation of carriers. If we want to store

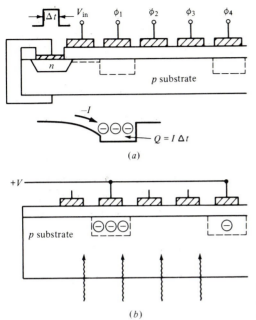

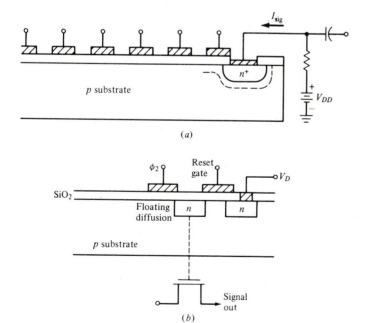

FIGURE 13-12
Charge injection with (*a*) a *pn* junction and (*b*) a light source.

FIGURE 13-13
Charge-detection methods: (*a*) a reverse-biased *pn* junction collector and (*b*) a charge-sensing floating-gate amplifier.

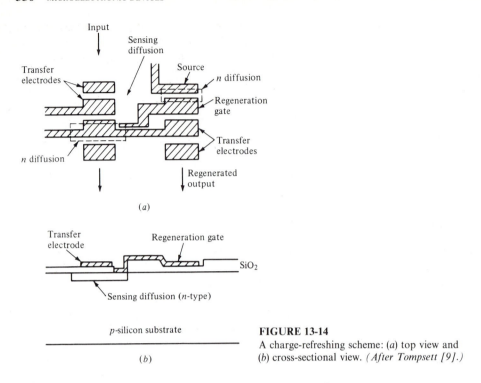

(a)

(b)

p-silicon substrate

FIGURE 13-14
A charge-refreshing scheme: (a) top view and
(b) cross-sectional view. (*After Tompsett [9].*)

information longer than the thermal-relaxation time, the charge packet must be regenerated or refreshed periodically. In the case of binary bits, a circuit can be set-up to measure the charge packets at the end of a transfer channel and to compare them with a given threshold value. For each packet that exceeds the threshold, a new full-sized packet is injected to replace the old one. A refreshing scheme is shown in Fig. 13-14 in its simplified form. Note that the two CCD transfer channels are oriented vertically rather than horizontally, as in previous CCD diagrams. A sensing *n* diffusion is fabricated under a transfer electrode in the first channel. Since the silicon surface is always depleted in charge-coupled operation, this diffusion region will adopt the surface potential beneath the electrode. The surface potential can now be measured by connecting the sensing diffusion to the gate of an MOSFET which serves as the input stage of the second transfer channel. The drain of the MOSFET is under the first transfer electrode, and the source is either grounded or biased positively. When a full charge packet (logic ONE) arrives at the sensing diffusion, the surface potential ψ_s is at its lowest value (Fig. 13-3c). Therefore, the gate voltage is below the threshold voltage of the transistor, and the drain registers a logic ZERO. On the other hand, the absence of a charge packet in the sensing diffusion produces the highest ψ_s (Fig. 13-3b) so that the MOSFET is turned on and a charge packet is injected from the source to the drain, delivering a logic ONE. In either case an inverse bit is regenerated.

An efficient charge transfer requires the charge packets to be confined to a narrow channel to prevent charge leaking and to reduce charge trapping by surface states. This is accomplished by either a channel-stop diffusion or an oxide step, as shown in Fig. 13-15. The p^+ channel-stop diffusion in Fig. 13-15a is the same p^+ diffusion shown in Fig. 13-2. The use of an oxide step requires that the threshold voltage of the thick oxide area outside the channel be such that the surface is never depleted. It is compatible with p-channel devices on (111) wafers with a low-resistivity n substrate because the fixed positive oxide charge increases the threshold voltage of the thick oxide area. The diffusion channel-stop method is most widely used and is applicable to both n- and p-channel CCDs.

13-6 BULK- (BURIED-) CHANNEL CCDs [10]

The devices described so far store and transfer the charge in potential wells at the silicon surface under the silicon dioxide and are known as *surface-channel* charge-coupled devices (SCCDs). As discussed earlier, surface states may have a strong influence on the transfer loss and noise, particularly if the signal level is low. These difficulties can be alleviated if the channel is moved away from the Si–SiO$_2$ interface. This results in a *bulk-* or *buried-channel* CCD (BCCD), as shown in Fig. 13-16. The channel is formed by a thin epitaxial or diffused n layer on a p substrate. The energy-band diagram under thermal equilibrium is drawn in Fig. 13-17a. Now if a large positive voltage is applied to the channel via the input and output diodes, the majority carriers in the channel will be completely depleted. A depletion layer is formed on both the reverse-biased pn junction and under the surface of the MOS capacitors. The energy-band diagram in Fig. 13-17b illustrates the potential well in the n-type layer. The corresponding channel is shown as the dashed lines in Fig. 13-16. A mobile signal charge (electron) injected from the source will stay in the potential well to produce a flat portion, as depicted in Fig. 13-17c. This well

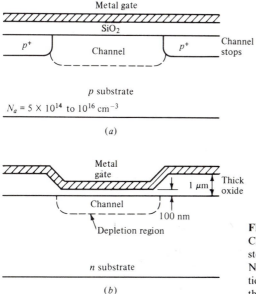

FIGURE 13-15

Channel-confinement techniques: (a) channel-stop diffusion and (b) oxide-step structure. Note that the channel is oriented in the direction normal to the paper, i.e., a 90° turn from that of Fig. 13-2.

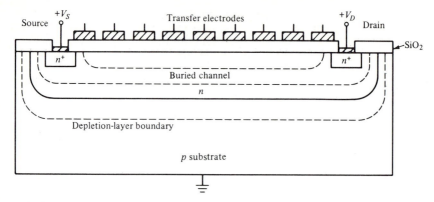

FIGURE 13-16
The structure and biasing of a buried-channel charge-coupled device (BCCD).

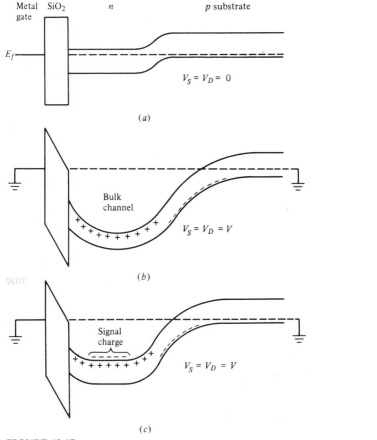

FIGURE 13-17
Energy-band diagram of a BCCD (*a*) at thermal equilibrium, (*b*) under complete depletion of all mobile carriers in the channel by reverse bias, and (*c*) with additional signal charge.

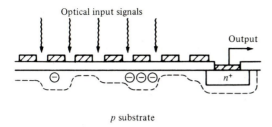

FIGURE 13-18

Carrier generation in a CCI by front illumination.

p substrate

can store and transfer charge by applying appropriate clock pulses to the electrodes, just as in an SCCD.

The advantages of the BCCD include the elimination of the surface-state trapping effect and increased mobility. Therefore, a transfer efficiency of 99.99 percent or higher is realizable without fat zero at room temperature. Furthermore, these devices can be constructed with large fringing fields since the bulk channel is farther away from the electrodes. As a result, a high clock frequency in the range of 100 MHz is obtainable. The only drawback is the added processing complexity and smaller capacitance that reduces signal-handling capability.

13-7 CHARGE-COUPLED IMAGER [12]

An imaging device converts an optical image into electric signals. In a standard Vidicon TV camera system, the optical image is converted into charge pulses by a photodiode matrix upon exposure [13]. The collected charge in the diode matrix is then discharged by an electron beam which scans the individual diodes and produces discharging current pulses. A *charge-coupled imager* (CCI) is a self-scanning system that does away with the electron beam. It is a device that will undoubtedly find more applications in the future.

To convert an optical signal into electric pulses, we can use either the front illumination shown in Fig. 13-18 or the back illumination shown in Fig. 13-12*b*. Electron-hole pairs are generated by the light source. If we introduce storage potential wells at all the CCD stages by applying appropriate clock pulses, the photogenerated minority carriers will be collected in these wells for a time called the *optical integration time*. The collected charge packets are shifted down the CCD register and are converted into current or voltage pulses at the output terminal. Usually, the optical integration time is much longer than the total shifting time of all stored charge packets to avoid smearing the image. Presently, two types of CCD imagers are available commercially, the *line imager* and *area imager*.

A practical CCD line imager is shown in Fig. 13-19. The shaded areas represent photosensing elements with potential wells for optical charge integration. After charge packets are accumulated, they are first transferred into two parallel CCD shift registers; then they are shifted to the output, following the directions of the arrows. This is the basic feature of a 256-stage buried-channel line imager produced by Fairchild. Line imagers with 1500 elements or more have been reported in the literature.

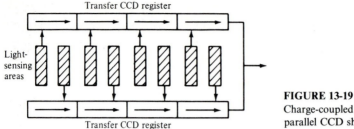

Transfer CCD register

Light-
sensing
areas

Transfer CCD register

FIGURE 13-19
Charge-coupled line imager with two
parallel CCD shift registers.

The two types of CCD area imaging systems are distinguished by the method of transfer, *interline transfer* and *frame transfer*, shown in Fig. 13-20. In essence, the interline-transfer CCI can be visualized as stacking line imagers in parallel. Charge packets are transferred into parallel shift-register lines and sequentially shifted down to the output register, which is read by transferring the charge horizontally.

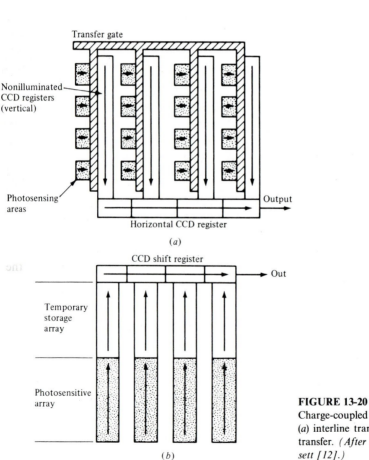

Transfer gate

Nonilluminated
CCD registers
(vertical)

Photosensing
areas

Output

Horizontal CCD register

(a)

CCD shift register

Out

Temporary
storage
array

Photosensitive
array

(b)

FIGURE 13-20
Charge-coupled area imagers with
(a) interline transfer and (b) frame
transfer. (*After Séquin and Tompsett [12].*)

The frame-transfer system has an opaque temporary storage array with the same number of elements as the photosensing array. The charge packets under the photosensing array are transferred over to the temporary storage array as a frame of a picture. Subsequently, the information in the temporary storage array is shifted down one by one to the output shift register and transferred out horizontally.

The limitations of a CCI result from transfer inefficiency and dark current. These undesirable effects introduce noise and distortion of the input video signal. The dark current is produced by thermally generated minority carriers in the absence of light. This current should be a small fraction of the expected video current, particularly if it is nonuniform over the imaging area. There are three principal dark-current components: one arising from carriers generated in the bulk depletion region, one from the neutral bulk, and one from the semiconductor surface. For quality bulk material with long lifetimes, say $100\,\mu s$, the dark current is essentially surface-dominated and is on the order of $30\,nA/cm^2$.

13-8 NONVOLATILE MIS MEMORY DEVICES

Before discussing nonvolatile devices, let us first explain two semiconductor memory systems: the read-only memory (ROM) and read-write random access memory (RAM). In an ROM, "programmed" information is stored in the memory, and only the *read* operation is performed. Examples of ROM applications include look-up tables and code-conversion systems. The RAM is an array of memory cells that stores information in binary form in which information can be randomly written into, or read out of, each cell as required. In other words, the RAM is a *read-write memory.* Most semiconductor memory systems utilize the bistable flip-flop or capacitive-charge-storage circuit as their basic memory cell in that the stored information is lost when the power supply is interrupted. It is often desirable to have nonvolatile semiconductor memories so that they can stand power failure or be stored and shipped without being energized all the time. In the following sections we introduce nonvolatile memory devices which make use of the MIS structure. Nonvolatility of an MIS device is usually achieved by storing charge in the gate structure. The stored charge alters the threshold voltage of the MIS transistor so that two stable states can be detected electrically. Presently, nonvolatile memory devices are used in ROMs only because the writing speed is too slow. Undoubtedly, as the switching speed is improved, nonvolatile devices will also be used in other forms of memory systems.

Among the nonvolatile MIS devices, we limit our discussion to the *metal-nitride-oxide-silicon,* (MNOS), the *floating-gate avalanche-injection MOS* (FAMOS), and the *floating-gate tunnel-oxide* (FLOTOX) structures. Used as a ROM, the FAMOS is electrically programmable but can be erased only with ultra-violet light. It can be reprogrammed and is known as the *electrically programmable ROM* (EPROM). An MNOS device can be electrically programmed and erased so that it is capable of functioning as a non-volatile read-write memory, but the speed in writing is rather slow (in microseconds). The FLOTOX makes use of attractive features in both MNOS and FAMOS devices, and has emerged as the industrial standard EEPROM.

13-9 MNOS TRANSISTOR

A cross section of a p-channel MNOS transistor is shown in Fig. 13-21. It is a conventional MOSFET in which the oxide is replaced by a double layer of nitride and oxide. The oxide layer is very thin, typically 5 nm thick, and the nitride layer is about 50 nm thick.

For low values of negative voltage applied to the gate, the MNOS transistor behaves like a conventional p-channel MOSFET. Upon application of a sufficiently high positive charging voltage V_C to the gate, as shown in Fig. 13-22a, electrons will tunnel from the silicon conduction band to reach traps in the nitride oxide interface and nitride layer, resulting in negative charge accumulation there. The stored charge as a function of time is given by

$$\frac{dQ}{dt} = J_o - J_n \tag{13-17}$$

where J is the current density and the subscripts denote the corresponding layers. The voltages across the oxide and nitride layers are, respectively, V_o and V_n, and they establish an electric field $\mathscr{E}_o$ in the oxide and $\mathscr{E}_n$ in the nitride. The relationship between the potential drops across the structure is

$$V_C = V_o + V_n = \mathscr{E}_o x_o + \mathscr{E}_n x_n \tag{13-18}$$

If the charge Q can be considered to have been stored at or near the interface only, the continuity of the electric flux leads to

$$K_o \varepsilon_o \mathscr{E}_o = K_n \varepsilon_o \mathscr{E}_n + Q \tag{13-19}$$

The stored charge as a function of time can be found by solving Eqs. (13-17) to (13-19) provided that the currents as functions of the electric field are known. Since the derivation of the transport equations through the oxide and nitride is beyond the scope of this text, we shall reproduce the current-field relationships here to help us understand their physical significance.

The transport across the thin oxide is by means of quantum-mechanical tunneling. As illustrated in Fig. 13-22a, the electrons in the conduction band of the

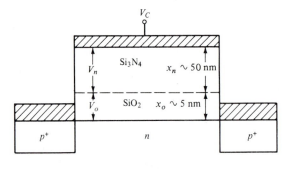

FIGURE 13-21
The MNOS transistor.

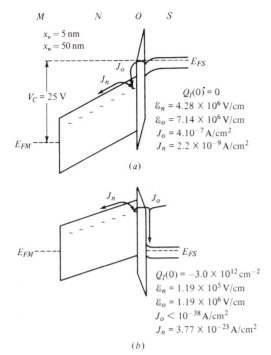

$M \qquad N \qquad O \qquad S$

$x_o = 5$ nm
$x_n = 50$ nm

E_{FS}

$V_C = 25$ V

$Q_I(0) = 0$
$\mathscr{E}_n = 4.28 \times 10^6$ V/cm
$\mathscr{E}_o = 7.14 \times 10^6$ V/cm
$J_o = 4.10^{-7}$ A/cm^2
$J_n = 2.2 \times 10^{-9}$ A/cm^2

E_{FM}

(a)

$J_n \qquad J_o$

E_{FM}

E_{FS}

$Q_I(0) = -3.0 \times 10^{12}$ cm^{-2}
$\mathscr{E}_n = 1.19 \times 10^5$ V/cm
$\mathscr{E}_o = 1.19 \times 10^6$ V/cm
$J_o < 10^{-38}$ A/cm^2
$J_n = 3.77 \times 10^{-23}$ A/cm^2

(b)

FIGURE 13-22
Energy-band diagram for an MNOS transistor (a) in the charging mode under an applied positive voltage $V_C = +25$ V and (b) at storage after voltage removal. (*After Bentchkowsky [14].*)

semiconductor may penetrate a triangular potential barrier, giving rise to a current

$$J_o = C_1 \mathscr{E}_o^2 \exp\left(-\frac{\beta}{\mathscr{E}_o}\right) \tag{13-20}$$

where $C_1 = 9.625 \times 10^{-7}$ A/V^2 and $\beta = 2.765 \times 10^8$ V/cm. This is known as *Fowler-Nordheim tunneling*, and the current depends strongly on the electric field [15].

The transport of carriers in the silicon nitride is very different. As illustrated in the insert in Fig. 13-23, electrons passing through the bulk of the nitride are captured and emitted by the numerous traps that exist there. This mechanism is known as *Frankel-Poole transport*, and the current-field relationship is given by

$$J_n = C_2 \mathscr{E}_n \exp\left(-\frac{\phi_B - \sqrt{q\mathscr{E}_n/\pi K_n \varepsilon_o}}{\phi_T}\right) \tag{13-21}$$

where ϕ_B represents the trap depth in energy and C_2 is a constant. Experimental values of ϕ_B and C_2 are estimated to be 0.825 eV and 1.1×10^{-9} $(\Omega \cdot \text{cm})^{-1}$, respectively. The barrier height is determined by a current vs. $1/T$ plot, and C_2 is estimated from the straight-line segment of Fig. 13-23. These are obtained at an electric field of 5.3×10^6 V/cm. At low electric field, the metal-insulator interface barrier becomes important as indicated by the bias-polarity effect and nonexponential plot of current in Fig. 13-23. In other words, Eq. (13-21) is applicable to the transport through the bulk nitride, and it should not be used for the low field region.

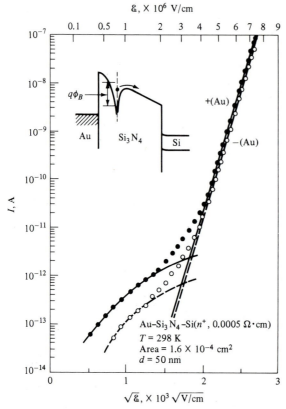

FIGURE 13-23

Current-voltage characteristics of Au–Si$_3$N$_4$–Si diode at room temperature. Insert shows Frenkel-Poole emission from trapped electrons. *(After Sze [16].)*

Since the majority of presently available MNOS devices have an oxide conduction current considerably larger than the nitride conduction current, we assume that $J_o \gg J_n$. As a result, electrons tunneling through the oxide will accumulate in the interface traps until the electric field in the oxide is reduced to zero. Therefore, Eq. (13-18) reduces to

$$V_C = \mathscr{E}_n x_n \qquad (13\text{-}22)$$

and the stored charge is given by

$$Q = -K_n \varepsilon_o \mathscr{E}_n = -\frac{K_n \varepsilon_o}{x_n} V_C = -C_n V_C \qquad (13\text{-}23)$$

where $C_n = K_n \varepsilon_o / x_n$. With a stored charge Q at the interface, the threshold voltage is shifted by

$$\Delta V_T = -\frac{Q}{C_n} = V_C \qquad (13\text{-}24)$$

and the corresponding turn-on characteristics are shown in Fig. 13-24, where V_{T_1} represents the original threshold.

On the other hand, if the interface charge does not reduce $\mathscr{E}_o$ by a significant amount, and if the oxide current as a function of the electric field is known, we can write the stored charge as

$$-Q = \int J_o \, dt \approx J_o \, \Delta t \tag{13-25}$$

Thus, the corresponding change in threshold voltage is given by

$$\Delta V_T \approx \frac{J_o \, \Delta t}{C_n} \tag{13-26}$$

Further understanding of the charge-storage effect can be gained by working out Prob. 13-11. Equation (13-26) is a reasonable approximation if ΔV_T is small compared with V_C.

To switch to the low-threshold voltage, a large negative voltage pulse is applied to the gate, and electrons are driven out of the interface traps and returned to the silicon substrate. It is important to note that while the tunneling current through the oxide is high at high electric fields during write and erase operations, there is essentially no current conduction through the oxide at low electric fields during the read or storage period. In other words, the charge stored in the nitride traps is considered permanent.

Besides calculating the change in the threshold voltage, one would like to be able to estimate the switching speed and charge-retention time. Unfortunately, it is difficult to obtain a reasonable estimate because of the lack of thorough understanding of the electrical properties of nitride traps, e.g., energy levels, densities, and capture cross sections. It has been found experimentally that the switching speed increases as the oxide thickness is reduced or the nitride-trap density is increased or both. But as the switching speed is increased, the ability to keep the stored charge, known as the *retentivity*, of the device is decreased. This tradeoff between retentivity and switching speed generally holds for all nonvolatile semiconductor memory devices. With MNOS structures, a switching speed on the order of nanoseconds can be achieved if a short retention time is tolerable. On the other hand, devices having a retention time estimate at tens of years can be fabricated if a slow switching speed is acceptable.

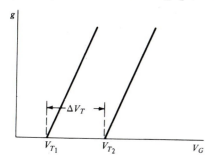

FIGURE 13-24
The shift of V_T after charging gate.

13-10 THE EPROM TRANSISTOR

A floating-gate avalanche-injection MOS transistor is shown in Fig. 13.25a. Like the MNOS device, the floating-gate structure is also a charge-storage transistor, but the charge is stored in a polycrystalline silicon floating gate rather than in traps at the nitride oxide interface. When the channel length becomes very short, i.e., about a micrometer or less, hot electrons are generated in the drain region by impact ionization without reaching the avalanche breakdown. For this reason, recent EPROMs are not really FAMOS. Hot-electron effects have been discussed in Sec. 11-5. Before we present the operation principle of the EPROM, let us consider the floating-gate potential as a function of drain bias.

When a voltage is applied at the drain, a potential is established at the floating gate by means of capacitive coupling, as shown in Fig. 13-25b. The potential drops across the structure, as represented by the equivalent circuit in Fig. 13-25c, are related by

$$V_D = V_1 + V_2 \tag{13-27}$$

Assuming there is no charge in the floating gate, Gauss' law may be written as

$$K_o \varepsilon_o \mathscr{E}_1 = K_o \varepsilon_o \mathscr{E}_2 \tag{13-28}$$

Using the relations $\mathscr{E} = V/x$ and $C = K_o \varepsilon_o / x$, Eq. (13-28) can be written with C_1

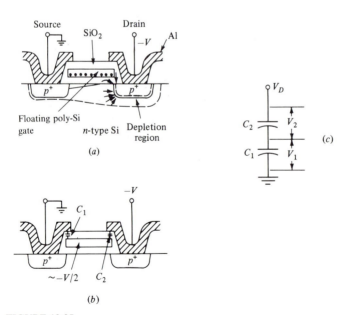

FIGURE 13-25
The FAMOS device: (a) injection of electrons into the floating gate, (b) potential of the floating gate due to capacitive voltage-divider action, and (c) an equivalent circuit. (*After Chang [17].*)

and C_2 as illustrated in Fig. 13-25c:

$$C_1 V_1 = C_2 V_2 \tag{13-29}$$

Solving Eqs. (13-27) and (13-29) yields

$$V_1 = \frac{C_2}{C_1 + C_2} V_D \tag{13-30}$$

In practical EPROM, we have $C_1 \approx C_2$ and $V_D = -V$. Therefore, we have $V_1 \approx -V/2$. as illustrated in Fig. 13-25b.

The typical oxide thickness is between 30 and 50 nm, so injection of charge by tunneling is not possible. To program the floating gate of the EPROM, the drain junction is reverse-biased to realize impact ionization. Electrons are accelerated in the depletion region, and hot electrons are injected from the silicon substrate into the silicon dioxide. These electrons are attracted to the floating gate which has a more positive voltage than the drain through capacitive coupling. The negative charge accumulated at the floating gate creates a p channel and gives rise to a low threshold voltage.

After programming, electrons are trapped in the floating gate, as depicted in Fig. 13-26. Since the barrier at the oxide-silicon interface is greater than 3 eV, electrons cannot easily get out and the EPROM has a very long retention time. Erasing the charge requires an ultraviolet light with enough energy to excite the trapped electrons into the conduction band of the gate dielectric.

There are two major disadvantages of the EPROM or FAMOS. First, it needs an ultraviolet light for erasing which requires a package with a quartz window. The device has to be removed from the circuit board and put under a special ultraviolet eraser. Second, it requires a high voltage for programming. Such a high power supply is not available in the integrated circuit, and a programming setup has to be provided. The advantages of the EPROM include a single transistor cell which occupies a small area, typically $20 \, \mu m^2$ or less, as well as lower cost in comparison with EEPROMs.

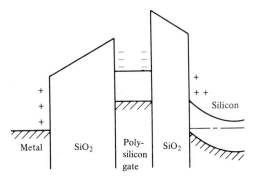

FIGURE 13-26
Energy-band diagram of the FAMOS device with charge stored in the silicon gate.

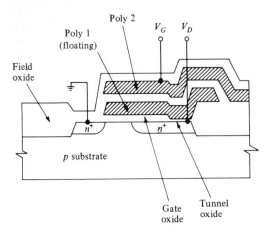

Poly 2

Poly 1
(floating)

V_G V_D

Field
oxide

n^+ n^+

p substrate

Gate   Tunnel
oxide   oxide

FIGURE 13-27
FLOTOX device structure. *(After Lai and Dham [15].)*

13-11   THE FLOTOX TRANSISTOR [15, 18]

It is desirable in some applications to have an electrically erasable programmable read-only memory (EEPROM) that requires neither ultraviolet light erasing nor a special power supply. The floating-gate tunnel-oxide MOS transistor (FLOTOX) satisfies these requirements and is a popular EEPROM in the market. The basic structure of the FLOTOX transistor is shown in Fig. 13-27. The key difference between the EPROM and the FLOTOX is that a small area of thin oxide is fabricated between the floating gate and drain where carriers may tunnel through. This transport mechanism follows the Fowler-Nordheim equation of (13-20), and allows electrons to conduct between the drain and the floating gate. The device operation may be explained by the idealized energy-band diagram shown in Fig. 13-28.

A section of the device with the thin tunneling oxide is shown in Fig. 13-28*a*, and the energy-band diagram under thermal equilibrium is depicted in Fig. 13-28*b*. To program the cell means to charge up the floating gate with electrons. This is accomplished by applying a high positive gate voltage at poly 2 with the source and the drain set at ground potential. A fraction of the positive gate voltage is developed in the floating gate through capacitive coupling so that a voltage is established across the thin tunnel oxide. The energy-band diagram is changed to Fig. 13-28*c*, where electrons in the conduction band of the n^+ drain can tunnel through the oxide to reach the floating gate. The current-voltage characteristics are shown in Fig. 13-29 for different oxide thicknesses, and they are in close agreement with Eq. (13-20). The most important part of these curves is the portion with the steepest slope in which the FLOTOX is operated. It should be noted that as electrons build up on the floating gate, it lowers the electric field and reduces the electron flow. Thus, the electron current is self-limiting, and the charge does not continue to grow.

In the storage mode, the gate voltage is removed, and the energy-band

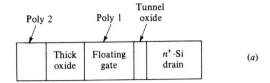

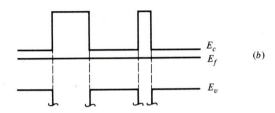

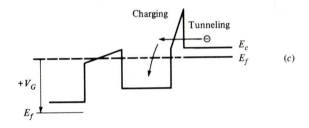

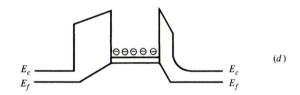

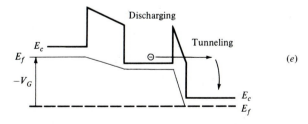

FIGURE 13-28

The FLOTOX transistor: (a) section with tunnel oxide, (b) equilibrium energy-band diagram, (c) conduction band with $+V_G$, programming mode, (d) storage mode, and (e) erasing mode with $-V_G$.

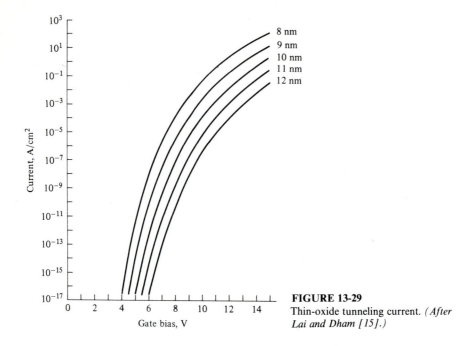

FIGURE 13-29
Thin-oxide tunneling current. *(After Lai and Dham [15].)*

diagram is shown in Fig. 13-28*d*. Since the system is not in equilibrium, the Fermi level does not line up. Note that the Fermi level in the floating gate has moved closer to the conduction band in comparison with Fig. 13-28*b*, indicating an increased electron concentration. The floating-gate charge induces an electric field in the surrounding oxide as illustrate to satisfy Gauss's law.

In the erasing mode, the top gate is grounded and the source is floating. A high positive voltage is applied to the drain, and the energy-band diagram is depicted in Fig. 13-28*e*. A fraction of the drain voltage is coupled capacitively to the floating gate, and electrons tunnel from the floating gate to the drain as illustrated.

The floating gate also acts as the gate of an *n*-channel MOS transistor. In the programmed state, the negative charge on the floating gate gives rise to a shift of the threshold voltage in the positive-voltage direction. After the stored charge is removed by erasing, the floating gate has a net positive charge, and the threshold voltage is shifted to a negative voltage.

The floating-gate transistor is the memory element for the EEPROM, and it requires a second transistor, called the *select transistor*, to form a cell. Without the select transistor, a high voltage applied to one drain will appear on the drain of other cells in the same memory column in a two-dimensional array. This leads to erasing all other cells that are not selected. The select transistor is used as an ON/OFF switch to disconnect the nonselected cells. The select transistor is necessary, and increases the area of the basic memory cell. For this reason, the typical area of a FLOTOX memory cell is about $100 \, \mu m^2$, a factor of 5 bigger than the EPROM, which limits the scaling of the memory cell for high density and results in higher cost. This is the weak point of the EEPROM, which is why the EPROM remains competitive today.

Example. A FLOTOX has a tunnel oxide of 10 nm with an area one-tenth of the floating gate. The gate oxide is 30 nm between poly 1 and the silicon substrate. Find the charge density at the floating gate after applying a programming current of $10^{-3} \, A/cm^2$ for 6 ms. What is the incremental change of the threshold voltage?

Solution. The total number of charge per unit area supplied by the tunneling current in the thin-oxide region is

$$Q = \frac{Jtq}{q} = \frac{6 \times 10^{-6} \, q}{1.6 \times 10^{-19}} = 3.75 \times 10^{13} q \, cm^{-2}$$

Since the actual floating-gate area is 10 times the tunnel-oxide area, the charge density is

$$Q_{FG} = \frac{Q}{10} = 3.75 \times 10^{12} q \, cm^{-2}$$

if the charge in the floating gate is assumed to be uniformly spread out.
The capacitance between the floating gate and the substrate is

$$C_{FG} = \frac{K_o \varepsilon_o}{x_{o1}} = 1.15 \times 10^{-7} \, F/cm^2$$

The incremental change of the threshold voltage at the floating gate is

$$\Delta V_T = \frac{Q_{FG}}{C_{FG}} = 5.2 \, V$$

A more accurate calculation involves the replacement of C_{FG} by the total capacitance defined in the next section. The analysis is given in Prob. 13-14.

13-12 CHARGE COUPLING DURING PROGRAMMING AND ERASING

The principle of operation of the FLOTOX transistor can be understood by considering the programming and erasing conditions or the read condition. In the read condition, the transistor is acting like an EPROM with the threshold voltage specified by the floating-gate charge. Because of the thin tunnel oxide, however, the drain voltage couples strongly to the floating gate, and the threshold voltage is sensitive to the drain voltage. This is different from that of the EPROM.

During programming and erasing, the transistor may be considered as a network of capacitors tied to the floating gate, as illustrated in Fig. 13-30 where

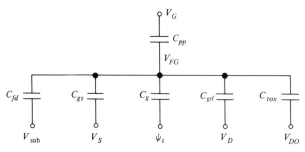

FIGURE 13-30
Equivalent circuit to represent capacitances for charge-coupling calculation. *(After Lai and Dham [15].)*

the capacitors are defined. The coupling of voltage through the capacitors to the floating gate can be developed by the same method in deriving Eq. (13-30) except that the total capacitance C_T must be used to replace $C_1 + C_2$. Therefore, the floating-gate voltage can be written as

$$V_{FG} = \frac{Q_{FG}}{C_T} + V_G \frac{C_{pp}}{C_T} + V_B \frac{C_{fd}}{C_T} + V_S \frac{C_{gs}}{C_T}$$

$$+ V_D \frac{C_{gd}}{C_T} + V_{DO} \frac{C_{tox}}{C_T} + (V_{FB} + \psi_s) \frac{C_g}{C_T} \qquad (13\text{-}31)$$

where

$$C_T = C_{pp} + C_{fd} + C_{gs} + C_{gd} + C_{tox} + C_g \qquad (13\text{-}32)$$

and C_{pp} is the poly-to-poly capacitance, C_{fd} is the floating-gate over field-oxide capacitance, C_{gs} is the floating-gate-to-source capacitance, C_{gd} is the floating-gate-to-drain capacitance excluding the tunneling area, C_{tox} is the tunnel-oxide capacitance, and C_g is the floating-gate over gate-oxide capacitance. In addition, Q_{FG} is the floating-gate charge, V_B, V_S, V_D, and V_{DO} are the voltages of the body, source, drain, and effective drain under the tunnel oxide, respectively. The inversion-layer charge may be written as

$$Q_I = C_g(V_{FG} - V_{FB} - \psi_s) \qquad (13\text{-}33)$$

Equations (13-31) to (13-33) have been used to predict accurately the device characteristics. An exception occurs during erasing when the surface charge changes from accumulation to inversion. The charge inversion leads to an MOS capacitance given by Eq. (9-17) and Fig. 9-6 (dashed line) so that the gate-oxide capacitance decreases with time. It gives rise to a lower coupling ratio of voltages to the floating gate as a function of time.

An important design parameter is the program coupling ratio, defined by

$$\text{C.R.} = \frac{V_{FG}}{V_g} \times 100\% \qquad (13\text{-}34)$$

The effect of the coupling ratio and tunnel-oxide thickness on the change in threshold voltage during programming is illustrated in Fig. 13-31. For a given

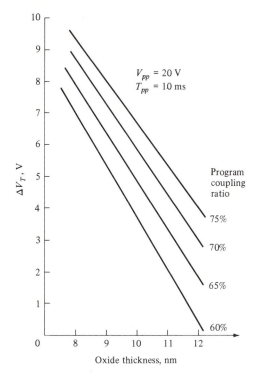

FIGURE 13-31
Threshold-voltage dependence on program coupling ratio and oxide thickness. *(After Lai and Dham [15].)*

ΔV_T, a lower coupling ratio is required for a thinner oxide. It turns out that a lower coupling ratio has smaller cell area, but a thinner oxide results in lower yield and reliability. In trading off the oxide thickness and the cell area, the coupling ratio is typically set around 70 percent.

REFERENCES

1. Amelio, G. F., W. J. Bertram, Jr., and M. F. Tompsett: Charge-Coupled Imaging Devices: Design Considerations, *IEEE Trans. Electron Devices*, **ED-18**:986 (1971).
2. Kim, C-K., and E. H. Snow: *p*-Channel Charge Coupled Devices with Resistive Gate Structure, *Appl. Phys. Lett.*, **20**:514 (1972).
3. Bertram, W. J., et al.: A Three-Level Metallization Three-Phase CCD, *IEEE Trans. Electron Devices*, **ED-21**:758 (1974).
4. Kosonocky, W. F., and J. E. Carnes: Design and Performance of Two Phase CCDs with Overlapping Polysilicon and Aluminum Gates, *Int. Electron Device Meet., Washington, 1973, Tech. Dig.*, p. 123.
5. Kim, C-K.: Two-Phase Charge Coupled Linear Imaging Devices with Self-aligned Barrier, *Int. Electron Device Meet., Washington, 1974, Tech. Dig.*, p. 55.
6. Kim, C-K, and M. Lenzlinger: Charge Transfer in Charge Coupled Devices. *J. Appl. Phys.*, **42**: 3856 (1971).
7. Carnes, J. E., W. E. Kosonocky, and E. G. Ramberg: Drift-Aiding Fringing Fields in Charge-Coupled Devices, *IEEE J. Solid-State Circuits*, **SC-6**:322 (1971).
8. Joyce, W. B., and W. J. Bertram: Linearized Dispersion Relation and Green's Function for Discrete Charge Transfer Devices with Incomplete Transfer, *Bell Syst. Tech. J.*, **50**:1741 (1971).

9. Tompsett, M. F.: A Simple Charge Regenerator for Use with Charge-Transfer Devices and the Design of Functional Logic-Arrays, *IEEE J. Solid-State Circuits*, **SC-7**:237 (1972).

10. Kim, C-K.: Design and Operation of Buried Channel Charge Coupled Devices, *CCD Appl. Conf. Proc. Nav. Electron. Lab., San Diego, Calif., September 1973.*

11. Sangster, F. L. J.: Integrated MOS and Bipolar Analog Delay Line Using Bucket-Brigade Capacitor Storage, *IEEE Solid-State Circuits Conf., 1970, Dig.*, p. 74.

12. Séquin, C. H., and M. F. Tompsett: "Charge Transfer Devices," chap. 5, Academic, New York, 1975.

13. Crowell, M. H., and E. F. Labuda: The Silicon Diode Array Camera Tube, *Bell Syst. Tech. J.*, **48**:1481 (1969).

14. Bentchkowsky, D. Frohman: The Metal-Nitride-Oxide-Silicon (MNOS) Transistor: Characteristics and Applications, *Proc. IEEE*, **58**:1207 (1970).

15. Lai, S. K., and V. K. Dham: VLSI Electrically Erasable Programmable Read Only Memory (EEPROM), in N. G. Einspruch (ed.), "VLSI Handbook," vol. 13, Academic, New York, 1985, p. 167.

16. Sze, S. M.: Current Transport and Maximum Dielectric Strength of Silicon Nitride Films, *J. Appl. Phys.*, **38**:2951 (1967).

17. Chang, J. J.: Nonvolatile Semiconductor Memory Devices, *Proc. IEEE*, **64**:1039 (1976).

18. Lai, S. K., and V. K. Dham: Comparison and Trends in Today's Dominant E^2 Technologies, *Tech. Proc. Int. Electron Device Meet.*, **580** (1986).

PROBLEMS

13-1. Sketch the energy-band diagram and charge distribution of an MOS capacitor fabricated on an n substrate under the condition of (a) deep depletion and (b) thermal equilibrium.

13-2. Verify Eq. (13-7).

13-3. An MOS capacitor has the following parameters: substrate with $N_a = 10^{15}$ cm^{-3}, $V_{FB} = 2$ V, $x_o = 100$ nm, electrode area of 10 by 20 μm. Calculate (a) the oxide capacitance, (b) the surface potential at $V_G = 10$ V, (c) the depletion-layer depth, and (d) the depletion-layer charge.

13-4. Verify Eq. (13-10).

13-5. (a) Calculate the capacitance between the electrode and the substrate of the MOS capacitor of Prob. 13-3 at $V_G = 10$ and $Q_{sig} = 0$.

(b) A signal-charge packet is injected into the potential well, and the resulting C_{GS} is found to double the value obtained in (a). Find the total number of injected electrons and the electron density.

13-6. A CCD is fabricated on a p-type substrate with $N_a = 2 \times 10^{14}$ cm^{-3}. The oxide thickness is 150 nm, and the electrode area is 10 by 20 μm.

(a) Calculate the surface potential and depletion-layer depth for two adjacent electrodes biased at $V_G = 10$ and 20 V, respectively. Assume $V_{FB} = 0$ and $Q_{sig} = 0$.

(b) Repeat (a) after 10^6 electrons are introduced into the cell.

(c) Sketch the potential-well diagram for (b).

13-7. (a) What is the fringing field at the electrode boundary in Prob. 13-6a if the interelectrode spacing is 3 μm?

(b) Assume that before the charge transfer, 10^6 electrons are uniformly distributed in a potential well with $V_G = 10$ V. Estimate the time required to transfer all the electrons to an adjacent well with $V_G = 20$ V by means of the fringing-field current. Assume $\mu_n = 650$ cm^2/V·s.

13-8. (a) Sketch the cross-sectional diagram of a CCD two-stage shift register together with the necessary charge-generation and sensing devices. Assume an inter-electrode area of 10 by 50 μm.

(b) Find the maximum clock rate of the device for both n- and p-channel structures if carriers are assumed to be transferred by diffusion. Use $\mu_p = 200 \text{ cm}^2/\text{V} \cdot \text{s}$ and $\mu_n = 650 \text{ cm}^2/\text{V} \cdot \text{s}$.

13-9. A three-phase CCD analog delay line has 1000 stages and is operated at a clock frequency of 1 MHz. The transfer inefficiency is 10^{-4}. Determine the attenuation and phase-shift degradation for a 100-kHz signal and for its third harmonic.

13-10. Assume an interface trap density (Q_{ss}) of 10^{11} cm^{-2} in Fig. P13-10. (a) Sketch the potential diagram for the CCD cell shown with $V_o = -2 \text{ V}$. (b) What is the required V_o to facilitate charge transfer?

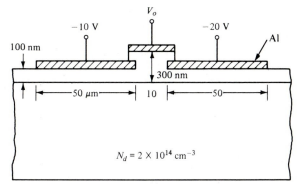

FIGURE P13-10

13-11. Assume that the insulating layer above the floating gate in an n-channel MNOS memory transistor has $K_n = 40$ and $x_n = 100 \text{ nm}$, and the lower insulator has $K_o = 4$ and $x_o = 10 \text{ nm}$. If the conductive properties of both insulators are the same and given by $J = \sigma \mathscr{E}$, where $\sigma = 10^{-7} \text{ mho/cm}$, find the approximate threshold-voltage shift of the transistor induced by a voltage of 10 V applied to the external gate for 1 μs.

13-12. Explain the conditions (physical parameters of insulators) under which an ideal nonvolatile MIS memory transistor would be achieved.

13-13. The concept of capacitive coupling may be derived in a circuit with two capacitors C_1 and C_2 connected in series (see Fig. P13-13). Show that

$$V_{fg} = \frac{C_1}{C_1 + C_2} V_g$$

by using small-signal analysis.

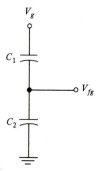

FIGURE P13-13

13-14. For the example given in Sec. 13-11, the oxide between poly 1 and poly 2 is 50 nm thick. Find the threshold voltage change at the floating gate for a programming time of 10 ms. Explain why C_T is used.

13-15. Repeat the example in Prob. 13-14 if the tunnel oxide is 8 nm with an area one-twentieth of the floating gate.

CHAPTER
14

SOLAR CELLS AND PHOTODETECTORS

In this chapter, we present the principle of light absorption and its applications in energy conversion and photodetection. Optical energy can be absorbed in a semiconductor if the photon energy is greater than the band-gap energy. The absorbed photons generate electron-hole pairs, which produce a photocurrent in a semiconductor or a *pn* junction. In the case of a junction, a potential difference is established across the space-charge layer. This process of converting optical energy into electric energy is known as the *photovoltaic effect*. The converter is known as the *solar cell*.

The high sensitivity to light stimulation makes a semiconductor ideal for photodetection. However, the requirements of a photodetector differ considerably from that of a solar cell. Instead of focusing on the conversion efficiency of a broadband solar spectrum, the photodetector must have the appropriate speed, sensitivity, and gain in a narrow range of photon energy. We shall first introduce the theory and characteristics of the solar cell, which will be followed by a discussion of the important issues in cell design. Then, we shall present the basic device physics of the photoconductor, *pin* diode, and avalanche photodiode.

14-1   OPTICAL ABSORPTION IN A SEMICONDUCTOR

The unit energy of light, called a *photon*, is hv, where v is the light frequency and h is Planck's constant. The wavelength of light λ is related to the frequency by

$$\lambda = \frac{c}{v} = \frac{hc}{E_{ph}} = \frac{1.24}{E_{ph}} \ \mu m \tag{14-1}$$

where E_{ph} is the photon energy hv in electronvolts and c is the speed of light, that is, 3×10^{10} cm/s. When a semiconductor is illuminated, photons may or may not be absorbed, depending on the photon energy and the band-gap energy E_g. Photons with energy smaller than E_g are not readily absorbed by the semiconductor because there is no energy state available in the forbidden gap to accommodate an electron (Fig. 14-1a). Thus, light is transmitted through and the material appears transparent. If $E_{ph} = E_g$, photons are absorbed to create electron-hole pairs, as shown in Fig. 14-1b. When the photon energy is greater than E_g, an electron-hole pair is generated and, in addition, the excess energy $E_{ph} - E_g$ is dissipated as heat (Fig. 14-1c).

Let us consider the nature of absorption for a semiconductor shown in Fig. 14-2. The optical source provides monochromatic light with $hv > E_g$ and a flux F_{ph} or F in photons per square centimeter per second. As the light beam penetrates the crystal, the fraction of the photons absorbed is proportional to the intensity of the flux $F(x)$. Therefore, the absorbed photons within Δx are

$$\alpha F(x) \, \Delta x$$

where α is a proportional constant called the *absorption coefficient*. From the continuity of light in Fig. 14-2, we find

$$F(x + \Delta x) - F(x) = \frac{dF(x)}{dx} \, \Delta x = -\alpha F(x) \, \Delta x$$

or

$$-\frac{dF(x)}{dx} = \alpha F(x) \tag{14-2}$$

The negative sign indicates decreasing intensity of the flux along x due to absorption. With the boundary condition $F = F_{ph}$ at $x = 0$, we obtain the solution of Eq. (14-2) as

$$F(x) = F_{ph} e^{-\alpha x} \tag{14-3}$$

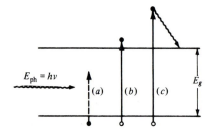

$E_{ph} = hv$

(a) (b) (c)

E_g

FIGURE 14-1
Optically generated electron-hole pairs in a semiconductor.

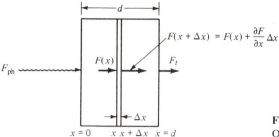

FIGURE 14-2
Optical absorption in a semiconductor.

Therefore, the fraction of light transmitted through the semiconductor is

$$F_t = F(d) = F_{ph} e^{-\alpha d} \qquad (14\text{-}4)$$

where d is the thickness of the semiconductor. Since the absorption is determined by the photon energy, it is conceivable that α is a function of $h\nu$. Figure 14-3 shows the absorption spectra of a few semiconductors that are used for optoelectronic applications. In this figure, the band-gap energies are indicated, along with the wavelength. We notice that the absorption coefficient drops off sharply at the band-gap energy, indicating negligible absorption for photons with energy smaller than E_g. Thus, silicon absorbs photons with $\lambda \leqslant 1.1 \; \mu m$, and GaAs absorbs photons with $\lambda \leqslant 0.9 \; \mu m$.

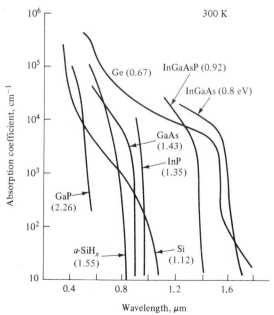

FIGURE 14-3
Absorption coefficients of some important optoelectronic semiconductors.

14-2 PHOTOVOLTAIC EFFECT AND SOLAR-CELL EFFICIENCY

The process of converting optical energy into electric energy in a *pn* junction involves the following basic steps: (1) Photons are absorbed, so electron-hole pairs are generated in both the *p* and *n* sides of the junction (Fig. 14-4*a*). (2) By diffusion, the electrons and holes generated within a diffusion length from the junction will be able to reach the space-charge region (Fig. 14-4*b*), since the diffusion length is defined as the average distance a carrier diffuses before recombination. (3) Electron-hole pairs are then separated by the strong electric field; thus, electrons in the *p* side slide down the potential to move to the *n* side and holes go in the opposite direction (Fig. 14-4*c*). (4) If the *pn* diode is open-circuited, the accumulation of electrons and holes on the two sides of the junction produces an *open-circuit voltage* (Fig. 14-4*d*). If a load is connected to the diode, a current will conduct in the circuit (Fig. 14-4*a*). The maximum current is realized when an electrical short is placed across the diode terminals, and this is called the *short-circuit current*.

The current produced by light in the load is in the same direction as the reverse saturation current of the *pn* junction. Therefore, the total diode current under illumination is given by

$$I = I_L + I_o(1 - e^{V/\phi_T}) \tag{14-5}$$

where I_L is the *light-generated current* and the second term on the right-hand side

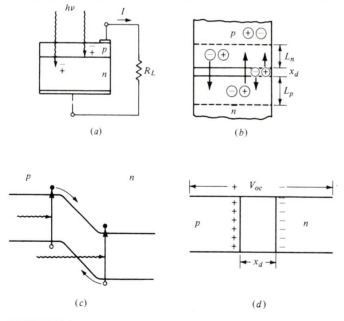

(*a*)   (*b*)

(*c*)   (*d*)

FIGURE 14-4
Conversion of optical energy into electric energy: (*a*) a solar cell with load resistor, (*b*) diffusion of electrons and holes that produces current, (*c*) energy-band diagram of (*b*), and (*d*) establishment of the open-circuit voltage (schematic representation).

is the reverse diode current, i.e., the negative of Eq. (4-24). For uniform absorption throughout the device, I_L is given by (Prob. 14-4)

$$I_L = qG_L(L_n + L_p)A \qquad (14\text{-}6)$$

where G_L is the generation rate. The current-voltage characteristics expressed by Eq. (14-5) are depicted in Fig. 14-5 for an experimental device with the light intensity as the variable. Data were taken under air-mass 1 (AM1) illumination, defined as the sun at the zenith and the test device at sea level under a clear sky. The energy reaching the solar cell under the AM1 condition is slightly higher than $100\ \mathrm{mW/cm^2}$. The solar spectrum just outside the atmosphere is known as AM0, where the sun energy is $135\ \mathrm{mW/cm^2}$. It is noted that I_L is the current at zero voltage in Fig. 14-5 and thus the short-circuit current. Setting $I = 0$ in Eq. (14-5), we obtain the open-circuit voltage as

$$V_{oc} = \phi_T \ln\left(1 + \frac{I_L}{I_o}\right) \qquad (14\text{-}7)$$

In Eq. (14-7) we have an open-circuit voltage or electric power source which supplies a current if an external load is connected across it. The conversion of optical to electric energy is thus realized.

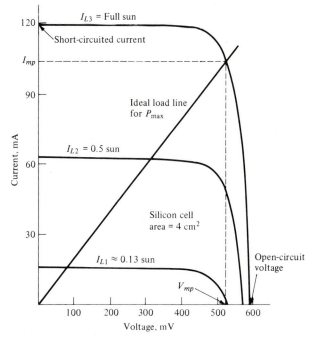

FIGURE 14-5
Current-voltage characteristics of a typical solar cell under air-mass 1 (AM1) illumination, i.e., sun energy at sea level under clear sky with sun at zenith.

An equivalent circuit of the solar-cell characteristic expressed by Eq. (14-5) can be drawn as shown in Fig. 14-6. The power delivered to the load is given by

$$P = IV = I_L V - I_o V(e^{V/\phi_T} - 1) \tag{14-8}$$

Example. Calculate the open-circuit voltage of a silicon n^+p cell for substrate dopings between 10^{15} and 10^{18} cm^{-3}. Assume that $L_n = 100 \, \mu$m, $D_n = 36$ cm^2/s, and $I_L/A = 35$ mA/cm^2 and that these values are independent of the doping concentration.

Solution. Since $N_d \gg N_a$ in an n^+p junction, the saturation current [Eq. (4-25)], can be simplified to

$$I_o = qAn_i^2 \left(\frac{D_p}{L_p N_d} + \frac{D_n}{L_n N_a} \right) \approx \frac{qAn_i^2 D_n}{L_n N_a} = \frac{1.3 \times 10^5 \, A}{N_a} \quad \text{A/cm}^2$$

Therefore, Eq. (14-7) becomes

$$V_{oc} = 26 \ln \left(1 + \frac{35 \times 10^{-3} \, N_a}{1.3 \times 10^5} \right) \quad \text{mV}$$

Thus

N_a, cm^{-3}	10^{15}	10^{16}	10^{17}	10^{18}
V_{oc}, mV	504	565	625	684

In the foregoing example, the open-circuit voltage is found to increase with the substrate doping. Since the light-generated current is relatively independent of doping level, the output power of the solar cell should increase with substrate-doping concentration monotonically. In practice, however, the open-circuit voltage and output power reach a maximum in a doping less than 10^{17} cm^{-3}. Beyond this point, the space-charge-recombination dark current becomes significant, and the open-circuit voltage decreases with further increase of substrate doping [2]. The voltage corresponding to the maximum power delivery V_{mp} is obtained by taking $\partial P/\partial V = 0$. Therefore, we derive

$$\left(1 + \frac{V_{mp}}{\phi_T} \right) e^{V_{mp}/\phi_T} = 1 + \frac{I_L}{I_o} \tag{14-9}$$

The current at maximum power is denoted by I_{mp}, as indicated in Fig. 14-5. The efficiency of the solar cell is therefore

$$\eta = \frac{I_{mp} V_{mp}}{P_{in}} \times 100\% \tag{14-10}$$

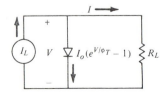

FIGURE 14-6
Equivalent circuit of the solar cell shown in Fig. 14-4.

where P_{in} is the input light power. To optimize the solar-cell efficiency, we should have large I_{mp} and V_{mp}. The maximum current and maximum voltage achievable in a solar cell are I_L and V_{oc}, respectively. Therefore, the ratio $V_{mp}I_{mp}/I_L V_{oc}$ is useful as a measure of the realizable power from the I–V curve. This is called the *fill factor*, and it is between 0.7 and 0.8 for a well-made cell. Typical efficiency for a commercially available silicon cell is 12 to 15 percent.

14-3 LIGHT-GENERATED CURRENT AND COLLECTION EFFICIENCY

In the preceding section, we presented the simple theory of the solar cell. The light-generated current was obtained by assuming uniform absorption throughout the device, and the effect of photon energy on absorption was neglected. It is, however, necessary to examine the nature of the light-generated current to gain further understanding of the solar cell. Let us consider an incident photon flux F_{ph} striking the surface of a *p*-on-*n* structure. The effect of surface reflection is ignored for the moment. By using Eq. (14-3) and assuming that each absorbed photon generates an electron-hole pair, the generation rate of electron-hole pairs as a function of surface penetration is obtained:

$$G_L = \alpha F_{ph} e^{-\alpha x} \qquad (14\text{-}11)$$

Adding Eq. (14-11) to the hole-continuity equation expressed in Eq. (4-8) leads to the following equation for holes in the *n* side of the junction under steady-state conditions:

$$D_p \frac{d^2 p_n}{dx^2} - \frac{p_n - p_{no}}{\tau_p} + \alpha F_{ph} e^{-\alpha x} = 0 \qquad (14\text{-}12a)$$

Here the generation term has a positive sign because it counteracts the recombination term. Similarly, the steady-state expression describing electrons in the *p* side is

$$D_n \frac{d^2 n_p}{dx^2} - \frac{n_p - n_{po}}{\tau_n} + \alpha F_{ph} e^{-\alpha x} = 0 \qquad (14\text{-}12b)$$

The electron- and hole-current components at the junction per unit area are given by

$$J_p = -qD_p \frac{dp_n}{dx}\bigg|_{x=x_j} \qquad (14\text{-}13a)$$

$$J_n = qD_n \frac{dn_p}{dx}\bigg|_{x=x_j} \qquad (14\text{-}13b)$$

The photon-collection efficiency is defined as

$$\eta_{col} = \frac{J_p + J_n}{qF_{ph}} \qquad (14\text{-}14)$$

Equations (14-12) to (14-14) can be solved if the boundary conditions are given.

Example. Derive expressions of light-generated minority-carrier density and current in the n side of the p^+n cell shown in Fig. 14-7. Assume the surface recombination velocity at the back contact is S and the incoming light is monochromatic. Absorption in the p^+ layer is neglected.

Solution. Consider the band diagram and carrier motions in Fig. 14-7b. The boundary conditions are

$$p_n - p_{no} = 0 \qquad \text{at } x = 0$$

$$S(p_n - p_{no}) = -D_p \frac{dp_n}{dx} \qquad \text{at } x = W_n$$

The general solution of Eq. (14-12a) is

$$p_n - p_{no} = K_1 e^{x/L_p} + K_2 e^{-x/L_p} - \frac{\alpha F_{ph} \tau_p}{\alpha^2 L_p^2 - 1} e^{-\alpha x}$$

Substituting the boundary conditions into the foregoing equation yields

$$p_n - p_{no} = \frac{\alpha F_{ph} \tau_p}{\alpha^2 L_p^2 - 1} \left[\cosh \frac{x}{L_p} - e^{-\alpha x} \right.$$
$$\left. - \frac{S\left(\cosh \dfrac{W_n}{L_p}\right) + \dfrac{D_p}{L_p}\left(\sinh \dfrac{W_n}{L_p}\right) - (\alpha D_p - S)e^{-\alpha W_n}}{S\left(\sinh \dfrac{W_n}{L_p}\right) + \dfrac{D_p}{L_p}\cosh \dfrac{W_n}{L_p}} \sinh \frac{x}{L_p} \right] \quad (14\text{-}15)$$

The hole current flowing from the n side to the p^+ side is obtained using Eq. (14-13a):

$$J_p = \frac{q F_{ph} \alpha L_p}{\alpha^2 L_p^2 - 1} \left[\frac{S\left(\cosh \dfrac{W_n}{L_p}\right) + \dfrac{D_p}{L_p}\left(\sinh \dfrac{W_n}{L_p}\right) + (\alpha D_p - S)e^{-\alpha W_n}}{S\left(\sinh \dfrac{W_n}{L_p}\right) + \dfrac{D_p}{L_p}\cosh \dfrac{W_n}{L_p}} - \alpha L_p \right] \quad (14\text{-}16)$$

The electron current flowing from the p^+ side to the n side can be found in the same way.

Let us present some results based on the complete solution to gain some physical insight into photon collection under different wavelengths. At short wavelengths, the absorption coefficient α obtained from Fig. 14-3 is large. Therefore, the absorption of photons expressed in Eq. (14-3) decays in a short distance from the surface. In other words, most photons are converted into electron-hole pairs in a narrow layer near the surface for a short λ (550 nm). At a longer wavelength (900 nm), α is small, and absorption takes place mostly in the n side of the junction. The resulting minority-carrier distributions are illustrated in Figure 14-8. If we consider that the incoming photon flux is monochromatic and the number of photons per unit area per second is given, we can obtain the collection efficiency

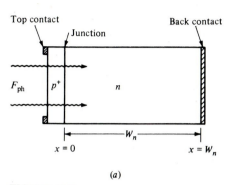
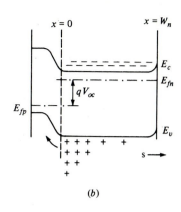

FIGURE 14-7
A p-on-n cell.

in the n side for each wavelength by substituting Eq. (14-16) into Eq. (14-14). The theoretical collection efficiency at different wavelengths is calculated and plotted in Fig. 14-9. The components resulting from the absorption in the n side and p side are separated to show the individual effects.

The collection efficiency is influenced by the minority-carrier diffusion length and the absorption coefficient. The diffusion length should be as long as possible to collect all light-generated carriers. In some solar cells, a built-in field is established by impurity gradient to improve carrier collection. As to the effect of the absorption coefficient, a large α leads to heavy absorption near the surface, resulting in a strong collection in the *skin layer*. A small α allows deep penetration of photons

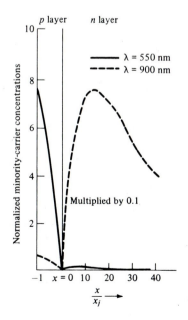

FIGURE 14-8
Normalized minority-carrier distributions for incident radiation at $\lambda = 550$ and 900 nm. Device parameters are $x_j = 2.8\ \mu m$, $W_n = 20$ mils, $\tau_p = 4.2\ \mu s$, $\tau_n = 10$ ns, and $S_n = 1000$ cm/s. *(After Wolf [3].)*

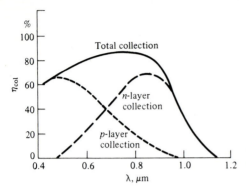

FIGURE 14-9
Collection efficiency vs. wavelength for solar cell of Fig. 14-8. *(After Wolf [3].)*

so that the *base* of the solar cell becomes more important in carrier collection. A typical GaAs or *a*-Si solar cell is of the former type, and the silicon cell belongs to the latter type.

14-4 MATERIAL SELECTION AND DESIGN CONSIDERATIONS

In the preceding derivation, we have obtained the ideal conversion efficiency with which a photon generates an electron-hole pair producing a current without any loss of energy in the process. In a practical solar cell, since various factors limit the device performance, it is necessary to consider these limiting factors in solar-cell design and material selection.

Spectral Considerations

Let us consider the matching of the solar spectrum and the solar-cell absorption characteristic for terrestrial applications. The solar spectrum at sea level under a clear sky (AM1) is given in Fig. 14-10. The predominant portion of the energy in the sunlight is in the visible region, and the total power reaching the earth at sea level is approximately $100\,\text{mW/cm}^2$. Since only the portion of energy greater than E_g can be absorbed, photons with energy lower than E_g are not utilized. The number of photons available for electron-hole pair generation is obtained by integrating Fig. 14-10 from E_g to the maximum energy. It is found that the flux density of solar photons at sea level is $4.8 \times 10^{17}\,\text{cm}^{-2}\,\text{s}^{-1}$, and the maximum number of photons that can be absorbed is $3.7 \times 10^{17}\,\text{cm}^{-2}\,\text{s}^{-1}$ in silicon and $2.5 \times 10^{17}\,\text{cm}^{-2}\,\text{s}^{-1}$ in GaAs. Therefore, the fraction of solar photons available for absorption in silicon is roughly 77 percent.

From the foregoing consideration, silicon would be a better material than GaAs. However, a large number of absorbed photons in silicon have energy greater than E_g. The excess energy $E_{ph} - E_g$ is dissipated as heat instead of generating more electrons and holes. For example, an energy of 1.1 eV is needed to produce an electron-hole pair in silicon. If a solar photon with $E_{ph} = 2.2\,\text{eV}$ is absorbed by

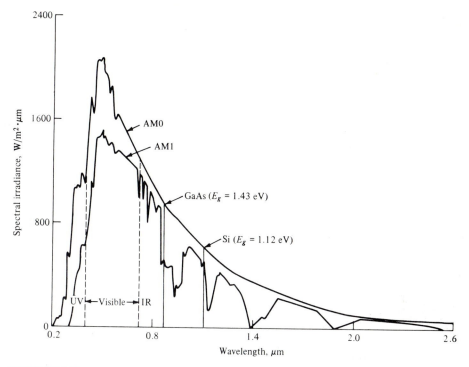

FIGURE 14-10
Solar spectra at AM0 and AM1 conditions with energy-cutoff points in GaAs and Si.

silicon, half the photon energy is dissipated as heat instead of producing electricity. It turns out that if all solar photons with energy greater than 1.1 eV are considered, the total energy loss of the absorbed photons is 43 percent. This partial utilization of absorbed photon energy must be taken into account in the overall assessment of a material. In general, the smaller the energy gap, the more power is wasted near the peak of the solar spectrum. As a consequence, we find that silicon and GaAs are comparable as far as matching the solar spectrum is concerned.

Maximum-Power Considerations

The maximum power output of a solar cell is determined by the open-circuit voltage and the short-circuit current. From spectral considerations, it is found that I_L decreases with increasng E_g. The open-circuit voltage is expressed by Eq. (14-7), in which V_{oc} is inversely related to the reverse saturation current I_o. Using Eqs. (1-19), (1-21), and (4-25), we find

$$I_0 \propto e^{-E_g/kT} \tag{14-17}$$

Substituting Eq. (14-17) into (14-7) yields

$$V_{oc} \propto E_g \tag{14-18}$$

which states that the open-circuit voltage is proportional to E_g. Since I_L decreases and V_{oc} increases with increasing E_g, the product $V_{oc}I_L$ exhibits a maximum. Using Fig. 14-10 and available semiconductor parameters, we plot the maximum conversion efficiency vs. E_g in Fig. 14-11 for different temperatures. It is seen that Si and GaAs are among the semiconductors most suitable for solar-cell applications.

Series-Resistance Considerations

The series resistance, the sum of the contact and sheet resistance, modifies the current-voltage characteristics as shown in Fig. 14-12. It increases the internal power dissipation and reduces the fill factor. The effect of the shunt resistance is found to be of little significance. The contact resistance can be reduced to a negligible value, but the choice of the sheet resistance is less straightforward. A small sheet resistance corresponds to a heavily doped surface layer, which reduces the carrier lifetime and diffusion length of the surface layer. A compromise of doping level and junction depth is therefore necessary to arrive at the optimum design.

In addition, a small series resistance requires a large metallic contact area, which limits the area of light absorption. A practical contact in the form of a grid is shown in Fig. 14-13. This structure allows a large area of exposure but at the same time keeps the series resistance to a reasonable value.

Surface Reflection

The number of photons penetrating the surface is less than that of the incident photons because of reflection at the surface. The percentage of light reflected is determined by the angle of incidence and the dielectric constant of the material.

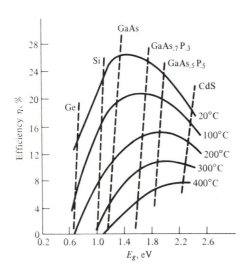

FIGURE 14-11
Maximum theoretical conversion efficiency vs. energy gap. *(After Wysocki and Rappaport [4].)*

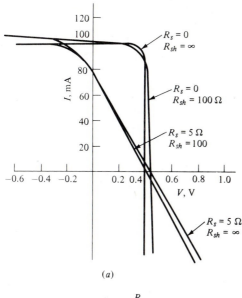

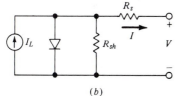

(b)

FIGURE 14-12

(a) Effects of series and shunt resistances on the *I-V* curve and (b) the equivalent circuit including parasitic resistances.

If we assume normal incidence, the reflectance is given by the following law of optics [5]:

$$R = \frac{(n-1)^2 + (\lambda\alpha/4\pi)^2}{(n+1)^2 + (\lambda\alpha/4\pi)^2} \tag{14-19}$$

where $n = n_2/n_1$ and n_1 and n_2 are the refractive indices[1] of the air and semiconductor, respectively. In addition, α is the absorption coefficient of the semiconductor. In a silicon cell, the term $(\lambda\alpha/4\pi)^2$ is negligible, and $n = 3.5$. Therefore, the reflected light amounts to about 30 percent. To reduce the reflectance, we can coat the semiconductor surface with a material having a refractive index between n_1 and n_2. A practical antireflective coating for Si cells is a silicon oxide layer, which has a refractive index of 1.9. The ideal antireflective coating material should have a refractive index of $\sqrt{n_1 n_2}$. When a (100) silicon surface is etched by an anisotropic etchant, a "texturized" surface is obtained as shown in Fig. 14-14.

---

[1] Refractive index is given by the square root of the dielectric constant. Thus, $n = \sqrt{K_s \varepsilon_0/\varepsilon_0} = \sqrt{K_s}$, where K_s is the dielectric constant of the semiconductor.

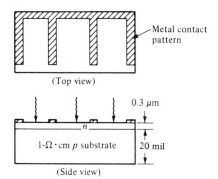

(Top view)

0.3 μm

n

1-Ω·cm *p* substrate

20 mil

(Side view)

FIGURE 14-13
Top and side views of a diffused *n*-on-*p* silicon cell.

The incident light may either be absorbed directly or after reflection, as depicted. Along with antireflection coating, surface-light reflection can be reduced to less than 5 percent.

Cost Considerations

Among the various limiting factors mentioned in the preceding paragraphs, materials synthesis to match the solar spectrum should provide the greatest improvement as far as efficiency is concerned. However, to capture a significant share of the energy market, cost of solar cells will be the dominant factor. Present research effort is leaning heavily toward methods for low-cost material preparation and device fabrication. Recently quoted prices for complete solar-cell panels were below $5 per peak watt. The term *peak watt* specifies a panel that produces 1 W at the peak solar energy of $100 \, \text{mW/cm}^2$. To compete with existing sources of electric energy, the price of a solar cell must be reduced to less than 50 cents per peak watt, with the ultimate goal of 10 cents per peak watt by the year 2000. This figure is unlikely to be attained in crystalline cells, but it could conceivably be realized by using amorphous-silicon thin-film devices to be described in a later section.

The total cost in making a solar cell consists of crystal growth and wafer preparation, device fabrication, and packaging. Packaging cost is least likely to be reduced, although automation could help somewhat. Conventional silicon crystals are grown in cylindrical rods, and they have to be sliced and polished before a

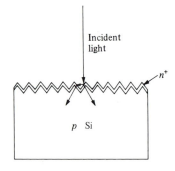

Incident
light

n⁺

p Si

FIGURE 14-14
Surface texturization to reduce light reflection.

junction is diffused. If crystals are grown in the form of a sheet or film, the wafer-preparation steps can be simplified. Newly proposed techniques include the edge-defined film-fed growth (EFG), the dendritic web growth, and edge-stabilized ribbon (ESR) growth. These semiconductor films are mostly polycrystalline, and the resulting solar cells have low efficiency [6]. Although these film-growing methods are suitable for mass production at a very low cost, significant improvement in film quality is necessary before they can be employed. As for the device-fabrication technology, the use of ion implantation or Schottky barriers to replace or supplement the solid-state diffusion could be of practical importance in the near future. The principle of Schottky-barrier cells will be discussed in the next section.

Besides silicon cells, the use of other semiconductors such as copper indium diselenide, cadmium telluride, and zinc phosphide provides another dimension in cost reduction. Solar cells fabricated on these materials are usually of the Schottky-barrier or heterojunction type.

Light Concentration

A different approach to cost cutting is making use of a light concentrator. In this method, a large optical lens is employed to concentrate sunlight to a small area of solar cells. The light intensity can be increased up to a few hundred times. The lens can be made of plastic materials, which are much cheaper than high-quality silicon, thus lowering the overall system cost.

By putting both contacts on one side of the wafer, one can eliminate metal shadowing and take full advantage of the light concentration on a texturized surface. Such a device has been made, and its structure is shown in Fig. 14-15. For

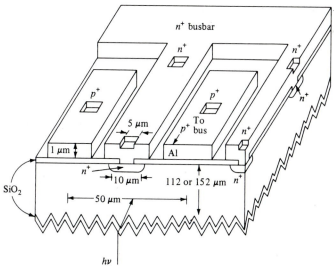

FIGURE 14-15
A cross section of a region in the solar cell near one of the Al busbars. *(After Sinton et al. [7].)*

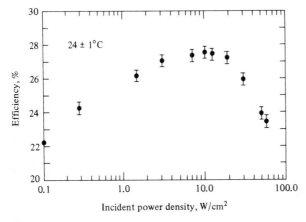

FIGURE 14-16
The efficiency vs. incident power density for a texturized solar cell. *(After Sinton et al. [7].)*

various sunlight-concentration intensities, the cell efficiency is plotted in Fig. 14-16. The efficiency at AM1 is seen to be 22 percent at 0.1 W/cm², and the maximum efficiency reaches 27 percent at 100 suns. This represents the highest efficiency of a silicon cell reported in the literature, and it is even beyond the theoretically predicted value for silicon in Fig. 14-11. The improvement comes mainly from the reduction in surface reflection and metal shadowing.

14-5   SCHOTTKY-BARRIER AND MIS SOLAR CELLS

One of the low-cost methods in fabricating a solar cell is using a Schottky barrier to replace the *pn* junction. The cost reduction is due to fabrication simplicity. In a typical process, a thin semitransparent metal film (5 to 10 nm thick) is evaporated onto the semiconductor, and the top contact with thick metal grid is then deposited. To minimize surface reflection at the metal-air interface, an antireflective coating is added in most devices to arrive at the final structure shown in Fig. 14-17a.

Using the energy-band diagram illustrated in Fig. 14-17b, we find two different modes of converting optical energy into electric energy. If the incoming photon energy is greater than the barrier height but is smaller than the band-gap energy, that is, $E_g > hv > q\phi_b$, electrons in the metal can be excited to overcome the barrier height, resulting in a current flow. This is not a very efficient process, however, because of the requirement of momentum conservation across the M-S barrier. If the photon energy is greater than E_g, electron-hole pairs will be generated in both the depletion and bulk regions of the semiconductor. As a consequence, holes will move toward the metal and electrons toward the semiconductor, leading to a current flow. Since the bulk absorbs most of the photons, the light-generated current is primarily constituted by hole flow from the semiconductor to the metal. This second mode of operation is similar to that in a *pn* junction cell and provides a good conversion efficiency.

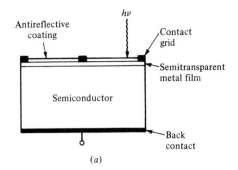

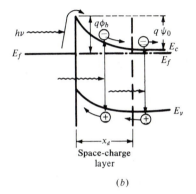

FIGURE 14-17
Schottky-barrier solar cell: (*a*) device structure and (*b*) energy-band diagram.

In comparison with the *pn* junction cell, the Schottky-barrier cell has a lower open-circuit voltage, resulting in a lower efficiency. According to Eq. (14-7), V_{oc} is inversely proportional to I_o. Since I_o in a Schottky-barrier diode is a few orders of magnitude higher than that of the *pn* junction (Sec. 7-2), the open-circuit voltage of the Schottky cell is significantly less than that of the *pn* junction cell. It is noted that I_o represents the majority-carrier thermionic current which opposes the light-generated current. The thermionic current I_o can be reduced if a thin insulating layer is inserted between the metal and the semiconductor, and V_{oc} can thereby be increased [8]. The new structure is shown in Fig. 14-18, along with the energy-band diagram. In this MIS device, current conduction results from tunneling of carriers through the thin insulating layer. An efficiency of 12 percent for Au–Si cells and 15 percent for Au–GaAs cells has been obtained with this structure.

14-6 AMORPHOUS-SILICON SOLAR CELL [9, 10]

A major development in photovoltaics in the last few years is the commercialization of thin-film amorphous-silicon solar cells. In 1985, one-third of all solar cells shipped were amorphous-silicon (*a*-Si) devices, with most applications in consumer electronics such as calculators and cameras. Thin *a*-Si films are deposited by glow-discharge decomposition of silane (SiH_4), which produces *a*-Si with approximately 10 percent

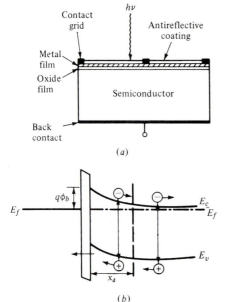

FIGURE 14-18
Metal-insulator-semiconductor solar cell: (a) device structure and (b) energy-band diagram.

of hydrogen. It is found that hydrogen atoms help to saturate dangling bonds so that the hydrogenated a-Si becomes a better material for electronic applications.

In a crystalline silicon, electrons and holes move freely in the conduction and valence bands, respectively. The energy states are considered to be "extended" or nonlocalized. In an amorphous silicon, inherent atomic disorder increases carrier scattering so that the mean free path becomes quite short, in the order of atomic dimension. Consequently, a carrier may be confined or "localized" as if it were attached to the host atom. Since the mean free path depends on the carrier's energy, one may define a *critical energy* at which point the mean free path is equal to the spacing between neighboring atoms. The critical energy becomes the boundary between localized and nonlocalized or extended states. In Fig. 14-19, E_c and E_v represent the critical energy for the conduction band and the valence band, respectively. In contrast with crystalline silicon, the energy-band diagram contains a large number of states between E_c and E_v. These are the localized states where carrier transport may take place by means of "hopping." Transport by the hopping mechanism is very slow with a typical mobility between 10^{-6} and 10^{-3} cm^2/V · s. On the other hand, the mobility in the extended states above E_c and below E_v is between 1 and 10 cm^2/V · s. Consequently, conduction through the band-gap states is insignificant because of the low mobility. The energy gap is called the *mobility gap*, and E_c and E_v are known as the *mobility edges*. The band gap can be controlled by impurity concentration or by alloying selected elements. Adding germanium or tin narrows the band gap, while introducing nitrogen or carbon widens it. At the present, hydrogenated silicon (a-SiH$_x$) with a 1.7-eV band gap produces the highest efficiency.

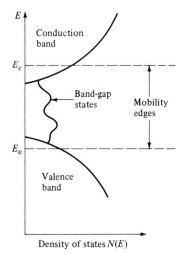

FIGURE 14-19

Energy-band diagram of amorphous silicon.

As shown in Fig. 14-3, the absorption coefficient of a-SiH$_x$ is much higher than the crystalline silicon at $hv = 1.7$ eV. Therefore, most of the sunlight is absorbed within 1 μm of the surface so that a thin a-Si layer is sufficient to convert the solar energy into electric current. Thus, a very little amount of silicon is needed to make a cell. Another advantage of a-SiH$_x$ is that a large area of film can be deposited; for example, a substrate of 40 cm wide and a few meters long is now available. In contrast, the minimum thickness of crystalline silicon cells is 50 μm, and the cell area is limited to a 5-in diameter at the present.

The most widely used structure for amorphous solar cells is the *pin* device shown in Fig. 14-20a. The a-SiH$_x$ is deposited on a tin-oxide coated-glass substrate. The p^+ and n^+ layers are about 10 nm, and the intrinsic layer is between 500 nm and 1 μm. Tin oxide or indium tin oxide is a transparent conducting film used as the front contact. Aluminum or silver is used as the back contact as well as a light reflector to trap the light. The energy-band diagram of the *pin* diode is shown in Fig. 14-20b, where we assume p^+ and n^+ are heavily doped regions so that the Fermi level is pinned at E_c and E_v at the two ends. This results in a high electric field inside the i layer. Carriers generated in the i region are separated by the electric field to produce a photo current (Fig. 14-20c). At the time of this writing, the maximum achievable open-circuit voltage is 1 V and the short-circuit current is 18 mA/cm². With a fill factor of 0.8, the best efficiency is about 11 percent. Most commercially available cells have an efficiency of 6 percent. In comparison, the corresponding values for crystalline silicon cells are 22 and 12 percent.

14-7 LIGHT-WAVE COMMUNICATION [11]

In the past decade, a fundamental change has taken place in communication systems as we observe the startling success of digital-information transmission via optical fibers. The basic fiber-optics system has a light source and a detector

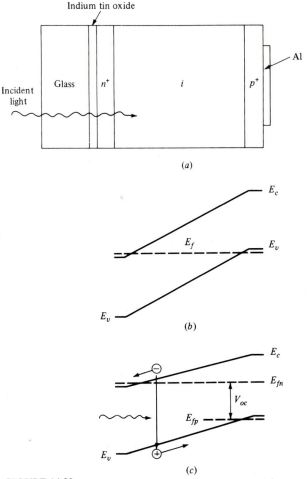

Indium tin oxide

Incident light

Glass

n^+

i

p^+

Al

(a)

E_c

E_f

E_v

E_v

(b)

E_c

E_{fn}

V_{oc}

E_{fp}

E_v

(c)

FIGURE 14-20
An amorphous-silicon *pin* solar cell: (a) structure, (b) energy-band diagram at equilibrium, and (c) energy-band diagram under illumination.

connected by an optical-fiber link, as shown in Fig. 14-21. It promises to have almost unlimited capacity with high reliablity, immunity to electric interference, and low costs. The characteristics of the best-quality fibers are shown in Fig. 14-22, where the attenuation and dispersion are plotted as a function of the wavelength. In curve (b), there are two minima located at 1.2 and 1.55 μm. The corresponding attenuations are 0.65 and 0.35 dB/km, which represent an amplitude reduction of 7 and 4 percent, respectively. Thus, attenuation means that the fiber transparency is less than 100 percent. The chromatic dispersion describes the phenomenon that different wavelengths of light travel through the fiber at slightly different velocities. As a result, a time-delay difference of as much as 100 ns may exist for a wavelength difference of 1 nm after traversing 1 km of fiber at a wavelength of 0.8 μm in (a). Thus, an input optical pulse will be seen at the output as distorted and spread

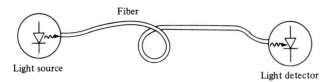

FIGURE 14-21
A fiber-optics link between a transmitter and a receiver.

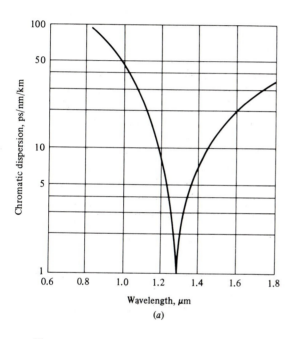

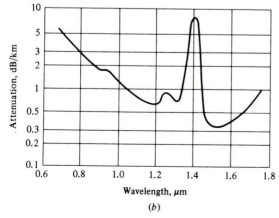

FIGURE 14-22
(*a*) Chromatic dispersion varies with wavelength, reaching zero near 1.3 μm. (*b*) Attenuation varies with wavelength and is least at 1.5 to 1.6 μm.

out. The chromatic dispersion at 1.3 μm is zero, so an optical pulse can be reproduced faithfully at this wavelength. Alternatively, the effect of dispersion becomes unimportant if the light source is a pure single-frequency device. In any case, minimum attenuation and dispersion are desirable for practical design of optical communication systems.

Semiconductor light sources will be described in the next chapter. We shall take up the subject of the photodetector here. There are three types of devices that are commonly employed for optical detection: the photoconductor, the *pin* photodiode, and the avalanche photodiode. The basic operation principles of these detectors are presented in the following sections.

14-8 PHOTOCONDUCTORS

The simplest device structure is a *photoconductor* which consists of a slab of semiconductor with two ohmic contacts, as shown in Fig. 14-23. When the energy of the incident photons is greater than the forbidden gap of the semiconductor, electron-hole pairs are generated so that the conductivity is increased. Making use of Eq. (2-34), the conductivity produced by the light is

$$\sigma = q(\mu_n + \mu_p)\Delta p \tag{14-20}$$

where $\Delta n = \Delta p$ is used. Let the incident optical power be P_L and the photon energy be hv. The carrier generation rate per unit volume is then given by

$$G_L = \frac{P_L}{hv}\frac{1}{v_o}\eta \tag{14-21}$$

where η is the quantum efficiency defined as the number of electrons generated per photon, P_L/hv represents the incoming photons per second, and the total volume v_o is wdl. The steady-state generation rate is related to the excess carrier density by

$$G_L = \frac{\Delta p}{\tau_p} \tag{14-22}$$

where Eq. (2-32) has been used for $t = 0$. Assuming uniform generation throughout the sample, the photocurrent is given by

$$I_{ph} = q\Delta p(\mu_n + \mu_p)\mathscr{E}wd \tag{14-23}$$

where wd is the cross-sectional area. By eliminating Δp in Eqs. (14-22) and (14-23), we obtain

$$I_{ph} = qG_L(\mu_n + \mu_p)\mathscr{E}\tau_p wd \tag{14-24}$$

Since $\mu_n\mathscr{E}$ is the electron velocity, the time required for the electron to traverse the detector, known as the *transit time*, is

$$t_n = \frac{l}{\mu_n\mathscr{E}} = \frac{l^2}{\mu_n V} \tag{14-25}$$

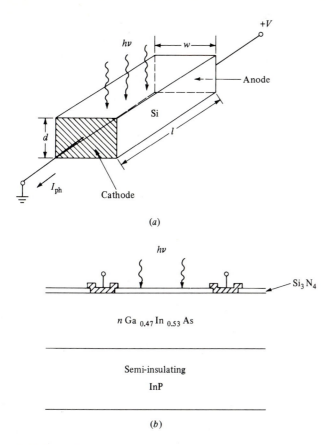

FIGURE 14-23
(a) Schematic diagram of a silicon photoconductor. (b) Schematic cross-sectional view of a GaInAs photoconductor for $\lambda = 1.3\,\mu$m. *(After Chen et al. [12].)*

where V is the applied voltage and $\mathscr{E} = V/l$ has been used. We have

$$I_{\mathrm{ph}} = qG_L \left(\frac{\tau_p}{t_n}\right)\left(1 + \frac{\mu_p}{\mu_n}\right)v_o \qquad (14\text{-}26)$$

The gain of the photoconductor, Γ, is defined as the ratio of carriers collected by the contacts and the carrier generated per unit time:

$$\Gamma = \frac{I_{\mathrm{ph}}}{qG_L v_o} = \frac{\tau_p}{t_n}\left(1 + \frac{\mu_p}{\mu_n}\right)$$

$$= \frac{\tau_p(\mu_n + \mu_p)V}{l^2} \qquad (14\text{-}27)$$

For a large carrier lifetime and small device length, a gain of 100 or more is realizable.

Example. A silicon photoconductor has the following parameters: $l = 100\,\mu m$, $w = 1$ mm, $d = 10\,\mu m$, and $\tau_p = 10^{-4}$ s. The incident photons have an energy of 1.2 eV, and the optical power is 1 μW. Find the photocurrent and gain if the quantum efficiency is 0.8 and the applied voltage is 10 V.

Solution. The generaton rate is

$$G_L = \frac{10^{-6}}{1.2 \times 1.6 \times 10^{-19}} \frac{0.8}{10^{-6}} = 4.3 \times 10^{18}\ \text{cm}^{-3} \cdot \text{s}^{-1}$$

The photocurrent is

$$I_{ph} = 1.6 \times 10^{-19} \times 4.3 \times 10^{18}\ (1350 + 450) \times 10^3 \times 10^{-4} \times 10^{-1} \times 10^{-3}$$

$$= 12.4\ \text{mA}$$

The gain is

$$\Gamma = \frac{10^{-4}(1350 + 450) \times 10}{10^{-4}} = 1.8 \times 10^4$$

The gain in a simple bar of semiconductor comes as a surprise, and it deserves further explanation. Let us consider what happens physically after the generation of an electron-hole pair. The electron is collected almost immediately by the positively biased anode because of its high drift velocity, leaving behind a slow-moving hole. The hole, being positively charge, attracts another electron from the cathode into the semiconductor which in turn is accelerated toward the anode. The process repeats itself so that electrons continue to be drawn into the conducting layer until the hole is removed by the cathode or by recombination. For this reason, the gain is simply the ratio of the hole lifetime to the electron transit time. At the end of the optical pulse, the photocurrent persists until the disappearance of the hole. Therefore, the actual switching speed is limited by the hole's lifetime. The frequency response or bandwidth is inversely proportional to the carrier lifetime.

From the gain equation, a large lifetime is desirable. But a large τ_p lowers the speed or the bandwidth. Thus, the gain and bandwidth are physically linked and cannot be maximized at the same time. For a given transit time, a large bandwidth is obtainable at the expense of a lower gain, or vice versa. For silicon photoconductors, a gain of 1000 or more is achievable with a time response between a microsecond and a millisecond. On the other hand, InGaAs photoconductors (Fig. 14-23b) have a gain of less than 100 but a speed of a nanosecond. For high-speed fiber-optics applications InGaAs appears to be more attractive.

Another limitation of the photoconductor is the noise derived from the dark current. The dark current results from random motions of carriers in the dark which could overwhelm the optical signal. The magnitude of the noise is directly proportional to temperature and dark conductance. Thus, low-temperature operations with lighter doping are desirable. In general, the overall performance of the photoconductor is inferior to the *pin* or avalanche photodiodes.

14-9 THE *pin* DIODE

The structure of a typical *pin* diode is shown in Fig. 14-24a. It has a thick near-intrinsic layer sandwiched between heavily doped n^+ and p^+ layers. A reverse bias is applied across the device so that the *i* region is completely depleted during its operation. The incoming photon flux is designated by F_{ph}, and RF_{ph} represents the reflected light. In the following analysis, we assume $R = 0$ to simplify the expressions. Carriers generated near or inside the depletion layer will be separated by the high electric field, giving rise to a photocurrent (Fig. 14-24b). In most photodiodes, the surface *p* layer is very thin so that carriers generated there are negligible. The photocurrent derived from electron-hole generation inside the depletion layer can easily be found as

$$J_1 = q \int_0^W G(x)\, dx = q \int_0^W F_{ph} \alpha e^{-\alpha x}\, dx = q F_{ph}(1 - e^{-\alpha W}) \qquad (14\text{-}28)$$

Since recombination within the reverse-biased depletion layer is negligible, essentially all the generated electrons and holes are collected and the efficiency is very close to 100 percent.

The current produced by carrier generation in the n^+ layer can be derived (Prob. 14-13):

$$J_2 = q F_{ph} \frac{\alpha L_p}{1 + \alpha L_p} e^{-\alpha W} \qquad (14\text{-}29)$$

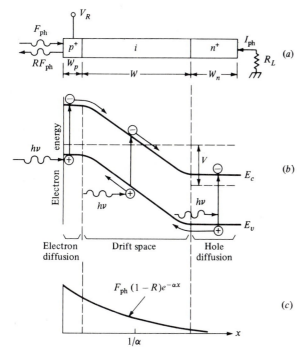

FIGURE 14-24
Operation of photodiode. (*a*) Cross-sectional view of *pin* diode. (*b*) Energy-band diagram under reverse bias. (*c*) Carrier generation characteristics. (*After Melchior* [13].)

Within the neutral n region, some of the excess carriers recombine so that the collection efficiency is less than 100%.

The total photocurrent is, therefore,

$$J_{ph} = J_1 + J_2 = qF_{ph}\left(1 - \frac{e^{-\alpha W}}{1 + \alpha L_p}\right) \tag{14-30}$$

The quantum efficiency is defined as

$$\eta = \frac{J_{ph}/q}{F_{ph}} = 1 - \frac{e^{-\alpha W}}{1 + \alpha L_p} \tag{14-31}$$

where $F_{ph} = P_L/hv$ and P_L is the optical power after the surface reflection is taken into account. A figure of merit for a photodiode is known as the *responsivity*, which is defined by

$$\mathscr{R} = \frac{J_{ph}}{P_L} = \frac{q\eta}{hv} = \frac{\eta\lambda}{1.24} \quad \text{A/W} \tag{14-32}$$

where λ is expressed in micrometers. A higher responsivity is preferred since it represents a higher photocurrent for a given incident light power.

The speed of a *pin* diode is limited by carrier transport. Since carriers drift at the saturation velocity across the depletion layer under a high electric field, the slower process is caused by carrier diffusion in the n region. For this reason, it is preferred to have all carriers generated inside the intrinsic region. The selection of the width of the i layer is constrained by conflicting requirements. A wide i layer increases the transit time which slows down the switching speed. But an excessively thin i layer has a large junction capacitance, giving rise to the resistance-capacitance time-constant limitation. A practical compromise is to have a transit time equal to one-half the modulation period T of the optical signal. Thus,

$$\tau_t = \frac{T}{2} = \frac{1}{2f} \tag{14-33}$$

where f is the modulation frequency.

Example. The intrinsic-layer thickness of a silicon *pin* diode is 10 μm. Calculate the transit time and the ideal modulation frequency.

Solution. In silicon, the saturation velocity is 10^7 cm/s. The transit time is

$$\tau_t = \frac{10 \times 10^{-4}}{10^7} = 100 \text{ ps}$$

The ideal modulation frequency has a period of $2\tau_t$. Thus,

$$f = \frac{1}{2\tau_t} = \frac{1}{200 \times 10^{-12}} = 5 \text{ GHz}$$

In a *pin* photodiode, there is no gain and the maximum efficiency is unity. Although the intrinsic speed of the device is specified by the transit time, the actual bandwidth may be limited by the extrinsic resistance-capacitance time constant. In practice, a bandwidth of 10 GHz is obtainable. Frequently, the choice of a photodetector depends on its noise performance. A *pin* photodiode has much lower dark current (i.e., current flow under bias but without illumination) so that the noise is substantially lower than that of a photoconductor. For this reason, the *pin* diode is the popular photodetector in practical fiber-optic telecommunication systems.

14-10 THE AVALANCHE PHOTODIODE (APD)

An avalanche photodiode is similar to the *pin* diode except that the bias voltage is sufficiently large to cause avalanche multiplication of carriers. The quantum efficiency and speed considerations are the same as the *pin* photodiode, but the APD has a current gain introduced by impact ionization and carrier multiplication. The multiplication factor is given by the empirical relation

$$M_{ph} = \left[1 - \left(\frac{V_R - IR}{V_B} \right)^n \right]^{-1} \qquad (14\text{-}34)$$

where the ohmic drop IR is included and V_B is the breakdown voltage. The exponent n is a material constant; for example, n is 4 for nSi and 6 for pSi. The increase in gain makes it possible to achieve a gain-bandwidth product of 100 GHz or more. However, the gain improvement is not without its disadvantage. Since the current multiplication comes from impact ionization which is random, it gives rise to current fluctuation and noise. It is found that, for materials where the impact ionization is initiated by both electrons and holes, the secondary carriers increase the randomness of carrier generation and thus the noise level. Therefore, the best material for APD should have a high impact-ionization rate for one type of carriers only. In silicon, the impact ionization by electrons is about 50 times greater than that of holes, and it results in excellent signal-to-noise ratios in silicon avalanche

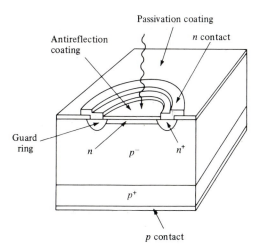

FIGURE 14-25
Planar silicon avalanche photodiode.

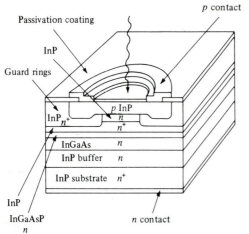

Passivation coating

InP

Guard rings

InP n^+

InP

InGaAsP
n

p contact

p InP
n
n^+

InGaAs n

InP buffer n

InP substrate n^+

n contact

FIGURE 14-26
Light absorbed in InGaAs and multiplication
takes place at the upper InP *pn* junction.

photodiodes. A typical APD structure is shown in Fig. 14-25. Note that a guard ring is used to eliminate the surface leakage and dark current.

A new structure for long-wavelength applications is known as the SAM-APD (separate absorption-and-multiplication region APD). In materials with a low band-gap energy, the leakage dark current is very large so that high-quality APD cannot be fabricated. Let us consider an APD with two regions. The first region is a semiconductor with a narrow band-gap energy where the incident light is absorbed to generate electron-hole pairs. These generated carriers are then introduced to the second region where a reverse-bias junction is made of a large band-gap semiconductor. Within the second region, the carriers experience avalanche multiplication by impact ionization. In the device shown in Fig. 14-26, the incident light passes through the antireflection coating and the high E_g InP *pn* junction and is then absorbed by the InGaAs layer which has a low E_g. A special layer of InGaAsP is inserted between the InP and InGaAs to reduce carrier trapping at the direct InP–InGaAs interface. The InGaAsP layer has a band gap between the InP and InGaAs so that the large band discontinuity or offset is reduced. An alternative is to use a graded band gap at the interface. Notice that a guard ring is used for the InP junction to reduce the electric field at the edges so as to minimize the dark current.

REFERENCES

1. Melchior, H.: Demodulation and Photodetection Techniques, in E. T. Arecchi and E. O. Schulz-Dubois (eds.), "Laser Handbook," vol. 1, North-Holland, A.nsterdam, 1972, p. 725.
2. Rittner, E. S.: An Improved Theory of the Silicon *pn* Junction Solar Cell, *Tech. Dig., Int. Electron Device Meet.*, 69 (Dec. 1976).
3. Wolf, M.: Limitations and Possibilities for Improvements of Photovoltaic Solar Energy Converters, *Proc. IRE*, **48**:1246 (1960).
4. Wysocki, J. J., and P. Rappaport: Effect of Temperature on Photovoltaic Solar Energy Conversion, *J. Appl. Phys.*, **31**:571 (1961).

5. Pankove, J. I.: "Optical Processes in Semiconductors," Prentice-Hall, Englewood Cliffs, N.J., 1971.
6. Card, H. C., and E. S. Yang: Electronic Process at Grain Boundaries in Polycrystalline Semiconductors under Optical Illumination, *IEEE Trans. Electron Devices*, **ED-24**:397 (1977).
7. Sinton, R. A., Y. Kwark, J. Y. Gan, and R. M. Swanson: 27.5% Silicon Concentrator Solar Cells, *IEEE Electron Device Lett.*, **EDL-7**:567 (1986).
8. Card, H. C., and E. S. Yang: MIS-Schottky Theory under Conditions of Optical Carrier Generation in Solar Cells, *Appl. Phys. Lett.*, **29**:51 (1976).
9. Carlson, D. E.: Amorphous Silicon Solar Cells, *IEEE Trans. Electron Devices*, **ED-24**:449 (1977).
10. Madan, A.: Amorphous Silicon: From Promise to Practice, *IEEE Spectrum*, 38 (September 1986).
11. Li, T.: Lightwave Communication, *Phys. Today*, 24 (May 1985).
12. Chen, C. Y., et al.: Modulation-Doped GaInAs/AlInAs Planar Photoconductive Detectors for 1.0–1.55 micron Applications, *Appl. Phys. Lett.*, **43**: 308 (1983).
13. Melchior, H.: Detector for Lightwave Communication, *Phys. Today*, 32 (Nov 1977).
14. Melchior, H., A. R. Hartman, D. P. Schinke, and T. E. Seidel: Planar Epitaxial Silicon Avalanche Photodiode, *Bell Syst. Tech. J.*, **57**:1791 (1978).
15. Campbell, J. C., et al.: High Speed InP/InGaAsP/InGaAs Avalanche Photodiode, *Tech. Dig. Int. Electron Device Meet.*, 464 (1983).

PROBLEMS

14-1. (a) Calculate the maximum wavelength λ of the light source that generates electron-hole pairs in Ge, Si, and GaAs.

(b) What is the photon energy for the light source with wavelengths of 550 and 680 nm?

14-2. A 0.46-μm-thick sample of GaAs is illuminated with a monochromatic light source of $hv = 2$ eV. The absorption coefficient α is 5×10^4 cm^{-1}, and the incident power of the sample is 10 mW.

(a) Calculate the total energy absorbed by the sample in joules per second.

(b) Find the rate of excess thermal energy given up by the electrons to the lattice before recombination in joules per second.

(c) Calculate the number of photons per second given off from recombination.

14-3. Assume that a p^+n diode is uniformly illuminated by a light source to produce an electron-hole generation rate of G_L. Solve the diffusion equation in the n side of the diode to show that

$$\Delta p_n = \left[p_{no}(e^{V/\phi_T} - 1) - G_L \frac{L_p^2}{D_p} \right] e^{-x/L_p} + \frac{G_L L_p^2}{D_p}$$

14-4. Use the result of Prob. 14-3 to derive Eq. (14-6).

14-5. (a) Carry out the derivation of Eq. (14-9).

(b) Assume that the dark current is 1.5 nA and the light-generated short-circuit current is 100 mA. Plot the I-V curve and find the load resistance graphically for maximum output power. What is the fill factor?

14-6. (a) Show that the current I_o in Eq. (14-5) for an n^+p cell is given by

$$I_o = \frac{qAn_{po}D_n}{L_n} \frac{S \cosh(W_p/L_n) + (D_n/L_n) \sinh(W_p/L_n)}{(D_n/L_n) \cosh(W_p/L_n) + S \sinh(W_p/L_n)}$$

where S is the surface-recombination velocity at the ohmic contact, W_p is the width of the p region, and other symbols designate minority-carrier parameters.

(b) Show that

$$
I_o = \begin{cases} \dfrac{qAn_{po}D_n}{L_n} \tanh \dfrac{W_p}{L_n} & \text{for } S \ll \dfrac{D_n}{L_n} \\[3ex] \dfrac{qAn_{po}D_n}{L_n} \coth \dfrac{W_p}{L_n} & \text{for } S \gg \dfrac{D_n}{L_n} \end{cases}
$$

Which is the preferred condition for solar-cell application?

14-7. Assuming $J_L = 40\,\text{mA/cm}^2$, plot the open-circuit voltage as a function of acceptor concentration in an n^+p GaAs cell with $J_o = I_o/A$ given by Prob. 14-6b for $S \ll D_n/L_n$ and $W_p = L_n = 5\,\mu\text{m}$.

14-8. The number of photons per square centimeter per second averaged over a small bandwidth $\Delta\lambda$ that has entered silicon is $Q(\lambda)$.

(a) Derive an expression to represent the light-generated current $\Delta J_L(\lambda)$ at wavelength λ as a function of the reflection coefficient at the back contact, the total cell thickness, and the absorption coefficient (neglecting the front reflection and recombination).

(b) Estimate the light-generated current at $\lambda = 900\,\text{nm}$. Assume that $Q(\lambda)$ is equal to 50 percent of the solar spectrum for a bandwidth of $\pm 50\,\text{nm}$. The average absorption coefficient is $500\,\text{cm}^{-1}$, and the reflection coefficient at the back contact is 0.8. The cell thickness is $10\,\mu\text{m}$.

14-9. The maximum theoretical conversion efficiency shown in Fig. 14-11 is based on a flat solar-cell surface. If the surface is texturized so that only 2 percent of the light is reflected, what will be the efficiency for Ge, Si, and GaAs cells at 20°C?

14-10. The absorption coefficient in a-SiH$_x$ is about $10^4\,\text{cm}^{-1}$ at 1.7 eV and $10^5\,\text{cm}^{-1}$ at 2 eV. What is the required cell thickness to capture 85 percent of the photons at these energies?

14-11. Derive an expression for the short-circuit current as a function of the i-layer thickness in a pin amorphous cell.

14-12 An InGaAs photoconductor has the following parameters: $\tau_p = 10^{-7}\,\text{s}$, $\mu_n = 10{,}000\,\text{cm}^2/\text{V} \cdot \text{s}$, $\mu_p = 200\,\text{cm}^2/\text{V} \cdot \text{s}$, and $E_g = 0.96\,\text{eV}$. Design a photoconductive structure to have a gain of 10^4. What is the speed of the device?

14-13. Derive the carrier distribution and photocurrent in the n layer of a pin diode. Show the result in the form of Eq. (14-29).

14-14. For a silicon pin diode with an i layer of 1, 10, or $100\,\mu\text{m}$, calculate the transit time and RC time-constant limit if $N_d = 10^{14}\,\text{cm}^{-3}$ and $A = 10$ square mils in a 50-Ω system. Let $V_R = 10\,\text{V}$.

14-15. Sketch the energy-band diagram and indicate carrier generation and multiplication regions in your diagram for the devices in Figs. 14-25 and 14-26.

CHAPTER
15

LIGHT-EMITTING AND LASER DIODES

In the last chapter, the process of converting optical energy into electric energy was described. It was shown that photons can be used to generate electron-hole pairs, resulting in an electric current. By applying a current through a *pn* junction, photons can be produced, giving rise to light emission. This inverse mechanism is called *electroluminescence*. The device is known as the *light-emitting diode* (LED). A typical emission spectrum of the LED has a bandwidth of wavelengths between 30 to 40 nm. If the LED structure is modified and the operating condition is changed, the device can function in a different mode where the emission bandwidth is less than 0.1 nm. This new device is known as the *laser diode*. In this chapter, the basic physics of the LED and the laser are presented. In addition, material consideration and applications are introduced. The coverage is mainly qualitative and elementary, and it may be used in an introductory course.

15-1 GENERATION OF LIGHT WITH A *pn* JUNCTION: THE LED

A light-emitting diode (LED) is a solid-state *pn* junction device that emits light upon the application of a forward-biasing current. It is different from the incandescent light bulb, in which light is generated by heating a filament to a very high temperature. An LED is a cold lamp which converts electric energy directly into optical energy without the intermediate step of thermal conversion. This luminescent mechanism, called *electroluminescence*, has an emission wavelength in the visible or infrared region. In the literature, these diodes are called *electroluminescent diodes*, and the acronym LED is reserved for devices emitting in the visible region only. In this chapter, however, the word LED is used to cover both visible and infrared emitters. LEDs are operated at low voltages and currents, typically 1.5 V and 10 mA, respectively; they can be made very small, so it is reasonable to consider them as point sources of light. These characteristics make LEDs attractive for optical displays. In addition, the emission spectrum of LEDs is relatively narrow, and they can be switched on and off in the order of 10 ns. These properties are suitable for applications in optical data communication.

When a forward bias is applied across a *pn* junction, carriers are injected across the junction to establish excess carriers above their thermal equilibrium values. The excess carriers recombine, and energy is released in the form of heat (phonons) or light (photons). In photon emission we derive optical energy from the biasing electric energy. The *pn* junction electroluminescence is illustrated graphically in Fig. 15-1. The injected electrons in the *p* side make a downward transition from the conduction band to recombine with holes in the valence band, emitting photons with an energy E_g. The corresponding emission wavelength is

$$\lambda = \frac{hc}{E_g} = \frac{1.24}{E_g} \, \mu m \tag{15-1}$$

where E_g is in electronvolts. For example, the emission wavelength of GaAs at room temperature is 890 nm, corresponding to a band-gap energy of 1.4 eV. Light emission in the *n* side of the junction follows the same pattern except that holes are now the excess carriers. In the following sections, the injection and recombination mechanisms are explained in more detail.

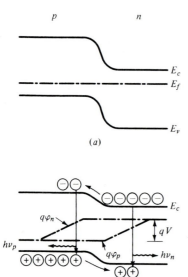

(a)

(b)

FIGURE 15-1
An electroluminescent *pn* junction under (*a*) zero bias and (*b*) forward-bias voltage *V*.

15-2 MINORITY-CARRIER INJECTION AND INJECTION EFFICIENCY

Depending on both the impurity profile and the external applied voltage, we can identify four current components in an LED under the forward-bias condition: (1) the electron diffusion current, (2) the hole diffusion current, (3) the space-charge-layer recombination current, and (4) the tunneling current. The tunneling current is important only in a heavily doped *pn* junction at a small forward bias; thus its effect is negligible in most LEDs at light-emitting current level. The other current components have been described in Chap. 4 and are rewritten in the following:

$$I_n = \frac{qD_n n_i^2}{L_n N_a}(e^{qV/kT} - 1) \tag{15-2}$$

$$I_p = \frac{qD_p n_i^2}{L_p N_d}(e^{qV/kT} - 1) \tag{15-3}$$

$$I_{\text{rec}} = \frac{qn_i x_d}{2\tau}e^{qV/2kT} \tag{15-4}$$

Recombination inside the space-charge layer is effective if trap levels exist near the center of the forbidden gap. This process is generally nonradiative, and the component I_{rec} does not contribute to the emission of light. Furthermore, luminescence originates from the electron diffusion current in the *p* side of the junction in most practical LEDs for reasons beyond the scope of the present text. Consequently, we can define the current-injection efficiency as

$$\gamma = \frac{I_n}{I_n + I_p + I_{\text{rec}}} \tag{15-5}$$

Frequently, the hole diffusion current is negligible because of the high electron-to-hole mobility ratio; e.g., in GaAs we have $\mu_n/\mu_p = 30$, and the foregoing equation can be simplified.

15-3 INTERNAL QUANTUM EFFICIENCY

The injection efficiency indicates the percentage of diode current that can produce radiative recombination in the *p* side of the junction. However, not all the electrons that reach the *p* side recombine radiatively. Electrons that survive the space-charge layer may recombine radiatively or nonradiatively, depending on the recombination paths in the *p* side of the junction, as shown in Fig. 15-2. The simplest recombination process is the band-to-band recombination R_1, in which a free electron and a free hole recombine directly. The second process R_2 involves a shallow impurity state, where an electron recombines with a hole trapped on a shallow acceptor state. Alternatively, the process may involve a shallow acceptor and a shallow donor state. The photon energy generated in this process is smaller than E_g. In the third

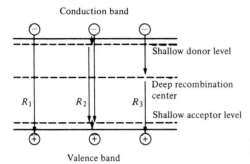

Conduction band

Shallow donor level

Deep recombination center

Shallow acceptor level

R_1 R_2 R_3

Valence band

FIGURE 15-2
Three possible recombination paths.

possibility R_3 via deep impurity states, photons may not be generated at all; even if photons are generated, their energy is much smaller than E_g so that it makes the emission intensity at E_g appear lower.

To simplify the complex picture, let us take an elementary case where a nonradiative recombination process via an intermediate state R_3 is competing with the band-to-band radiative recombination R_1. Following the notation in Appendix B, we define the *radiative* and *nonradiative recombination rates*, respectively, as (for the p-type region)

$$U_r \equiv \frac{\Delta n}{\tau_r} \tag{15-6}$$

$$U_{nr} \equiv \frac{\Delta n}{\tau_{nr}} \tag{15-7}$$

where Δn = excess-electron density
  τ_r = radiative recombination lifetime
  τ_{nr} = nonradiative lifetime

The *radiative efficiency* is defined as the percentage of electrons that recombine radiatively,

$$\eta \equiv \frac{U_r}{U_r + U_{nr}} = \frac{1}{1 + \tau_r/\tau_{nr}} = \frac{\tau}{\tau_r} \tag{15-8}$$

where τ is the effective lifetime, given by

$$\frac{1}{\tau} = \frac{1}{\tau_r} + \frac{1}{\tau_{nr}} \tag{15-9}$$

Using Eq. (2-26) along with Eq. (2-25) as the radiative lifetime and Eq. (2-29) as the nonradiative lifetime, we can write radiative efficiency in the p side of the junction as

$$\eta = \frac{1}{1 + c_n N_t/BN_a} \tag{15-10}$$

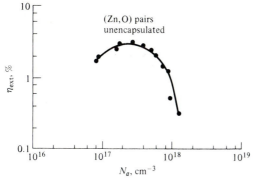

FIGURE 15-3
External quantum efficiency of red GaP diodes vs. net acceptor doping. *(After Bhargava [1].)*

where the approximation $n_o + p_o = N_a + N_a/n_i^2 \approx N_a$ has been employed. In some practical cases, e.g., in the red GaP LED, the recombination processes involve trapping effects with R_2 and R_3 as the competing mechanism. The radiative efficiency can be derived from the detailed balance of recombination and generation rates. The result is (Prob. 15-2)

$$\eta = \left[1 + \frac{N_t c_{p3} p}{N_a c_{n2} n} \exp\left(-\frac{E_t + E_a - 2E_f}{kT} \right) \right]^{-1} \tag{15-11}$$

The overall *internal quantum efficiency* can now be written as

$$\eta_i = \eta\gamma \tag{15-12}$$

The internal efficiency is influenced by the current-injection efficiency and radiative efficiency, and these two parameters depend strongly on the doping concentrations. In general, we can increase the doping concentration N_a in the p side to increase the radiative efficiency, as seen in Eqs. (15-10) and (15-11). A higher N_a also has the benefit of a smaller series resistance, thus reducing the forward-voltage drop and ohmic losses. However, very high concentration is not desirable because it increases crystal imperfections, which lead to an increase of the nonradiative centers N_t. In addition, a high doping in the p side reduces the injection efficiency. The foregoing consideration is confirmed by experimental data given in Fig. 15-3 for a GaP LED, where the externally measured efficiency peaks at $N_a = 2.5 \times 10^{17}$ cm^{-3}.

15-4 EXTERNAL QUANTUM EFFICIENCY

The most important parameter of an LED is the *external quantum efficiency*. It may be significantly smaller than the internal quantum efficiency because of internal absorption and reflection of light. After photons are generated at the *pn* junction, they must pass through the crystal to reach the surface. Some of the emitted photons are reabsorbed by the semiconductor. Furthermore, even after the photons have reached the surface, they may not be able to leave the semiconductor because of the large difference in the refractive indices of the semiconductor and air. According to

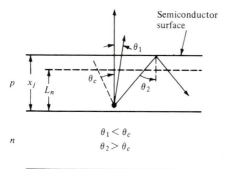

$\theta_1 < \theta_c$
$\theta_2 > \theta_c$

FIGURE 15-4
Internal reflection and critical angle in an LED.

the theory of optics, the *critical angle* θ_c (Fig. 15-4) at which total internal reflection occurs is given by the Fresnel equation

$$\sin \theta_c = \frac{1}{n} = \frac{n_1}{n_2} = \frac{1}{\sqrt{K_s}} \tag{15-13}$$

where n is the refractive index of the semiconductor with the air (n_1) as the external reference. All rays of light striking the surface at angles exceeding θ_c are reflected. Since n ranges between 3.3 to 3.8 for a typical LED material, θ_c is calculated to be between 15 and 18°. For the light striking within the critical angle, the portion that comes out is given approximately by the average transmissivity

$$T = \frac{4n}{(1 + n)^2} \tag{15-14}$$

Therefore, the total light emission within the solid angle θ_c is

$$\bar{T} = T \sin^2 \frac{\theta_c}{2} \tag{15-15}$$

A simple expression relating the external quantum efficiency to the internal quantum efficiency is given by [3]

$$\eta_{\text{ext}} = \frac{\eta_i}{1 + \bar{\alpha} v_o / A \bar{T}} = \frac{\eta_i}{1 + \bar{\alpha} x_j / \bar{T}} \tag{15-16}$$

where $\bar{\alpha}$ = average absorption coefficient
v_o = diode volume
A = emitting area

In an LED, the ratio v_o/A may be taken as the junction depth x_j from the emitting surface. Equation (15-16) indicates that the external quantum efficiency can be increased by reducing $\bar{\alpha}$ or x_j or by increasing $\bar{T}$.

Reducing the junction depth to less than a diffusion length from the surface introduces more minority carriers to the surface. Therefore, the surface recombination centers capture a larger portion of the injected carriers, thus reducing the

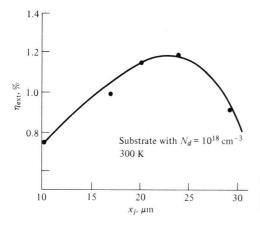

FIGURE 15-5
External efficiency vs. junction depth of GaAs LEDs. $L_p \approx 8\,\mu\text{m}$.

internal quantum efficiency. Experimental results of the junction-depth dependence of η_{ext} in a GaAs LED are shown in Fig. 15-5, where the optimized junction depth is between 15 and 25 μm.

The reduction of α can be achieved by generating luminescence with $h\nu < E_g$ (R_2 in Fig. 15-2), as illustrated in Fig. 15-6. A high efficiency is obtained since the emitted photons have an energy below E_g. Note that the absorption coefficient is very low at the emission peak but absorption is high at E_g. Alternatively, an optical window is used, as shown in Fig. 15-7. In this device, an additional AlGaAs layer is grown on top of the GaAs diode. Since the AlGaAs material has a band gap greater than GaAs, the emitted photons are not absorbed by this added layer. At the same time, the density of recombination centers at the AlGaAs–GaAs interface is significantly lower than that of the GaAs surface without the AlGaAs layer. Therefore, the depth of the junction from the interface can be made very small.

The reduction of the internal reflection can be achieved by using a dome-shaped diode geometry (Fig. 15-8a) so that most of the light emitted at the junction arrives within the critical angle at the semiconductor surface. The disadvantage of

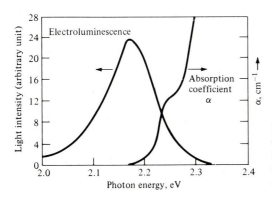

FIGURE 15-6
Comparison of a typical external electro-luminescent spectrum of a green GaP LED and the absorption coefficient of GaP. $E_g = 2.25\,\text{eV}$.

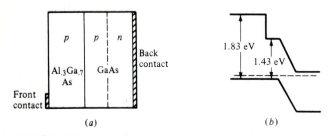

FIGURE 15-7
A GaAs LED with an AlGaAs layer as optical window: (*a*) structure and (*b*) energy-band diagram. $Al_{0.3}Ga_{0.7}As$ represents a ratio of 30% Al and 70% Ga in its composition.

this method is that a large amount of semiconductor material is needed and the machine work is not economical. A more practical technique makes use of an optical medium with a refractive index between that of air and the semiconductor, as shown in Fig. 15-8*b*. Hemispherical domes cast from epoxy or acrylic-polyester resin (with $n = 1.5$) are quite effective, increasing the external efficiency by a factor of 2 to 3.

15-5   EYE SENSITIVITY AND BRIGHTNESS

The response of the human eye, called *luminous efficiency*, is limited to wavelengths between 400 and 700 nm. The *standard-luminous-efficiency curve* is shown in Fig. 15-9. The eye is very sensitive to green or yellow color but is a poor detector in the red or violet region. Because of the large variation of eye sensitivity, the performance of an LED is appraised not only by its external quantum efficiency but also by the relative response of the eye at the wavelength of interest. An emission at 550 nm (2.23 eV) is most desirable as far as the luminous efficiency is concerned. For this reason, we define the brightness of an LED as a measure of the visual impact of the radiation,

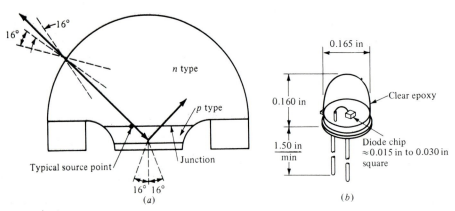

FIGURE 15-8
Dome-shaped LED structures: (*a*) with *n*-type semiconductor dome and (*b*) with clear epoxy dome.

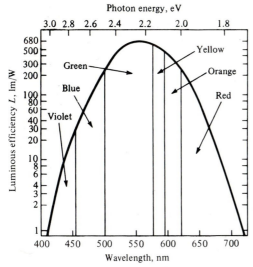

FIGURE 15-9
Luminous efficiency of eye vs. wavelength of incident light. $L_{\text{max}} = 680$ lm/W at 555 nm.

using

$$B = 1150L \frac{J}{\lambda} \frac{A_j}{A_s} \eta_{\text{ext}} \qquad \text{fL} \qquad (15\text{-}17)$$

where λ = emission wavelength, μm
$\quad J$ = current density, A/cm^2
$\quad L$ = luminous efficiency at λ, lm/W
$\quad A_j$ = pn junction area
$\quad A_s$ = observed emitting surface area

The unit of brightness is footlamberts (fL). To facilitate fair comparison of the performance of different types of LEDs, the brightness B is frequently normalized with respect to the current density. A brightness of 1500 fL at 10 A/cm^2 is typical for a commercial LED. By comparison, the brightness of a 40-W incandescent light bulb is about 7000 fL, but it is at a much higher current density.

15-6 LEDs FOR DISPLAY APPLICATIONS

The maximum possible energy for the emitted photons in a semiconductor is determined approximately by the energy gap. For a visible display, the wavelength of the emitted light should be between 0.4 and 0.72 μm. This requirement limits our materials to those having an energy gap between 1.7 and 3.0 eV.

GaP LEDs

With an energy gap of 2.3 eV, gallium-phosphide diodes emit red or green light, depending on the recombination mechanism involved. In general, recombination in an indirect semiconductor such as GaP tends to take place via impurity levels because this facilitates conservation of crystal momentum. We shall describe two types of GaP diodes separately, the red and the green.

RED GaP DIODE. The radiation mechanism in a red diode is through a donor-acceptor impurity pair. When both donors and acceptors are simultaneously present in the semiconductor, the donor and acceptor states are partly occupied. It is therefore possible for an electron in a donor state to make a downward transition, recombining with a hole in an acceptor state. The radiated photon energy is given by

$$hv = E_g - (E_d + E_a) + \frac{q^2}{4\pi K_s \varepsilon_0 r} \qquad (15\text{-}18)$$

where v = emitted frequency
E_d = donor energy
E_a = acceptor energy

The last term in Eq. (15-18) describes the Coulomb interaction energy between donors and acceptors where r is the donor-acceptor separation. In an oxygen-doped GaP diode, an emission near 690 nm is observed. It is believed that the emission is due to a transition from a deep donor caused by oxygen to a shallow acceptor such as zinc. The zinc level is 0.04 eV above the valence-band edge, and the oxygen level is 0.803 eV below the conduction-band edge. Since the donor-acceptor separation varies among different pairs, the emission band is quite broad, having a half-width of 87 nm. Typically, red-emitting diodes have external efficiencies of 2 to 3 percent at current levels of 10 A/cm². The light output does not increase linearly with the current. As shown in Fig. 15-10, the drop of efficiency in the red GaP diode at high current is caused by the saturation of impurity centers at high carrier-injection levels. Maximum efficiency in red diodes of 15 percent has been achieved in the laboratory; however, the maximum brightness is low because the emission output of these high-efficiency diodes saturates at a very low current. Presently, the GaP red LED is seldom used.

GREEN GaP DIODE. Green emission has been observed in GaP, and this has been attributed to recombination at a nitrogen atom on a phosphorus site. Because both nitrogen and phosphorus are in the same column of the periodic table, replacement of phosphorus by a nitrogen atom is described as *isoelectronic*. An isoelectronic center is a very localized potential well that can trap an electron, thus becoming charged. The resulting Coulomb field then attracts a hole which pairs with the trapped electron to form an *exciton*, i.e., a hydrogenlike bound electron-hole pair. The annihilation of this exciton by radiative recombination gives rise to green light with $\lambda = 570$ nm at room temperature. Because of competing nonradiative recombination and thermal activation processes, the internal efficiency of green emission in GaP is less than 1 percent. However, the self-absorption is very small since the emission is below the band-gap energy, and

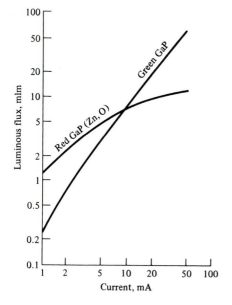

FIGURE 15-10
Light output vs. current in GaP LEDs showing saturation effect in the red diode. (*After Bhargava [1].*)

an external efficiency of 0.7 per cent has been realized. The green emission is nearly at the peak of the eye-sensitivity curve shown in Fig. 15-9. For this reason, the green GaP diode provides high brightness despite its low external quantum efficiency. In addition, its light output does not saturate with current, as seen in Fig. 15-10, where the red GaP diode is included for comparison. The current-voltage characteristic of a typical green diode is shown in Fig. 15-11.

$GaAs_{1-x}P_x$ LEDs

Material synthesis is a practical approach to obtaining a semiconductor with a particular desired energy-band gap. Such synthesis is desirable because efficient emission in direct-band-gap semiconductors such as GaAs is not visible to the

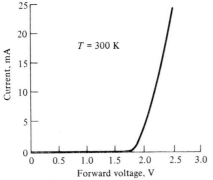

FIGURE 15-11
Current-voltage characteristic of a green GaP LED.

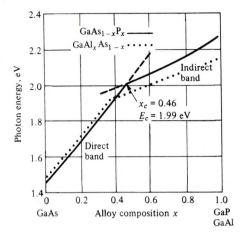

FIGURE 15-12
Band-gap energy E_g of GaAs$_{1-x}$P$_x$ as a function of alloy composition. *(After Casey and Trumbore [2].)*

human observer and because GaP was a material difficult to handle in its early development. When GaAs is mixed with GaP, a *ternary alloy* can be formed having an energy gap between that of GaAs and GaP. The mixed alloy is called GaAs$_{1-x}$P$_x$, where x specifies the alloy composition of phosphorus. Using x as a parameter, we obtain a different energy-band gap, as shown in Fig. 15-12. It is noted that in addition to an increased E_g, the crystal changes from direct gap to indirect gap at $x = 0.46$. Since the radiative recombination is more efficient in direct-gap materials, we expect a decreasing external quantum efficiency for increasing x and E_g, as shown in Fig. 15-13 for LEDs with and without nitrogen doping. Nitrogen doping is found to enhance the radiative-recombination process and thus improve the quantum efficiency. It is also shown that the epoxy encapsulation yields an improvement of $2\frac{1}{2}$ times. When the eye-sensitivity response is taken into account,

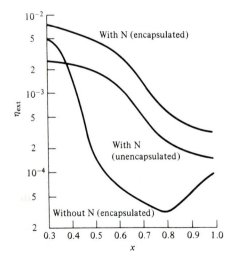

FIGURE 15-13
External quantum efficiency as a function of alloy composition for GaAs$_{1-x}$P$_x$ diodes with and without nitrogen doping. *(After Bhargava [1].)*

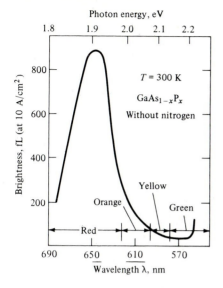

FIGURE 15-14
Brightness of GaAsP diodes vs. emission peak.
(After Bhargava [1].)

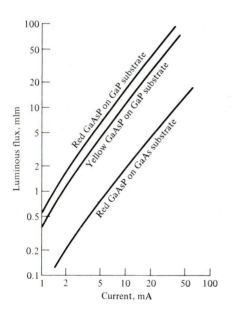

FIGURE 15-15
Effect of substrate material on GaAsP LED
performance. *(After Bhargava [1].)*

we find that the brightness of $GaAs_{1-x}P_x$ diodes peaks approximately at $x = 0.4$ and $E_g = 1.9$ eV (Fig. 15-14). The corresponding emission wavelength is 650 nm, which is in the red region. The external efficiency is greater than 1 percent for an encapsulated diode with a brightness of $300\,fL/A \cdot cm^2$. By increasing the GaP concentration, we can obtain orange emission, but with reduced efficiency and brightness.

Package physical dimensions

Orientation marks

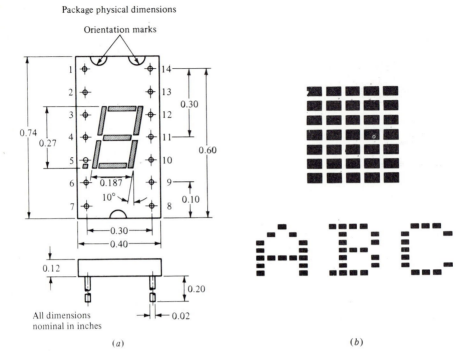

All dimensions
nominal in inches

(a)

(b)

FIGURE 15-16

(a) Seven-segment LED display and (b) alphanumeric 5 × 7 array. *(After Casey and Trumbore [2].)*

In most earlier commercial diodes, the GaAsP layer was grown on a GaAs substrate. Since GaAs has a lower band-gap energy, it absorbs the light emitted from the GaAsP junction and thus reduces the light output. Absorption by the substrate can be eliminated if the band-gap energy of the substrate is larger than the emitted-photon energy. For this reason, recent GaAsP LEDs are fabricated on a GaP substrate. The improved light-current characteristic is seen in Fig. 15-15. The output light power increases without saturation even at a high current similar to that of the GaAs diodes.

By using planar technology $GaAs_{0.6}P_{0.4}$ monolithic arrays can be fabricated for numeric and alphanumeric displays, as shown in Fig. 15-16. Individual segments are clearly visible because of the strong self-absorption, so that light does not penetrate deep into the inactive regions. With the help of a lens, $\frac{1}{2}$-in numerals can be obtained. Most commercially available red LEDs are GaAsP devices because of lower cost and ease of fabrication.

$GaAl_xAs_{1-x}$ LEDs

Making use of a heterojunction, a much brighter LED has been fabricated by growing thin $GaAl_xAs_{1-x}$ layers on a GaAs substrate. The device structure is shown in Fig. 15-17 along with the energy-band diagram. It should be pointed

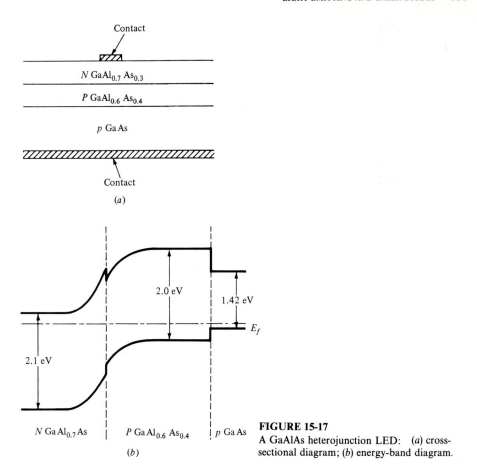

Contact

$N\,GaAl_{0.7}\,As_{0.3}$

$P\,GaAl_{0.6}\,As_{0.4}$

$p\,GaAs$

Contact

(a)

2.0 eV

1.42 eV

E_f

2.1 eV

$N\,Ga\,Al_{0.7}\,As$ $P\,Ga\,Al_{0.6}\,As_{0.4}$ $p\,Ga\,As$

(b)

FIGURE 15-17
A GaAlAs heterojunction LED: (a) cross-sectional diagram; (b) energy-band diagram.

out that the band-gap energy is a function of the aluminum concentration (Fig. 15-12). The different compositions across the heterojunction give rise to a band discontinuity or offset as explained in Sec. 7-11. Since the structure favors electron injection from the wider-band-gap N layer to the narrower P layer, the injection efficiency is unity for all practical purpose. The injected electrons in the P layer recombine radiatively with holes to produce an emission of 650 nm, corresponding to the band-gap energy E_{gP}. Since $E_{gP} < E_{gN}$, the photons can escape through the N layer with little absorption. In other words the N layer constitutes an excellent optical window, and the external quantum efficiency is enhanced. With improved material quality by liquid-phase or molecular-beam epitaxy, the $GaAl_xAs_{1-x}$ diode has achieved a brightness 10 times that of the typical $GaAs_{1-x}P_x$ LED. The efficiency can be improved further if the substrate GaAs is removed. This step eliminates the light absorption in the substrate but it requires complicated technology. With the latest development in the field, it will not be surprising if new applications, such as customized taillight designs in automobiles, should appear in the near future.

15-7 INFRARED LEDs FOR OPTICAL COMMUNICATION

For optical communication applications, the light source does not have to be visible. In fact, the lowest attenuation and dispersion in Fig. 14-22 are situated at wavelengths between 1.2 and 1.6 μm. In this range of wavelengths, the quaternary compound InGaAsP is found to be most suitable. For example, $In_{0.72}Ga_{0.28}As_{0.6}P_{0.4}$ has a wavelength of 1.25 μm, matching one of the minima in the attenuation curve. Heterojunction devices can be realized with InP as the substrate for near-perfect lattice matching.

Gallium arsenide is a direct-gap semiconductor with an energy gap of 1.4 eV at room temperature, which corresponds to an emission wavelength of 890 nm. A typical GaAs LED is made by solid-state impurity diffusion with zinc as the p-type impurity diffused into an n-type substrate doped with tin, tellurium, or silicon. To achieve high efficiency, the concentration of both types of dopants is of the order of 10^{18} cm^{-3}. The external efficiency at room temperature is typically 5 percent. The emission spectra of a GaAs diode are shown in Fig. 15-18. The spectral line width, i.e., the width of the half-power points, is typically less than 30 nm. A GaAs diode can also be fabricated by liquid-phase epitaxy with silicon as both its n and p dopants. If a silicon atom replaces a Ga atom, it provides one additional electron; thus the resulting GaAs is an n type. If a silicon atom replaces an arsenic atom, an electron is missing, and the resulting GaAs is a p type. Because of this special property, silicon in GaAs is called an *amphoteric dopant*. In a Si-doped GaAs diode, the emission peak shifts down to 1.32 eV. The self-absorption becomes much smaller, and an external quantum efficiency as high as 20 percent has been achieved in dome-shaped devices. Since the emission is in the infrared region, GaAs light sources are suitable for

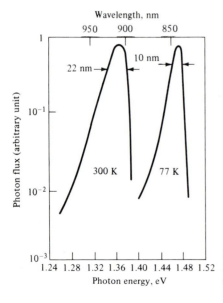

FIGURE 15-18
Experimental emission spectra of GaAs LED at 300 and 77 K. *(After Bergh and Dean [3].)*

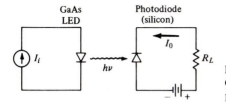

FIGURE 15-19
Optical isolator using a GaAs LED and a silicon photodiode.

applications such as the optical isolator illustrated in Fig. 15-19. The high switching speed, with a recovery time between 2 and 10 ns, makes them ideal for data transmission.

The disadvantages of the GaAs emitter are the emitted wavelength and the associated attenuation and dispersion. However, the GaAs LED can be constructed much easier than its competitor, the injection laser. In addition, the LED is more reliable, and its emitted power has a small temperature dependence. Its ouput changes by less than a factor of 2 if the temperature is raised from 25 to 100°C. As a result, the drive circuitry can be much simpler.

A critical issue of using an LED for fiber optics is the coupling of light from the semiconductor to the fiber. Because of the large refractive index of GaAs relative to air, the external efficiency of the LED can be quite low. A practical approach is to use epoxy resin in encapsulation, as shown in Fig. 15-20. The epoxy has a refractive index between that of the semiconductor and the air, thus enhancing the light coupling. As shown, the diode emits light through its top surface, and the LED is known as a *surface emitter*.

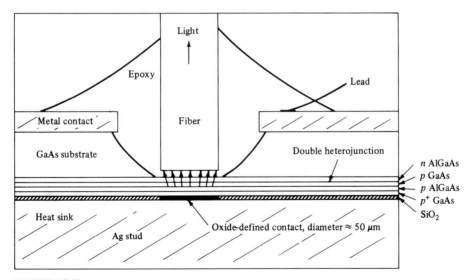

FIGURE 15-20
An etched-well surface-emitting LED designed for fiber optics, in a schematic cross section. (*After Burrus and Miller [4].*)

15-8 THE SEMICONDUCTOR LASER DIODE

The word *laser* stands for *l*ight *a*mplification by *s*timulated *e*mission of *r*adiation, and the laser comes in different sizes, speed, and power. In this section, we present an elementary description of the laser diode, a device that is extremely important in optical communication systems.

When an electron makes a transition from an upper to a lower energy level, it gives up an energy quantum which may be emitted in the form of a photon, as shown in Fig. 15-21a. This phenomenon has been discussed in the LED where light emission is *spontaneous* in that each photon acts independently. Suppose that a monochromatic light source impinges upon the semiconductor and the photon energy is the same as $E_2 - E_1$. The electron in the upper energy level E_2 will now act differently. While it will drop to E_1 and emit a photon as before, its timing is such that it synchronizes with the incoming photons, as shown in Fig. 15-21b. In fact, the presence of the external photons tends to increase the likelihood of the downward transition. For this reason, the electrons are *stimulated* to produce additional photons, resulting in optical gain or amplification. The radiation is considered to be *coherent* since all the emitted photons are *in phase*, just like a perfect line of marching soldiers.

Under thermal equilibrium, the electron distribution follows the Fermi-Dirac or Boltzmann statistics. In either case, the concentration of electrons in E_1 is always higher than that of E_2. The ratio of concentrations is given by

$$\frac{N_2}{N_1} = e^{-(E_2 - E_1)/kT} \tag{15-19}$$

To arrive at the lasing action, there must be more electrons in the upper state than in the lower state. This condition is known as *population inversion*, and it does not come naturally. Population inversion can be achieved by making use of a GaAs *pn* junction operated under forward bias.

A fundamental requirement for population inversion is that the doping levels on both sides of the junction must be degenerate so that the Fermi level lies within the conduction and valence bands. The energy-band diagram of such a junction is shown in Fig. 15-22a. Upon forward bias, electrons are injected from the *n* side into the *p* side, as depicted in Fig. 15-22b. It should be pointed out that the electron concentration is exaggerated to enhance the effect. Near the junction on the *p* side,

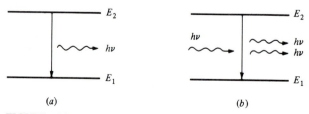

(a)   (b)

FIGURE 15-21
(a) Spontaneous emission and (b) stimulated emission.

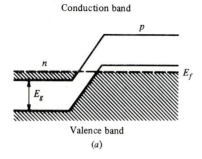

Conduction band

n

p

E_f

E_g

Valence band

(a)

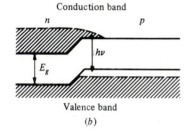

Conduction band

n

p

E_g

hv

Valence band

(b)

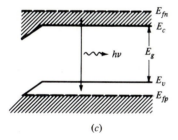

E_{fn}

E_c

hv

E_g

E_v

E_{fp}

(c)

FIGURE 15-22
A *pn* junction with heavy doping: (*a*) no bias, (*b*) forward bias, and (*c*) enlarged region near the depletion layer.

the energy-band diagram may be enlarged, as illustrated in Fig. 15-22c. Heavy injection of electrons has moved the quasi-Fermi level of electron into the conduction band, while the hole quasi-Fermi level remains inside the valence band. The significance of the picture is that many electrons are in the conduction band (the higher energy level) and that the lower energy states between E_v and E_{fp} are empty. This is the necessary condition of population inversion. The downward transition of electrons from occupied conduction-band states to empty valence-band states is accomplished by radiative recombination. From the diagram shown, the emitted photons are between the energies (Fig. 15-22c)

$$E_g \leqslant hv \leqslant E_{fn} - E_{fp} \tag{15-20}$$

To have emission, it is necessary to have $E_{fn} - E_{fp}$ greater than E_g. For this reason, population inversion can be achieved only with heavily doped semiconductors.

15-9 OPTICAL CAVITY AND GAIN

In addition to the population inversion, two other requirements are needed to reach the lasing threshold in a semiconductor diode laser. An optical cavity must be provided to build up the optical wave through positive feedback, and the optical gain within the device must be greater than optical losses. A practical implementation is shown in Fig. 15-23.

When a forward bias is applied across the diode, electrons are injected into the *p* side and are recombined with holes radiatively to emit photons with energy *hv*. To have stimulated emission, the emitted photons must be kept in the junction region to provide stimulation for subsequent emissions. This can be accomplished by using a resonant cavity consisting of two parallel mirrors, known as the *Fabry-Perot resonator*. Within this cavity, the optical wave bounces back and forth, i.e., making many passes through the region with an inverted population of electrons. The resonant cavity is usually established by cleaving the GaAs crystal along the (110) plane (Fig. 15-23). Since two (110) planes are perfectly in parallel, the cleaved surfaces form a pair of ideal mirrors at the two ends of a diode normal to the junction. For resonance, there must be an integral number of half-wavelengths between the end mirrors (Fig. 15-24*a*). Thus,

$$N\left(\frac{\lambda}{2}\right) = L \qquad (15\text{-}21)$$

where N is an integer and L is the mirror separation. There are many combinations of N and λ that satisfy the resonance condition as long as the wavelength is within

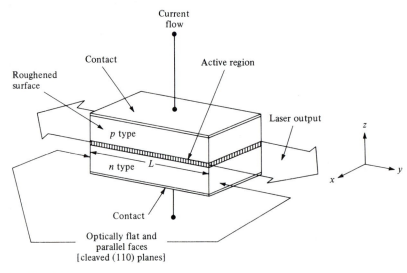

Current flow

Contact

Roughened surface

Active region

p type

n type   *L*

Laser output

Contact

Optically flat and parallel faces [cleaved (110) planes]

FIGURE 15-23
A broad-area *pn* junction laser diode. The two parallel planes (110) serve as the optical cavity in the *y* direction with a length of *L*. Roughened surfaces prevent optical positive feedback in the *x* direction.

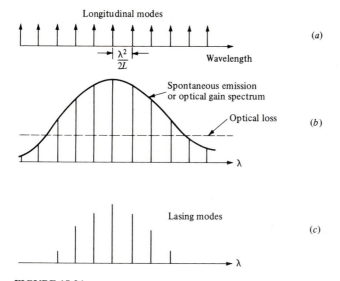

Longitudinal modes

$\dfrac{\lambda^2}{2L}$

Wavelength

(a)

Spontaneous emission or optical gain spectrum

Optical loss

(b)

λ

Lasing modes

(c)

λ

FIGURE 15-24
Graphical construction of laser emission modes: (*a*) resonant modes of laser cavity with L as cavity length, (*b*) spontaneous emission or gain curve, and (*c*) actual emission above optical loss.

the spontaneous emission spectrum (see, for example, Fig. 15-18). In Fig. 15-24*b*, the spontaneous-emission line shape is the optical gain curve, which is seen as the envelope upon which the discrete wavelengths satisfying the cavity resonator are shown. These emission lines are known as *longitudinal modes* with a wavelength separation of $\lambda^2/2L$ (Prob. 15-9). It should be noted that the wavelength separation between the longitudinal modes is very small compared with the laser-emission width. The lasing action begins when the optical gain is greater than the optical losses in Fig. 15-24*b*. Consequently, the lasing modes observed are illustrated in Fig. 15-24*c*. The physical picture of the longitudinal light wave traveling along the junction plane and the emitted modes are shown in Fig. 15-25.

The light wave can also resonate back and forth in the direction normal to the junction. This is known as the *transverse direction* which can reinforce a certain number of modes determined by the width of the optical cavity, resulting in *transverse* modes as shown in Fig. 15-26. The fundamental transverse mode shown is a sine-like standing wave with a half-length equal to the transverse cavity length. The higher-frequency transverse modes are undesirable because they give rise to nonuniform bright spots and should be eliminated.

In a Fabry-Perot cavity, optical gain is realized as a result of stimulated emission. But at the same time, the emitted light may be lost because of light leaking out or being absorbed. This loss mechanism is described by

$$\exp\left(-\alpha x\right) \tag{15-22}$$

where α is the optical loss coefficient per unit length, most of which is introduced by

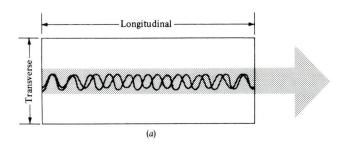

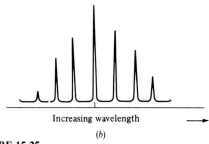

FIGURE 15-25
(a) Longitudinal traveling wave and (b) emitted longitudinal modes. *(After Bell [5].)*

absorption. Similarly, the optical gain follows the form of

$$\exp(gx) \qquad (15\text{-}23)$$

where g is the optical gain coefficient per unit length. At the onset of lasing, known as the *threshold*, the light wave must attain a gain of unity after traversing $2L$ where L is the cavity length. This condition may be expressed as

$$R_1 R_2 \exp(2g - 2\alpha)L = 1 \qquad (15\text{-}24)$$

where R_1 and R_2 are the reflectivity at the two end mirrors, which can be calculated

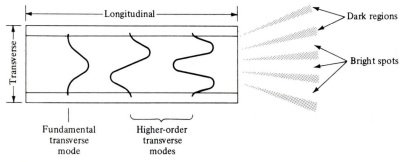

FIGURE 15-26
Transverse modes and nonuniform emission. *(After Bell [5].)*

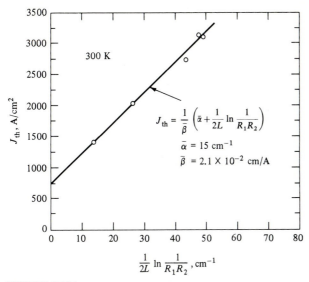

FIGURE 15-27
Threshold current density at 300 K as a function of the Fabry-Perot cavity end loss, $(1/2L)\ln(1/R_1/R_2)$ of a thin double-heterostructure laser. The facet reflectivity was varied with SiO coatings of various thicknesses. *(After Kressel [7].)*

by using Eq. (14-19). The gain at threshold may be written as

$$g_{th} = \alpha + \frac{1}{2L}\ln\frac{1}{R_1 R_2} \qquad (15\text{-}25)$$

The gain coefficient is found to be linearly proportional to the current. Thus, the diode current at threshold can be expressed as

$$J_{th} = \frac{1}{\beta}\left(\alpha + \frac{1}{2L}\ln\frac{1}{R_1 R_2}\right) \qquad (15\text{-}26)$$

A typical plot of J_{th} vs $(1/2L)\ln(1/R_1 R_2)$ is shown in Fig. 15-27, where α and β can be extracted.

15-10 THE DOUBLE-HETEROSTRUCTURE (DH) LASER

In a homojunction laser diode, the injected carriers diffuse away from the junction so that a higher current is required to reach population inversion. In addition, the emitted light may escape in the direction normal to the junction and become lost. Significant improvement can be made if the injected carriers and the emitted light can be kept near the junction's active region. These techniques are known as *carrier* and *optical confinement*, respectively, and they can be implemented by using two heterojunctions. The resulting structure is called the *double-heterostructure laser*, as shown in Fig. 15-28a. The basic concept involves three layers with a thin active

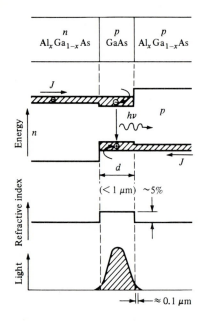

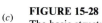

FIGURE 15-28

The basic structure of a GaAs–Al$_x$Ga$_{1-x}$ As DH laser shown schematically, with (*a*) the basic structure, (*b*) the band edge with forward bias, (*c*) a schematic representation of the refractive index change across the junction, and (*d*) the optical field distribution in the laser. (*After Casey and Panish [8].*)

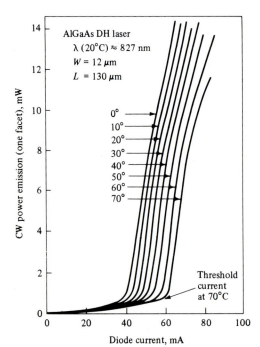

FIGURE 15-29

The power output as a function of current and heat-sink temperature of an oxide-defined GaAs–AlGaAs DH laser. (*After Kressel [9].*)

pGaAs sandwiched between a pAlGaAs and an nAlGaAs layer. The simplified energy-band diagram is shown in Fig 15-28b under forward bias. Electrons injected into the pGaAs are prevented from reaching the pAlGaAs layer by the conduction-band discontinuity. The potential step is about 0.28 eV for $Al_{0.3}Ga_{0.7}As$, which provides excellent carrier confinement. Population inversion can be easily reached, and radiative recombination is limited to the active layer. The optical confinement of the DH laser arises from the difference in the index of refraction for GaAs and AlGaAs, as illustrated in Fig. 15-28c. Although the refractive indices differ by only 5 percent or less, the optical confinement is quite remarkable, as shown in Fig. 15-28d. The threshold current of the DH laser is significantly lower so that continuous-wave (CW) operation at or above room temperature becomes possible.

Typical current-light characteristics of a DH laser diode are shown in Fig. 15-29. Note that the threshold current is defined as the break in the $I - L$ curve and I_{th} increases with temperature. The optical emission spectra are shown in Fig. 15-30 for

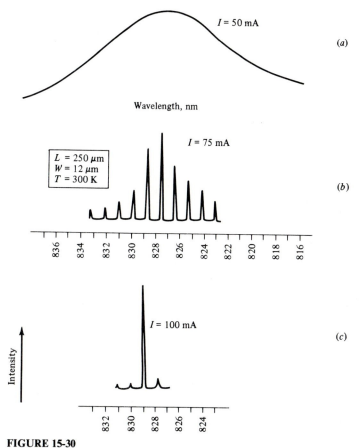

FIGURE 15-30
Typical CW lasing spectrum of an oxide-defined GaAs–AlGaAs DH laser with a cavity length of 250 μm. (*After Kressel [9].*)

different driving currents. At low bias, a relatively broad spectrum is observed which is the result of spontaneous emission (Fig. 15-30a). Above the threshold, peaks are seen corresponding to the longitudinal modes (Fig. 15-30b). At very high current, a single dominant mode is observed with an intensity many times those at or below the threshold (Fig. 15-30c).

Current research is directed toward new structures to lase at 1.3 or 1.55 μm. As shown in Fig. 14-22, these wavelengths have the lowest attenuation in a typical optical fiber. To avoid dispersion, lasers are also designed to produce an extremely narrow emission spectrum. These are known as single-frequency laser diodes. Detailed discussion of these structures are not covered here and the interested readers are referred to Bell [5], Boetz [6], and Dutta [10].

REFERENCES

1. Bhargava, R. N.: Recent Advances in Visible LEDs, *IEEE Trans. Electron Devices*, **ED-22**: 691 (1975).
2. Casey, H. C., Jr., and F. A. Trumbore: Single Crystal Electroluminescent Materials, *Mater. Sci. Eng.*, **6**:69 (1970).
3. Bergh, A., and P. Dean: Light-Emitting Diodes, *Proc. IEEE*, **60**:156 (1972).
4. Burrus, C. A., and B. I. Miller: Small-Area Double-Heterostructure AlGaAs Light Source for Optical-Fiber Transmission Lines, *Opt. Commun.*, **4**:307 (1971).
5. Bell, T. E.: Single Frequency Semiconductor Lasers, *IEEE Spectrum*, 38 (December 1983).
6. Boetz, D.: Laser Diodes Are Power-Packed, *IEEE Spectrum*, 43 (June 1985).
7. Kressel, H., et al: *RCA Rev.*, **32**: 393 (1971).
8. Casey, H. C., Jr., and M. B. Panish: "Heterostructure Lasers," parts A and B, Academic, New York, 1978.
9. Kressel, H., in M. K. Barnoski (ed.): "Fundamentals of Optical Fiber Communications," 2d ed., chap. 4, Academic, New York, 1981.
10. Dutta, N. K. et al.: Performance Comparison of InGaAsP, Lasers Emitting at 1.3 and 1.55 μm, *AT&T Tech. J.*, **64**:1857 (Oct. 1985).

ADDITIONAL READINGS

Bergh, A. A., and P. J. Dean: "Light-Emitting Diodes," Clarendon, Oxford, 1976.
Casey, H. C., Jr., and M. B. Panish: "Heterostructure Lasers," Academic, New York, 1978.
Gillessen, K., and W. Shairer: "Light Emitting Diodes," Prentice-Hall Int., New York, 1987.
Kressel, H., and J. K. Butler: "Semiconductor Lasers and Heterojunction LEDs," Academic, New York, 1977.

PROBLEMS

15-1. Show that the hole diffusion current is negligible in comparison with the electron current if $N_a \approx N_d$ in a GaAs LED. Use $\mu_n/\mu_p = 30$.

15-2. Assume $n_t = N_t \exp[-(E_t - E_f)/kT]$ and $p_a = N_a \exp[(E_a - E_f)/kT]$. Derive Eq. (15-11) by using Fig. P15-2 and the detailed balance of recombination and generation rates.

15-3. A GaAs infrared emitter has the following device parameters: $\eta_i = 80$ percent, $\bar{\alpha} = 10^3$ cm^{-1}, and $x_j = 10$ μm.

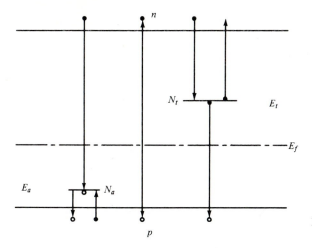

n

N_t E_t

E_f

E_a N_a

FIGURE P15-2

p

(a) Calculate the external quantum efficiency.

(b) Repeat (a) if a dome-shaped epoxy with a refractive index of 1.8 is used in the LED package. What is the ratio of improvement in the external quantum efficiency?

15-4. In GaAs the temperature dependence of the absorption coefficient can be approximated by $\bar{\alpha} = \alpha_o \exp(T/T_o)$, where α_o is α extrapolated to $T = 0\,\text{K}$ and T_o is approximately 100 K. The external quantum efficiency of a diode is 5 percent at 300 K. Other parameters are $x_j = 20\,\mu\text{m}$, $\bar{T} = 0.2$, and $\bar{\alpha}(300\,\text{K}) = 10^3\,\text{cm}^{-1}$.

(a) Calculate the internal quantum efficiency at 27°C.

(b) Assuming that the internal quantum efficiency is constant for the temperature range considered here, estimate the external quantum efficiency at -23 and 77°C.

15-5. Calculate the brightness of (a) a red GaP LED with $\eta_{\text{ext}} = 5$ percent at $10\,\text{A/cm}^2$, (b) a green GaP with $\eta_{\text{ext}} = 0.03$ percent at $10\,\text{A/cm}^2$, and (c) a red $\text{GaAs}_{0.6}\text{P}_{0.4}$ with $\eta_{\text{ext}} = 0.15$ percent at $20\,\text{A/cm}^2$. Assume that A_j/A_s is unity.

15-6. Estimate the range of the spacing of the donor-acceptor separation for the red GaP diode described in the text.

15-7. What are the minimum doping levels in a GaAs pn junction injection laser?

15-8. Show that the wavelength separation between the two adjacent longitudinal modes is $\lambda^2/2L$ by considering the difference of the two wavelengths associated with integers N and $N-1$. Let $N \gg 1$.

15-9. Assume the emission is equal to the band-gap energy. Find the longitudinal-mode separation in a GaAs laser with $L = 100\,\mu\text{m}$.

15-10. Determine the cavity length of a GaAs DH laser to achieve a threshold current density of $1000\,\text{A/cm}^2$ if it follows the curve in Fig. 15-27.

15-11. Assume $\Delta E_c = 0.3\,\text{eV}$ and $\Delta E_v = 0.15\,\text{eV}$ for a DH laser.

(a) Sketch the energy-band diagram accurately.

(b) Plot the minority-carrier concentration in p and P regions.

15-12. In Prob. 15-11, assume that the minority carriers cross the pGaAs–pAlGaAs interface by means of thermionic emission. Derive expressions for minority-carrier (electrons) distribution and current in the active layer.

APPENDIX

A

ATOMS, ELECTRONS AND ENERGY BANDS

This appendix has been prepared for students not previoulsy exposed to the theories of modern physics. It begins with the atomic model according to Bohr, which is followed by a discussion of the concept of wave-particle duality. Schrödinger's equation is presented, along with quantization of energy levels. Electronic structures of elements and the energy-band theory are then introduced.

A-1 THE BOHR ATOM

In the simplest model, the atom consists of a positively charged nucleus and negatively charged electrons. The total charge of all the electrons is equal to that of the nucleus, so that the atom as a whole is electrically neutral. Because the nucleus contains nearly all the mass of the atom, it is essentially immobile, whereas the electrons circle around in closed orbits.

Let us consider the case of hydrogen, which has only one electron per atom (Fig. A-1). Two forces are established between the nucleus and the electron. The

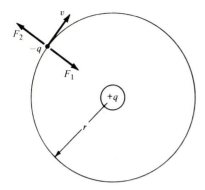

FIGURE A-1
The hydrogen atom according to the Bohr model.

first is the force of attraction described by Coulomb's law

$$F_1 = \frac{q^2}{4\pi\varepsilon_o r^2} \tag{A-1}$$

where q = electronic charge
ε_o = permittivity of free space
r = separation of charged particles

The second is the centrifugal force resulting from the orbital motion of the electron,

$$F_2 = \frac{mv^2}{r} \tag{A-2}$$

where m is the free-electron mass and v is the speed of the electron in its circular orbit. Since a stationary orbit is reached when the two forces are equal,

$$\frac{q^2}{4\pi\varepsilon_o r^2} = \frac{mv^2}{r} \tag{A-3}$$

By choosing the potential energy at infinity as zero reference, we find that the potential energy of the electron is

$$U = \int_\infty^r F_1 dr = -\frac{q^2}{4\pi\varepsilon_o r} \tag{A-4}$$

Since the kinetic energy of the electron is $mv^2/2$, the total energy is

$$E = \frac{mv^2}{2} - \frac{q^2}{4\pi\varepsilon_o r} = -\frac{q^2}{8\pi\varepsilon_o r} \tag{A-5}$$

The last term in this equation, obtained by making use of Eq. (A-3), demonstrates that the energy of the electron becomes smaller, i.e., more negative, as it approaches the nucleus.

According to Bohr, the electron may assume only certain discrete levels of energy in a stationary orbit. As long as the electron maintains a particular orbit, it cannot radiate or absorb energy. However, radiation does take place when the electron makes a transition from an orbit of higher energy E_2 to one of lower energy E_1. The radiation consists of a quantum of light, or *photon*, whose energy is given by

$$E_2 - E_1 = hv \qquad (A-6)$$

where h = Planck's constant

v = frequency of radiated energy

hv = photon energy

Bohr further postulated that a stationary orbit is determined by the condition in which the angular momentum of the electron is quantized and given by

$$mvr = \frac{nh}{2\pi} \qquad (A-7)$$

where n is an integer greater than or equal to 1.

Solving Eqs. (A-3), (A-5), and (A-7) to eliminate r and v, we obtain

$$E_n = -\frac{q^4 m}{8h^2 \varepsilon_o{}^2 n^2} \qquad (A-8)$$

For each value of n there is a corresponding energy E_n, which is discrete. Thus, the quantization of the angular momentum leads to discrete allowable energy levels. Substituting Eq. (A-8) into Eq. (A-6), we can find the frequencies of emitted photons corresponding to the transitions between different energy levels. For each frequency, the corresponding wavelength λ is given by c/v where c is the velocity of light.

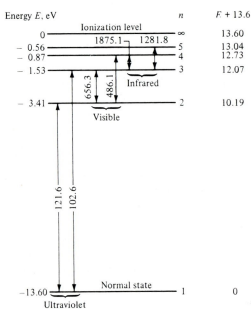

Energy E, eV

FIGURE A-2

The lowest five energy levels and the ionization energy of hydrogen. The spectral lines are expressed in nanometers.

The discrete energies and some of the transition photon wavelengths (in nanometer units) are shown in Fig. A-2. The energy level with $n = 1$, called the *ground* or *normal state*, is occupied by the electron when the temperature is absolute zero and there is no incident radiation. The higher energy levels correspond to n greater than 1 and are known as *excited states*. The excitation is accomplished by radiant energy, and the incident photon must have an energy exactly equal to the difference between the two energy levels involved. Thus,

$$E_2 = E_1 + hv \tag{A-9}$$

If this condition is not satisfied, the photon will not be absorbed.

If sufficient energy is imparted to the electron, it may reach the zero-total-energy level ($n = \infty$ in Fig. A-2) and the electron is said to be free from the influence of the nucleus. The required energy to reach this state, known as the *ionization energy*, is 13.6 eV for hydrogen.

A-2 THE WAVE PROPERTIES OF MATTER; SCHRÖDINGER'S EQUATION

The photon has the dual character of wave and particle. The wave properties are evidenced by the wavelength associated with the photon energy and are based on optical-interference experiments. On the other hand, when a photon is absorbed by an atom, the event takes place in a localized point of space so that the photon may be visualized as a particle. Since the energy of a photon is hv and its velocity is c, the momentum p is hv/c. Therefore, we have

$$p = \frac{hv}{c} = \frac{h}{\lambda} = \frac{hk}{2\pi} \tag{A-10}$$

where the relation $k = 2\pi/\lambda$ has been used to obtain the last expression and k is known as the *wave number*. Equation (A-10) is known as the *de Broglie relation*.

The de Broglie relation is found to be applicable to electrons and atoms, which also exhibit the dual character of wave and particle. A general theory describing the wave properties of matter is given by *Schrödinger's equation* as follows:

$$\nabla^2 \psi(x, y, z) + \frac{8\pi^2 m}{h^2} [E - U(x, y, z)] \psi(x, y, z) = 0 \tag{A-11}$$

with

$$\nabla^2 = \frac{\partial^2}{\partial x^2} + \frac{\partial^2}{\partial y^2} + \frac{\partial^2}{\partial z^2} \tag{A-12}$$

where $\psi(x, y, z)$ is known as the *wave function*, E is the total energy of the particles, and $U(x, y, z)$ is the potential energy. Physically, the absolute square of ψ defines the probability of finding the electron in a certain space. Thus, $|\psi|^2 \, dx \, dy \, dz$ represents the probability of finding the electron in the volume $dx \, dy \, dz$ around point (x, y, z) in space. Since the probability has a value ranging from 0 to 1, the wave function must be normalized so that

$$\int |\psi|^2 \, dx \, dy \, dz = 1 \tag{A-13}$$

The term $E - U$ represents the kinetic energy and is given by

$$E - U = \frac{p^2}{2m} = \frac{h^2 k^2}{8\pi^2 m} \tag{A-14}$$

where Eq. (A-10) has been used. For $U = 0$, we can plot E vs. k and obtain Fig. A-3, known as the E-k diagram. It expresses the energy-momentum relationship for a free particle.

As an example of the solution of Schrödinger's equation, let us consider a potential well in the form of a cube, each side of which has length L. Assume that the inside of the well is at zero potential and the potential is infinite outside the well so that no electrons can escape. Therefore, the probability of finding an electron outside the well must be zero; that is, $\psi = 0$ at $x, y, z < 0$ and $x, y, z > L$. For the one-dimensional case (Fig. A-4), we find, using Eqs. (A-11) and (A-14),

$$\frac{d^2\psi(x)}{dx^2} + k^2\psi(x) = 0 \tag{A-15}$$

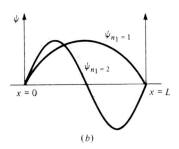

FIGURE A-3
The E–k diagram for a free particle.

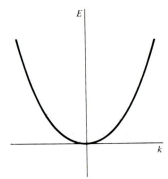

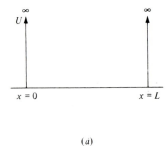

(a)   (b)

FIGURE A-4
(a) One-dimensional potential well and (b) the first two eigenfunctions from Eq. (A-22).

where

$$k^2 = \frac{8\pi^2 mE}{h^2} \tag{A-16}$$

The general solution of Eq. (A-15) is

$$\psi(x) = A \sin kx + B \cos kx \tag{A-17}$$

Since $\psi(x) = 0$ at $x = 0$, we have $B = 0$. Using the second boundary condition at L, we find

$$kL = n_1 \pi \qquad \text{or} \qquad k = \frac{n_1 \pi}{L} \tag{A-18}$$

where n_1 is a positive integer. Note that either $A = 0$ or $n_1 = 0$ represents the trivial solution in which ψ would vanish everywhere. Using Eq. (A-18), we can rewrite Eq. (A-16) as

$$E = \frac{h^2 k^2}{8\pi^2 m} = \frac{n_1^2 h^2}{8mL^2} \tag{A-19}$$

To complete solution of the Schrödinger equation, normalize ψ by rewriting (A-13) as

$$\int_0^L |\psi(x)|^2 \, dx = 1 \tag{A-20}$$

Thus, we have

$$\int_0^L A^2 \sin^2 \frac{n_1 \pi x}{L} \, dx = \frac{A^2 L}{2} = 1 \tag{A-21}$$

or $A = (2/L)^{1/2}$. Consequently, the solution, known as an *eigenfunction*, is

$$\psi = \left(\frac{2}{L}\right)^{1/2} \sin \frac{n_1 \pi x}{L} \tag{A-22}$$

The eigenfunctions for $n_1 = 1, 2$ are plotted in Fig. A-4b.

For the three-dimensional potential well, solution of the Schrödinger equation is obtained by the technique of separation of variables, and the result is

$$\psi = \left(\frac{8}{L^3}\right)^{1/2} \sin k_1 x \sin k_2 y \sin k_3 z \tag{A-23}$$

where

$$k_1 = \frac{n_1 \pi}{L} \qquad k_2 = \frac{n_2 \pi}{L} \qquad k_3 = \frac{n_3 \pi}{L} \tag{A-24}$$

and

$$k^2 = k_1^2 + k_2^2 + k_3^3 = \frac{8\pi^2 m}{h^2} E \tag{A-25}$$

In Eq. (A-25), each energy state is defined by three wave numbers, k_1, k_2, and k_3. Each set of wave numbers yields a different eigenfunction ψ. From other experimental evidence it is found that an eigenfunction has associated with it two electron states of opposite *spin* sign. Thus, four quantum numbers (n_1, n_2, n_3, s) are needed to define an energy state.

A-3   THE DENSITY OF STATES

In the calculation of electron concentration in a metal or semiconductor, it is necessary to know how many states there are between the energy levels E and $E + dE$. Let us consider the potential-box problem described in the preceding section and rewrite Eq. (A-25) as

$$E = \frac{h^2}{8mL^2}\,(n_1{}^2 + n_2{}^2 + n_3{}^2) = \frac{h^2 \mathbf{R}^2}{8mL^2} \tag{A-26}$$

where

$$\mathbf{R}^2 = n_1{}^2 + n_2{}^2 + n_3{}^2 \tag{A-27}$$

Each set of integers (n_1, n_2, n_3) determines both ψ and E, and $\mathbf{R}$ represents a vector to a point (n_1, n_2, n_3) in three-dimensional space. In this space, every unit cube $(L = 1)$ specifies a state so that the number of states in any volume is just equal to the numerical value of the volume. Thus, in a sphere of radius $\mathbf{R}$, the number of free electron states N is

$$N = \frac{4\pi |R|^3}{3} \tag{A-28}$$

Since $\mathbf{R}$ and E are related by Eq. (A-26), we can say that the corresponding number of states having a total energy less than E is

$$\frac{4\pi |R|^3}{3} = \frac{4\pi}{3}\left(\frac{8m}{h^2}\right)^{3/2} E^{3/2} \tag{A-29}$$

where we have set $L = 1$. Differentiating Eq. (A-29) with respect to E, we find that the number of states between E and $E + dE$ is

$$2\pi \left(\frac{8m}{h^2}\right)^{3/2} E^{1/2}\,dE \tag{A-30}$$

Since only positive integers are allowed for n_1, n_2, and n_3, we must divide Eq. (A-30) by a factor of 8; that is, each dimension can assume one-half of all integers. Furthermore, we must multiply the result by a factor of 2 to account for the two allowed values for the spin. Thus, we obtain

$$N(E)\,dE = \frac{4\pi(2m)^{3/2}}{h^3} E^{1/2}\,dE \tag{A-31}$$

This quantity is known as the *density-of-states function* per unit volume.

A-4 ELECTRONIC STRUCTURE OF THE ELEMENTS

Using the principle outlined in the previous section, we can solve Schrödinger's equation for hydrogen or any atom with multiple electrons. The solution for the hydrogen atom can be found in most texts in introductory quantum mechanics and is not given here. The potential energy expressed in Eq. (A-4) is first substituted into Eq. (A-11), and an appropriate coordinate system is chosen to simplify the mathematics. In general, it is found that four quantum numbers are required to define an energy state:

1. The principal quantum number $n = 1, 2, 3, \ldots$
2. The orbital angular-momentum quantum number $l = 0, 1, 2, \ldots, n-1$ for each n
3. The orbital magnetic quantum number $m = 0, \pm 1, \pm 2, \pm 3, \ldots, \pm l$ for each l
4. The electron spin $s = +\frac{1}{2}, -\frac{1}{2}$

Each set of quantum numbers specifies an energy state. According to the *Pauli exclusion principle*, no two electrons can have the same set of quantum numbers because each state can accommodate only one electron. However, different quantum states can be *degenerate*, i.e., have the same energy. For any atom with more than one electron, electrons will occupy different quantum states, and the electron configuration can be classified into *shells* and *subshells*.

All the electrons in an atom which have the same principal quantum numbr are said to belong to the same electron shell. The first four electron shells corresponding to $n = 1, 2, 3, 4$ are identified by the letters K, L, M, N, respectively. A shell is further divided into subshells, corresponding to the values of the orbital angular-momentum quantum number. The first four subshells corresponding to $l = 0, 1, 2, 3$ are identified as s, p, d, f, respectively. The distribution of electrons in an atom among the shells and subshells is tabulated in Table A-1 for the first four shells. To account for all the chemical elements given in the periodic table in Table A-2 (where the atomic number gives the number of electrons per atom), seven shells are required.

In Table A-1, we have two states in the K shell for $n = 1$ since $l = m = 0$ and there are two spins. These are called the $1s$ states. For the L shell with $n = 2$, we

TABLE A-1
Electron shells and subshells

Shell n	K 1	L 2		M 3			N 4			
l Subshell	0 s	0 s	1 p	0. s	1 p	2 d	0 s	1 p	2 d	3 f
Number of electrons	2	2	6	2	6	10	2	6	10	14
	2	8		18			32			

TABLE A-2
Periodic table of the elements†

Period	Group IA	Group IIA	Group IIIB	Group IVB	Group VB	Group VIB	Group VIIB	Group VIII			Group IB	Group IIB	Group IIIA	Group IVA	Group VA	Group VIA	Group VIIA	Inert gases
1	H 1 1.01																	He 2 4.00
2	Li 3 6.94	Be 4 9.01											B 5 10.81	C 6 12.01	N 7 14.01	O 8 16.00	F 9 19.00	Ne 10 20.18
3	Na 11 22.99	Mg 12 24.31											Al 13 26.98	Si 14 28.09	P 15 30.97	S 16 32.06	Cl 17 35.45	Ar 18 39.95
4	K 19 39.10	Ca 20 40.08	Sc 21 44.96	Ti 22 47.90	V 23 50.94	Cr 24 52.00	Mn 25 54.94	Fe 26 55.85	Co 27 58.93	Ni 28 58.71	Cu 29 63.54	Zn 30 65.37	Ga 31 69.72	Ge 32 72.59	As 33 74.92	Se 34 78.96	Br 35 79.91	Kr 36 83.80
5	Rb 37 85.47	Sr 38 87.62	Y 39 88.90	Zr 40 91.22	Nb 41 92.91	Mo 42 95.94	Tc 43 (99)	Ru 44 101.07	Rh 45 102.90	Pd 46 106.4	Ag 47 107.87	Cd 48 112.40	In 49 114.82	Sn 50 118.69	Sb 51 121.75	Te 52 127.60	I 53 126.90	Xe 54 131.30
6	Cs 55 132.90	Ba 56 137.34	La 57 138.91	Hf 72 178.49	Ta 73 180.95	W 74 183.85	Re 75 186.2	Os 76 190.2	Ir 77 192.2	Pt 78 195.09	Au 79 196.97	Hg 80 200.59	Tl 81 204.37	Pb 82 207.19	Bi 83 208.98	Po 84 (210)	At 85 (210)	Rn 86 (222)
7	Fr 87 (223)	Ra 88 (226)	Ac 89 (227)	Th 90 232.04	Pa 91 (231)	U 92 238.04	Np 93 (237)	Pu 94 (242)	Am 95 (243)	Cm 96 (247)	Bk 97 (247)	Cf 98 (251)	Es 99 (254)	Fm 100 (253)	Md 101 (256)	No 102 (254)	Lw 103 (257)	

The rare earths

Ce 58 140.12	Pr 59 140.91	Nd 60 144.24	Pm 61 (147)	Sm 62 150.35	Eu 63 151.96	Gd 64 157.25	Tb 65 158.92	Dy 66 162.50	Ho 67 164.93	Er 68 167.26	Tm 69 168.93	Yb 70 173.04	Lu 71 174.97

† The number to the right of the element symbol gives the atomic number. The number below the element symbol gives the atomic weight.

have $l = 0, 1$. In the s subshell corresponding to $l = 0$, there are again two states because of the two spins. This is known as the $2s$ subshell, where 2 stands for $n = 2$. In the p subshell, we have $l = 1$ and $m = 0, \pm 1$, which will make six states when the two spins are taken into account. This is known as the $2p$ subshell. The total number of energy states in the L shell is therefore $2 + 6 = 8$, as given in Table A-1.

The energy levels of an atom are filled from the lowest energy state to the highest energy state. The K shell has the lowest energy and is first occupied. As the atomic number increases, the next subshell will be filled and then the next shell. Thus, for carbon in column IVA in the periodic table with an atomic number of 6, the $1s$ and $2s$ subshells are filled, and only two states in the $2p$ subshell are occupied. The electronic structure of carbon can be designed by $1s^2 2s^2 2p^2$, and its schematic diagram is shown in Fig. A-5, where the shell structure of silicon is also shown. Note that both silicon and carbon have four electrons in the outermost shell, a characteristic common to all elements in the fourth column of the periodic table. These electrons are called the *valence* electrons, and carbon and silicon as well as germanium are known as *tetravalent* elements.

A-5 THE ENERGY-BAND THEORY OF CRYSTALS

In the previous sections we have considered the single, isolated atom, and the theory is applicable to gases in which individual atoms are sufficiently far apart. In a crystalline solid (see Sec. 1-1), the atoms are close enough to interact, and the potential energy $U(x, y, z)$ becomes a periodic function of space, as shown in Fig. A-6. The eigenfunctions of each atom obtained from the solution of Schrödinger's

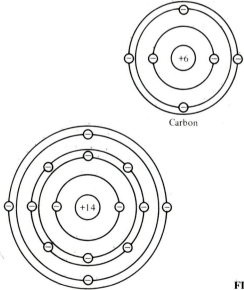

Carbon

Silicon

FIGURE A-5
Schematic diagrams of carbon and silicon atoms.

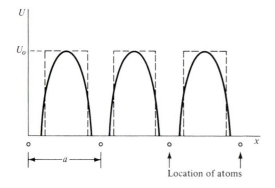

FIGURE A-6
Variation of the electron's potential energy in a one-dimensional crystal (solid curves) along with an idealized model (dashed lines) for analytical calculations.

equation overlap those of its neighbors because of the close proximity of atoms. Each energy level in the outermost shell now splits into a number of discrete levels. This phenomenon is similar to that when two *LC* tuning circuits with identical resonant frequencies are brought together, and the resonant curve becomes double-peaked instead of single-peaked.

Let us examine the carbon atom in further detail. The carbon atom has two shells. The first shell has two states which are filled. These are low-energy states, and electrons in these states are tightly bound to the nucleus. The second shell has eight states. Four of them are filled, and the other four are empty. If we were able to vary the atomic spacing between adjacent carbon atoms at will, we would obtain the energy-level distribution shown in Fig. A-7. At atomic spacing *d*, the atoms are sufficiently far apart to behave like isolated atoms. As the atoms are brought closer together at *c*, the two upper levels split into two bands, having two and six states, respectively. At atomic spacing *b*, the two bands overlap. As we reduce the atomic spacing further, two of the states from the upper band drop into the lower band, as indicated at atomic spacing *a*. It turns out that *a* represents the actual atomic spacing for carbon (as well as for silicon and germanium). At this atomic spacing, the four electrons fill the lower band, which is known as the *valence band*. The upper band is empty, and is known as the *conduction band*. In Fig. A-7, the *K* shell is shown as a discrete level. Actually, it is also split into a narrow band, but its interaction with neighboring atoms is small and can be neglected.

Mathematically, we can use the function $U(x) = U(x + a)$ to represent the periodic potential energy shown in Fig. A-6 and then solve Schrödinger's equation accordingly. This is known as the *Kronig-Penney model*. The solution, called the *Bloch wave function*, exhibits discrete bands of energy levels as depicted in the *E-k* diagram in Fig. A-8a. The solid curves represent *allowed energy levels* or *bands*. The discontinuities in the diagram indicate that the probability of finding an electron is zero in certain energy regions called *forbidden gaps*. The zones indicated in the diagram are called *Brillouin zones*; they can be simplified to the *reduced Brillouin zone* scheme shown in Fig. A-8b.

The Kronig-Penney model is the first step in understanding the energy-band theory of solids. In crystals such as silicon and gallium arsenide, the periodic potential energy is more complicated, leading to more complex energy-band

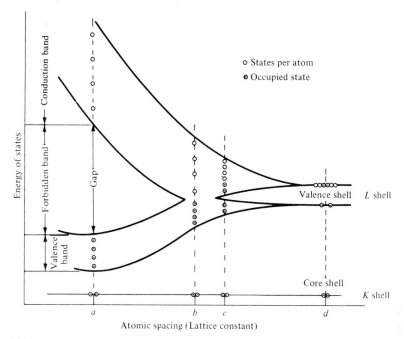

FIGURE A-7
The energy bands of the carbon lattice as a function of the atomic spacing at $T = 0$ K.

diagrams. Because these crystals are anisotropic, the E-k diagram along the principal axes is not necessarily the same and a three-dimensional picture is required to show all the features of the energy-band model. In practice, two-dimensional band diagrams are used to represent GaAs and Si (Fig. A-9). In these diagrams, a crystalline direction is designated along the k axis. Note that the minimum energy of the conduction band is *directly* above the maximum energy of the valence band in GaAs. Therefore, GaAs is known as a *direct-gap* semiconductor. In Si, the conduction-band minimum does not align with $k = 0$, and therefore it is known as an *indirect-gap* semiconductor. The energy-momentum relation shown in Fig. 1-12 is based on these diagrams.

One important piece of information we can obtain from the E-k diagrams is the *effective mass* (see Chap. 1) in a solid. Because of the influence of the periodic potential, the effective mass differs from the electronic mass in free space. By taking the partial derivative of E with respect to k twice in Eq. (A-14) we have

$$\frac{\partial^2 E}{\partial k^2} = \frac{h^2}{4\pi^2 m_e} \tag{A-32}$$

which can be rewritten as

$$m_e = \frac{h^2/4\pi^2}{\partial^2 E/\partial k^2} \tag{A-33}$$

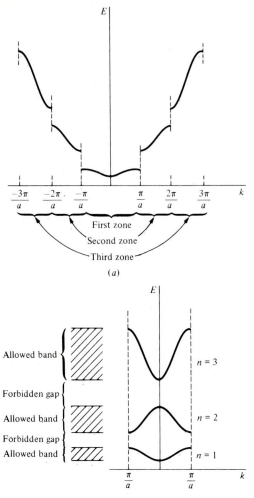

FIGURE A-8
(a) The energy as a function of wave number k for the Kronig-Penney model. The discontinuities in energy occur at $k = n\pi/a$, with $n = \pm 1,\ \pm 2,\ \pm 3,\ \ldots$. (b) The E–k diagram reduced to the first Brillouin zone. Allowed bands and forbidden gaps are shown.

By applying Eq. (A-33) to the conduction band in Fig. A-9, we obtain the effective mass of electrons in the conduction band. Similarly, the effective mass of *holes* (see text) in the valence band can be obtained. Note that the valence bands of Si and GaAs are both degenerate. The heavy-hole band has a smaller $\partial^2 E/\partial k^2$, which yields a larger effective mass from Eq. (A-33). The light-hole band has a larger $\partial^2 E/\partial k^2$ and a smaller effective mass. These finer features in the E-k diagram, though important in specifying the effective masses, are ignored in most device analyses.

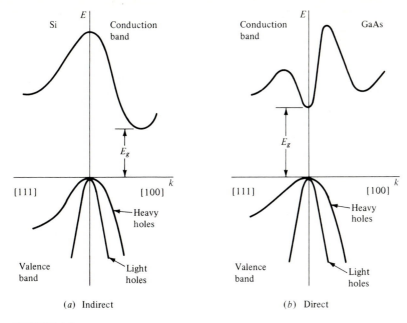

(a) Indirect   (b) Direct

FIGURE A-9
Energy-band diagrams for Si and GaAs, representing indirect and direct band-gap materials, respectively.

ADDITIONAL READINGS

Eisberg, R. M.: "Fundamentals of Modern Physics," Wiley, New York, 1961.
Sproull, R. L.: "Modern Physics," 2d ed., Wiley, New York, 1963.

APPENDIX

B

INDIRECT-RECOMBINATION STATISTICS

A schematic diagram showing the various steps involved in the recombination via an intermediate center is shown in Fig. 2-15, where the energy of these centers is E_t and the density of the centers is N_t. Four possible transition processes are shown: (1) an electron is captured by an empty center, (2) an electron is emitted from an occupied center, (3) an occupied center captures a hole, and (4) an empty center emits a hole. Since the probability of occupancy of a center follows the Fermi-Dirac function expressed by Eq. (1-10), we have

$$f_t^o = (e^{(E_t - E_f)/kT} + 1)^{-1} \tag{B-1}$$

The number of occupied centers is therefore $N_t f_t^o$, and the number of empty centers is $N_t(1 - f_t^o)$. The superscript o specifies the equilibrium condition.

By following the argument used in deriving the direct-recombination rate, we find the capture rate of an electron by an empty center to be

$$R_1 = c_n n N_t(1 - f_t) \tag{B-2}$$

where n is the electron density in the conduction band and c_n is the capture coefficient, which has a typical value of 10^{-8} cm^3/s. The electron capture coefficient is the product of the thermal velocity v_{th} and the electron-*capture cross section* σ_{cn}, where σ_{cn} is a measure of the closeness of an electron to an empty center and has a typical value of 10^{-15} cm^2. A hole-capture cross section is defined in the same

428

manner. The rate of emitting an electron to the condition band from an occupied center is given by

$$R_2 = e_n N_t f_t \tag{B-3}$$

where e_n is the emission coefficient. By analogy, the capture and emission rates of holes by the centers may be written as

$$R_3 = c_p p N_t f_t \tag{B-4}$$

$$R_4 = e_p N_t (1 - f_t) \tag{B-5}$$

where c_p and e_p are the hole capture and emission coefficient, respectively.

Under thermal equilibrium, the number of electrons emitted from the centers must be the same as that of the captured electrons, that is, $R_1 = R_2$. Equating Eqs. (B-2) and (B-3) and using Eq. (1-25), we obtain

$$c_n n_i e^{(E_f - E_i)/kT} N_t (1 - f_t) = e_n N_t f_t \tag{B-6}$$

Furthermore, since

$$\frac{1 - f_t}{f_t} = e^{(E_t - E_f)/kT} \tag{B-7}$$

we have

$$e_n = c_n n_i e^{(E_t - E_i)/kT} \tag{B-8}$$

Using the same procedure and Eq. (1-26), we find the hole emission rate

$$e_p = c_p n_i e^{(E_i - E_t)/kT} \tag{B-9}$$

Since the capture and emission probabilities are independent of equilibrium and nonequilibrium conditions, Eqs. (B-8) and (B-9) are valid at nonequilibrium although they have been derived under the thermal-equilibrium condition. However, the probability of occupancy expressed by Eq. (B-1) is not applicable at nonequilibrium because the Fermi level E_f is meaningful only at equilibrium.

Let us now consider the nonequilibrium case by applying an external source of energy, e.g., a light source, so that a generation rate G_L exists uniformly throughout the semiconductor. Under steady-state conditions, the electrons entering the leaving the conduction band in Fig. B-1 must be equal. This is called the *principle of detailed balance*, and it yields

$$G_L = R_1 - R_2 \tag{B-10}$$

Similarly, the detailed balance of holes in the valence band leads to

$$G_L = R_3 - R_4 \tag{B-11}$$

Equating Eqs. (B-10) and (B-11), we have

$$R_1 - R_2 = R_3 - R_4 \tag{B-12}$$

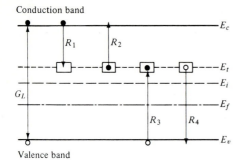

Conduction band

Valence band

FIGURE B-1
Generation and recombination processes under illumination.

We can now substitute Eqs. (B-2) to (B-5) into the foregoing expression to obtain

$$c_n n N_t(1 - f_t) - e_n N_t f_t = c_p p N_t f_t - e_p N_t(1 - f_t) \tag{B-13}$$

By assuming $c_n = c_p = c$ and using Eqs. (B-8) and (B-9), we can derive f_t from Eq. (B-13), yielding

$$f_t = \frac{n + n_i e^{(E_i - E_t)/kT}}{n + p + 2n_i \cosh\,[(E_t - E_i)/kT]} \tag{B-14}$$

Therefore, the net recombination rate is

$$U \equiv R_1 - R_2 = \frac{c N_t(pn - n_i^2)}{n + p + 2n_i \cosh\,[(E_t - E_i)/kT]} \tag{B-15}$$

From Eq. (B-15), we find that at equilibrium, that is, at $pn = n_i^2$, the net recombination rate is zero. It is also interesting to note that the maximum recombination rate occurs when $E_t = E_i$. In other words, the most efficient recombination centers are located at or near the center of the forbidden gap. Away from the level E_i, the centers would be less effective because it is more probable to capture one type of carrier but less probable to capture the other type.

APPENDIX

C

DERIVATION OF THE THERMIONIC-EMISSION CURRENT

We can rewrite Eq. (1-5) as

$$E = \frac{P^2}{2m} = \frac{1}{2m}(p_x^2 + p_y^2 + p_z^2) \tag{C-1}$$

where m is the free-electron mass. Differentiating (C-1) leads to

$$dE = \frac{P\, dP}{m} \tag{C-2}$$

The conversion to rectangular coordinates is obtained by considering the incremental volume in momentum space between a sphere of radius P and $P + dP$, and the relationship is

$$4\pi P^2\, dP = dp_x dp_y dp_z \tag{C-3}$$

Substituting Eqs. (C-1) to (C-3) into Eq. (7-8) yields

$$I = \frac{2qA}{mh^3} \int_{p_{xo}}^{\infty} \int_{p_y=-\infty}^{\infty} \int_{p_z=-\infty}^{\infty} p_x e^{-(p_x^2 + p_y^2 + p_z^2 - 2mE_f)/2mkT}\, dp_x dp_y dp_z$$

$$= \frac{2qA}{mh^3} \int_{p_{xo}}^{\infty} e^{-(p_x^2 - 2mE_f)/2mkT} p_x dp_x \int_{-\infty}^{\infty} e^{-p_y^2/2mkT}\, dp_y \int_{-\infty}^{\infty} e^{-p_z^2/2mkT}\, dp_z \tag{C-4}$$

where $p_{xo}^2 = 2m(E_f + q\phi_m)$. The last two integrals can be evaluated by using the definite integral

$$\int_{-\infty}^{\infty} e^{-ax^2}\,dx = \left(\frac{\pi}{a}\right)^{1/2} \tag{C-5}$$

and each integral yields $(2\pi mkT)^{1/2}$. The first integral is evaluated by setting

$$\frac{p_x^2 - 2mE_f}{2mkT} = u \tag{C-6}$$

Therefore we have

$$du = \frac{p_x dp_x}{mkT} \tag{C-7}$$

The lower limit of the first integral can be written as

$$\frac{2m(E_f + q\phi_m) - 2mE_f}{2mkT} = \frac{q\phi_m}{kT} \tag{C-8}$$

so that the first integral becomes

$$mkT \int_{q\phi_m/kT}^{\infty} e^{-u}\,du = mkTe^{-q\phi_m/kT} \tag{C-9}$$

Substituting the results of (C-5) and (C-9) into Eq. (C-4), we obtain

$$I = A\,\frac{4\pi mqk^2}{h^3}\,T^2 e^{-q\phi_m/kT} = AA^*T^2 e^{-q\phi_m/kT} \tag{C-10}$$

where $A^* = 4\pi mqk^2/h^3$ and is known as Richardson's constant.

$$\mathrm{erf}\ z \equiv \frac{2}{\sqrt{\pi}} \int_0^z e^{-a^2}\, da \qquad \mathrm{erfc}\ z \equiv 1 - \mathrm{erf}\ z$$

$$\mathrm{erf}\ 0 = 0 \qquad \mathrm{erf}\ \infty = 1$$

$$\mathrm{erf}\ z \approx \frac{2}{\sqrt{\pi}}\, z \qquad \text{for } z \ll 1$$

$$\mathrm{erf}\ z \approx \frac{1}{\sqrt{\pi}}\, \frac{e^{-z^2}}{z} \qquad \text{for } z \gg 1$$

$$\frac{d\ \mathrm{erf}\ z}{dz} = \frac{2}{\sqrt{\pi}}\, e^{z^2}$$

$$\int_0^z \mathrm{erfc}\ (z')\, dz' = z\ \mathrm{erfc}\ z + \frac{1}{\sqrt{\pi}}\, (1 - e^{-z^2})$$

$$\int_0^\infty \mathrm{erfc}\ (z)\, dz = \frac{1}{\sqrt{\pi}}$$

erfc z

z	erfc z	z	erfc z	z	erfc z	z	erfc z
0	1.000 00	1.00	0.157 30	2.00	0.004 68	3.00	0.000 022 09
0.10	0.887 54	1.10	0.119 80	2.10	0.002 98	3.10	0.000 011 65
0.20	0.777 30	1.20	0.089 69	2.20	0.001 86	3.20	0.000 006 03
0.30	0.671 37	1.30	0.065 99	2.30	0.001 14	3.30	0.000 003 06
0.40	0.571 61	1.40	0.047 72	2.40	0.000 689	3.40	0.000 001 52
0.50	0.479 50	1.50	0.033 90	2.50	0.000 407	3.50	0.000 000 743
0.60	0.396 14	1.60	0.023 65	2.60	0.000 236	3.60	0.000 000 356
0.70	0.322 20	1.70	0.016 21	2.70	0.000 134	3.70	0.000 000 167
0.80	0.257 90	1.80	0.010 91	2.80	0.000 075	3.80	0.000 000 77
0.90	0.203 09	1.90	0.007 21	2.90	0.000 041	3.90	0.000 00 35

INDEX

Page references in **boldface** include illustrations or tables; page references in *italic* refer to problems.